Quantitative Techniques

6th Edition

Terry Lucey

M.Soc.Sc., F.C.M.A., F.C.C.A., J.Dip.M.A.

Terry Lucey has been an accountant and consultant in industry and has had over twenty years examining and teaching experience at all levels of professional studies and for diploma and degree courses in business studies. He was previously Head of Department of Business Studies at the University of Wolverhampton and Visiting Fellow at Aston Business School, Aston University.

Amongst his other published works are:

Investment Appraisal: Evaluating Risk and Uncertainty, Accounting and Computer Systems (co-author), *Management Information Systems, Costing, Management Accounting*, and several study and revision manuals for Professional examinations.

THOMSON

Australia • Canada • Mexico • Singapore • Spain • United Kingdom • United States

THOMSON
™

Quantitative Techniques

Copyright © Terry Lucey 2002

The Thomson logo is a registered trademark used herein under licence.

For more information, contact Thomson Learning, High Holborn House, 50-51 Bedford Row, London WC1R 4LR or visit us on the World Wide Web at: http://www.thomsonlearning.co.uk

British Library Cataloguing-in-Publication Data
A catalogue record for this book is available from the British Library

ISBN 1-84480-106-3

First edition 1979
Second edition 1982
Third edition 1988
Fourth edition 1992
Fifth edition 1996
Sixth edition 2002 by Continuum
Reprinted 2003 by Thomson Learning

Typeset by YHT Ltd, London
Printed in the UK by TJ International, Padstow, Cornwall

The author would like to express thanks to the following for permission to reproduce past examination questions:
Chartered Association of Certified Accountants (ACCA)
Chartered Institute of Management Accountants (CIMA)
Chartered Institute of Public Finance and Accountancy (CIPFA)
Institute of Chartered Accountants (ICA)

Contents

Contents

Preface

Aims of the book

1. This book is designed to provide a sound understanding of Quantitative Techniques and is particularly relevant for:

    a) Students preparing themselves for professional examinations of the following bodies:

    - Chartered Association of Certified Accountants
    - Chartered Institute of Management Accountants
    - Chartered Institute of Public Finance and Accountancy
    - Institute of Chartered Accountants
    - Institute of Chartered Secretaries and Administrators
    - Institute of Data Processing Management
    - Institute of Company Accountants

    b) Students on BTEC Higher level courses and undergraduates reading Business Studies, Accounting and any course including Quantitative Techniques.

    c) Managers and others in industry, commerce, and local authorities who wish to obtain a working knowledge of quantitative techniques to aid them in their own work and to facilitate communication with accountants and operational research specialists.

Teaching approach

2. The book has been written in a standardised format with numbered paragraphs, end of chapter summaries, with review questions and examination questions at the end of each chapter. This approach has been tested and found effective by thousands of students and the book can be used for independent study or in conjunction with tuition at a college.

How to use the book effectively

3. For ease of study the book is divided into self-contained chapters with numbered paragraphs. Each chapter is followed by self review questions, cross referenced to appropriate paragraph(s). You should attempt to answer the self review questions unaided then check your answer with the text.

    In addition, most chapters contain a number of test exercises. The exercises are shorter and simpler than typical examination questions and will be found useful for practice and consolidation. The answers are provided after the exercises but students are strongly advised to work through the problems themselves before looking at the answers.

At appropriate points in the text (after Chapters 3, 9, 15, 21, 26 and 31) there are Assessment and Revision Sections. These contain carefully selected examination questions drawn from the most recent Professional examinations. The questions have been chosen not merely to repeat the material in the chapters but to extend knowledge and understanding. Fully worked answers are provided in an Appendix to the book but, again, you are recommended to attempt the questions yourself before looking at the answers.

Each Assessment and Revision Section also contains a further selection of examination questions, without answers. These can be used by lecturers for class work and assignments when the book is being used as a course text or as extra practice when the book is used for independent study. A separate answer guide is available free to lecturers who adopt the book as a course text.

Sequence of study

4.   The book should be studied in the sequence of the chapters. The sequence has been arranged so that there is a progressive accumulation of knowledge and any given chapter either includes all the principles necessary or draws upon a previous chapter(s).

Notes to the 6th edition

The responses to previous editions of this book have been extremely gratifying and I would like to thank both lecturers and students for their comments and constructive feedback.

Features of the 6th edition

a)   There have been numerous detailed revisions and extensions of coverage.

b)   Carefully selected questions have been included from the latest professional examinations.

c)   Each chapter has learning objectives, self-review questions and exercises with answers.

d)   The book has been reset throughout to improve the readability and appearance.

e)   A separate Lecturers' Supplement is available free to lecturers who adopt the book as a course text and can be obtained from the publishers. The Supplement contains answers to the examination questions in the Assessment and Revision Sections and OHP masters of key diagrams from the book.

Terry Lucey

2002

1 Introduction to quantitative techniques

Objectives

1. After studying this chapter you will
   - know the principles of a model-based approach;
   - be able to describe various types of models;
   - understand the importance of risk and uncertainty;
   - be able to describe the stages in an operational research study;
   - know that the book assumes a knowledge of Foundation Mathematics and Statistics.

Differences in terminology

2. The title used for this book is Quantitative Techniques. However, many other terms are also used for the concepts and techniques described in the book. These include – Decision Making Techniques – Operational Research or OR – Analytical Techniques – Quantitative Analysis – Business Mathematics and Management Science. Regardless of the name used it is important to realise that the same group of techniques are being described and, more importantly, the same approach is employed. This approach is brought out by the following British Standard definition of Operational Research.

Operational research definition

3. 'The attack of modern science on complex problems arising in the direction and management of large systems of men, machines, materials and money, in industry, business, government and defence. The distinctive approach is to develop a scientific model of the system incorporating measurements of factors such as change and risk, with which to predict and compare the outcome of alternative decisions, strategies or controls. The purpose is to help management determine its policy and action scientifically.'

Essential features of the OR approach

4. The above formal definition contains several essential features and these are:
   a) Application of a model-based scientific approach.
   b) Systems approach to organisations.
   c) The recognition of risk and uncertainty.
   d) Assistance to management decision making and control.

Application of a model-based approach

5.   The basis of the OR approach is that of constructing models of problems in an objective, factual manner and experimenting with these models to show the results of various possible courses of action. A model is any representation of reality and may be in graphical, physical or mathematical terms. The type of model most frequently used in OR is a *mathematical model*, i.e. one which tries to show the workings of the real world by means of mathematical symbols, equations and formulae.

An example of a simple mathematical model familiar to accountants could be the following equation to estimate the total overheads for a period. Assume that total overheads comprise fixed overheads and variable overheads which are directly related to the units produced, then the equation:

$$y = a + bx$$

is a model of the relationship of total overheads to the number of units produced where

y = Total overheads

a = Fixed overheads per period

b = Variable overheads per unit (assumed to be constant)

x = Number of units produced

The above model is obviously very simple and most practical models are of necessity much more complex. It is important to realise that however complex looking the model is and however many variables it contains, it still involves considerable simplification of reality and any results or predictions obtained from the model must therefore be used with caution and judgement.

Although mathematical or symbolic models are the more usual, other types of models, e.g. Iconic, Analogue, Simulation and Heuristic, sometimes have applicability.

Iconic models are visual models of the real object(s) they represent. They may be larger or smaller than reality. For example, a model steam engine is much smaller than the real thing whilst the familiar coloured plastic models of molecular structures are much larger. Both of these are iconic models and so are pictures, maps and diagrams although these latter are in a different form to the reality they represent. These models are often difficult to manipulate experimentally (but not always, e.g. a wind tunnel model for aircraft) so are not used greatly in OR.

Analogue models use one set of physical movements or properties to represent another set. For example, the movement of a piece of metal under stress can be represented in a more observable form by the movement of a gauge finger. Fathom lines on charts and lines on graphs are also analogues of the reality they represent.

Although analogue models are more versatile than iconic models they generally lack the flexibility of mathematical models.

Simulation models represent the behaviour of a real system. Perhaps the best known example is the flight simulator used in the training of Pilots. Simulation models are used where there is no suitable mathematical model, where the mathematical model is too complex, or where it is not possible to experiment upon a working system without causing serious disruption – as in the case of the training of Pilots! One application of simulation models in the management context is the study of the behaviour of people and objects in queues. With the development of relatively inexpensive computing facilities many business games are based on the simulation of the operation of the complete business. This area is dealt with more fully later in the book.

Heuristic models are models which use a set of intuitive rules which managers hope will produce at least a workable solution, and a better solution than methods currently being used. For example, a delivery van driver may be instructed to plan the day's deliveries using the following rule: 'After each delivery, go to the nearest customer whom you have not yet visited.' This will certainly give a 'good' solution early in the day, but it can lead to some long distances being travelled at the end of the day back to the depot. The driver has no way of knowing whether the route gives optimum time and distances, and any improvements can come only through testing other heuristic approaches.

Models may be further classified into *normative* and *descriptive*. Normative models are concerned with finding the best, optimum or ideal solution to a problem. Many mathematical models fall into this classification. Descriptive models, as their name implies, describe the behaviour of a system without attempting to find the best solution to any problem. For example, simulation tends to fall within this category.

Perhaps the most important point to appreciate is not so much what a model is called, but what it does in helping managers to attain the goals that they have set.

Systems approach to organisation

6. The primary aim of OR is to attempt to identify the best way of conducting the affairs of the organisation, i.e. the optimum. In studying problems the OR practitioner tries to optimise the operation of the organisation as a whole rather than narrow aspects of the business such as a single department or section. This is easier said than done and, because of the practical necessity of dealing with manageable areas of work and thereby producing simplified and incomplete models of operations, there may be a tendency to produce sub-optimal solutions, i.e. a solution which is optimal for a small section of the firm, but not optimal for the firm as a whole. This is another point which should be watched when considering the results of an OR investigation.

Recognition of risk and uncertainty

7. All business planning and decision making involves forecasting future activities. This cannot be done with any certainty and so to provide the maximum possible assistance to planners and decision makers a systematic analysis of the possible

extent of the risks and uncertainties involved is a vital part of any OR study. OR techniques do not of themselves remove the risks and uncertainties, but they are able to highlight their effects on the firm's operations. An important method to help with this process is what is termed *sensitivity analysis*. When a solution has been obtained using one of the OR techniques (such as linear programming, inventory control, investment appraisal) alterations are made to the factors of the problem, such as, sales, costs, amount of materials, to see what effect there is on the original solution. If the value of the original solution alters considerably with minor changes in the factor values it is said to be sensitive. In such circumstances the whole problem will need much deeper analysis and a particularly sensitive factor may cause decisions to be altered.

Assistance to management decision making and control

8. In general, OR practitioners do not make the business decisions. Their role is the provision of information to assist the planners and decision makers. The skill, experience and judgement of managers cannot be replaced by formal decision making techniques. The results of an OR investigation are but one input of information into the decision making process. There is a strong parallel between the OR practitioner supplying information for management decision making and that of other information specialists such as accountants. This is why a knowledge of OR for the accountant (and accountancy for the OR practitioner) can be very useful and can improve the quality of the information provided. There are strong reasons why the most effective OR teams contain people drawn from various backgrounds – economists, accountants, mathematicians, engineers, psychologists etc. In this way there is more chance that the numerous facets of business problems can be recognised and analysed.

Quantification of factors

9. Not all the factors involved in a decision making or planning situation can be quantified, but the most readily usable OR techniques are those based on quantifiable factors such as costs, revenues, number of units, etc. These techniques are the ones most commonly included in examinations and form the basis of the syllabuses of the major professional bodies. Hence the contents and title of this book.

Stages in OR study

10. An OR study can be separated into seven steps:
    a) Problem recognition and definition
    b) Model building
    c) Data collection
    d) Problem solution
    e) Interpretation of solution
    f) Implementation of final solution

g) Review and maintain

Problem recognition and definition

11. No technique can be applied or analysis undertaken until the problem has been recognised and carefully defined. Vague descriptions of problems are insufficient. What is required is a clear, detailed statement of what the problem really is. This is a difficult stage and thorough investigation is necessary. A superficial study may not identify the real problem. Close and friendly liaison with the people involved in the area will provide the best results.

Model building

12. Having defined the problem, a model must be developed which, it is hoped, can be solved by an appropriate technique such as the ones described in this book. Care must be taken not to apply a standard OR technique to a non-standard problem. One must ensure that the model incorporates the essential features of the problem being studied.

Data collection

13. Data will be collected at the two previous stages, but to solve the problem much more data will be required. This data will include costs, revenues, production and stock data, etc., and will be gained from past records and from estimates of the future. The data required will also include the quantification of factors not always quantified, such as risk and uncertainty.

Problem solution

14. For many problems, particularly those encountered in examinations, a solution can usually be obtained by standard mathematical means using a recognised OR technique. This does not, unfortunately, apply to all problems encountered in practice which may require the use of advanced mathematical analysis and non-standard techniques. Even where a solution can be obtained by the use of a basic model, practical problems require large amounts of data necessitating repeated calculations making the use of computers an economic proposition.

Interpretation of solution

15. Although many OR techniques produce optimal solutions, it must be realised that the model does not provide a solution to the actual real life problem, but to a simplified version of the problem represented by the model. Accordingly results must be treated with caution and there must be due regard to the problems involved in all OR studies which include:

- The appropriateness of the model to the real problem
- The accuracy of the data used and assumptions made
- The dangers of sub-optimisation

Sub-optimisation is where the objectives of the sub-system (e.g. a department) are pursued to the detriment of the overall organisational goals which must be paramount.

Implementation of final solution

16. After careful interpretation of the results of the OR study and modification where appropriate, the resulting solution would normally be implemented. Sometimes the solution may not be implemented because, although technically valid, management may consider that its implementation would not be cost effective or that it might cause too much disruption to operations and/or employees, customers, suppliers and the general public.

Review and maintain

17. After implementation the performance of the model should be carefully monitored to ensure that it actually does work and fulfills its objectives. The review process should be at regular intervals so that appropriate adjustments can be made to meet minor changes in conditions or to recognise promptly when major changes have occurred, which render the implemented solution inappropriate. This is, of course, a form of audit procedure familiar to accountants.

All organisations are subject to change. Consequently, no solution remains optimal for ever and changes in the system or in the environment of the system makes it essential continually to review the models used and the existing solutions to see if adjustments are required.

Foundation mathematics and statistics

18. The syllabuses at which this book is aimed assume a basic knowledge of mathematics and statistics which would normally be gained in Foundation Level studies. Accordingly this book does not cover foundation level mathematics and statistics, but because of the step-by-step teaching approach used and the inclusion of some revision material (e.g. probability concepts) it should be readily comprehensible to all students, including those whose basic mathematics and statistics are a little rusty.

Summary

19. a) Numerous names exist for the concepts and techniques covered in this book. More important than the name is the common approach adopted.

   b) This approach is characterised by a scientific, model-based approach to solving problems.

   c) Whatever the problem, there are common stages in all OR studies. These are Problem Recognition, Model building, Data Collection, Solution, Interpretation and Implementation.

   d) This book covers the full range of Quantitative Techniques normally encountered in college courses and professional examinations and assumes a knowledge of foundation level mathematics and statistics.

Points to note

20. a) OR should not be regarded as a compartmentalised area of study. The basic approach is applicable in virtually any situation, particularly those related to planning, control and decision making.

   b) Formal mathematical notation is reduced to the absolute minimum in this book and formal proofs are not given. This is in line with the philosophy adopted by the major professional bodies for their examinations which concentrate on the application of OR techniques to practical problems.

   c) There is considerable overlap between the work of the OR specialist and that of a statistician and, indeed, many OR specialists originally trained as statisticians. The ability to handle, analyse and interpret masses of data are key elements in both disciplines.

Self review questions *Numbers in brackets refer to paragraph numbers*

1. Define OR in your own words. Compare your definition with the B.S. definition. (2)

2. What are the essential features of the OR approach? (4)

3. How does OR assist management decision making? (8)

4. What are the stages in an OR study? (10)

5. Give reasons why the results of an OR study may not be implemented. (16)

2 Probability and decision making

Objectives

1. After studying this chapter you will
    - know the meaning of probability;
    - understand what is meant by Subjective Probability;
    - be able to use the Addition and Multiplication rules of probability;
    - understand the principles of Conditional probability;
    - know the principles of Bayes' Rule or Theorem;
    - be able to distinguish between Permutations and Combinations;
    - understand how to calculate and use Expected Value;
    - know how to value Perfect and Imperfect information;
    - be able to use various decision rules, including maximin, maximax, minimax regret and the Hurwicz rule.

Probability definition

2. For our purposes, probability can be considered as the quantification of uncertainty. Uncertainty may also be expressed as 'likelihood', 'chance' or 'risk'. Probability is represented by p, and can only take values ranging from 0, i.e. impossibility, to 1, i.e. certainty. For example, it is impossible to fly to the moon unaided and it is certain that one day we will die. This is expressed as

    p(flying to the moon unaided) = 0

    and **p(dying) = 1**

    More generally, we may define the probability of an event, E, as follows:

    $$p(E) = \frac{\text{Number of favourable outcomes}}{\text{Total number of possible outcomes}}$$

    where an event is some occurrence, e.g. spinning a coin to show a head, drawing an Ace from a pack of cards.

    Probabilities can thus be regarded as *relative frequencies*, as shown in the following examples.

    What is the probability of drawing an Ace from a shuffled pack of cards?

    $$p(\text{drawing an Ace}) = \frac{4}{52} = \frac{1}{13}$$

    The numerator is 4 because there are four Aces in a pack and the denominator is 52 because there are fifty two cards in a pack.

    What is the probability of throwing a 3 with a six-sided unbiased die?

$$p(\text{throwing a } 3) = \frac{1}{6}$$

The numerator is 1 because only one side depicts a three and the denominator is 6 because there are six sides on the die.

Objective and subjective probability

3.   Where the probability of an event is based on past data and the circumstances are repeatable by test, the probability is known as *statistical* or *objective probability*. For example, the probability of tossing a coin and a head showing is 50% or ½ or 0.5. This value can be shown to be correct by repeated trials. In most circumstances objective probabilities are not available in business, so that *subjective probabilities* must be used. These are quantifications of personal judgement, experience and expertise.

For example, the Sales Manager considers that there is a 40% chance (i.e. $p = 0.4$) of obtaining the order for which the firm has just quoted. Clearly this value cannot be tested by repeated trials. In spite of the undoubted shortcomings of subjective probabilities they are all that are normally available and so they are used to help in the decision-making process. It should be emphasised that the use of probabilities does not of itself make the decision. It merely provides more information on which a more informed decision can be taken.

Note: When there are several possible outcomes, each with an associated probability, the total of all the probabilities must equal 1.

Basic rules of probability

4.   Whether the probabilities are objective or subjective, once they are established they are used in the same way according to the basic rules of probability which are described below. Before considering these rules it is necessary to define two terms: mutually exclusive events and independent events.

Mutually exclusive events are events which cannot happen at the same time. If one happens the other(s) cannot occur. For example, if we are considering the classification of people the events 'male' and 'female' are mutually exclusive. If a person is 'Female' this automatically excludes the possibility of being 'Male'.

Independent events. Two or more events are independent if the occurrence or non-occurrence of any one event does not effect the occurrence or non-occurrence of the others. For example, the outcome of any throw of a die is independent of the outcome of any preceding or succeeding throws.

The rules of probability to be covered are the *addition* rule, the *multiplication* rule and the rule of *conditional probability*.

a)   *Addition rule (OR)*

$$P(X \text{ or } Y) = P(X) + P(Y)$$

This rule is used to calculate the probability of two or more mutually exclusive events. In such circumstances the probabilities of the separate events must be added.

Example 1

What is the probability of throwing a 3 *or* a 6 with a throw of a die?

$$P(\text{throwing a 3}) = \frac{1}{6} \text{ and } P(\text{throwing a 6}) = \frac{1}{6}$$

$$\therefore \ P(\text{throwing a 3 } or \text{ a 6}) = \frac{1}{6} + \frac{1}{6} = \frac{1}{3}$$

Note: Outcomes are mutually exclusive if, when one outcome occurs, the other(s) cannot happen.

b) *Multiplication rule (AND)*

$$P(X \text{ and } Y) = P(X) \times P(Y)$$

This rule is used when there is a string of independent events for which each individual probability is known and it is required to know the overall probability.

Example 2

What is the probability of throwing a 3 *and* a 6 with two throws of a die?

$$P(\text{throwing a 3}) = \frac{1}{6} \text{ and } P(\text{throwing a 6}) = \frac{1}{6}$$

$$\therefore \ P(\text{throwing a 3 } and \text{ a 6}) = \frac{1}{6} \times \frac{1}{6} = \frac{1}{36}$$

Note: The probability of $\frac{1}{36}$ given above is the probability for throwing a 3 followed by a 6. If the order is unimportant i.e. if a 3 followed by a 6 or a 6 followed by a 3 is acceptable then the probability is doubled, i.e.

$$\text{where } P \ (3 \text{ and } 6) = P(3 \text{ followed by } 6) + P \ (6 \text{ followed by } 3)$$

$$\text{then } P(3 \text{ and } 6) = \frac{1}{36} + \frac{1}{36} = \frac{1}{18}$$

c) *Conditional probability*

This is the probability associated with combinations of events given that some prior result has already been achieved with one of them.

When the events are independent of one another, then the conditional probability is the same as the probability of the remaining event.

Example 3

The probability of throwing a total of 10 with 2 dice, before the events, is $\frac{1}{12}$; i.e.

$$P(5 \text{ and } 5) = \frac{1}{6} \times \frac{1}{6} = \frac{1}{36}$$

$$P(6 \text{ and } 4) = \frac{1}{6} \times \frac{1}{6} = \frac{1}{36}$$

$$P(4 \text{ and } 6) = \frac{1}{6} \times \frac{1}{6} = \frac{1}{36}$$

$$\frac{1}{36} + \frac{1}{36} + \frac{1}{36} = \frac{3}{36}$$

$$= \frac{1}{12}$$

but if one die has been thrown and shows a 4 then the conditional probability is the probability of throwing a 6 with the other die, which is a probability of $\frac{1}{6}$).

Conditional probability is usually expressed in the form:

$$P(x \mid y)$$

which means

Probability of *x* *given* that *y* has occurred.

Example 4

From past experience it is known that a machine is set up correctly on 90% of occasions. If the machine is set up correctly then the conditional probability of a good part is 95% but if the machine is not set up correctly then the conditional probability of a good part is only 30%.

On a particular day the machine is set up and the first component produced and found to be good. What is the probability that the machine is set up correctly?

Solution

This is displayed in the form of a probability tree or diagram as follows:

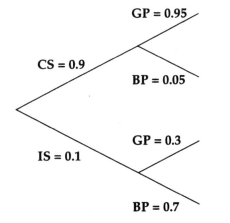

$\therefore$ CSGP $= 0.9 \times 0.95 = 0.855$
CSBP $= 0.9 \times 0.05 = 0.045$
ISGP $= 0.1 \times 0.3 \;\;\; = 0.03$
ISBP $= 0.1 \times 0.7 \;\;\; = \underline{0.07}$
$= \underline{1.00}$

CS $=$ Correct set up
IS $\;=$ Incorrect set up
GP $=$ Good Part
BP $=$ Bad Part

11

∴ Probability of getting a good part = **CSGP** + **ISGP** = 0.855 + 0.03 = 0.885

∴ Probability that machine is correctly set up after getting good part is

$$\frac{\textbf{CSGP}}{\textbf{CSGP} + \textbf{ISGP}} = \frac{0.855}{0.885} = 0.966$$

Good parts may be produced when the machine is correctly set up and also when it is incorrectly set up. In 1000 trials there will be 855 occasions when it is correctly set up and good parts are produced (shown as **CSGP** above) and 30 occasions when it is incorrectly set up yet good parts are produced (shown as **ISGP** above). Thus if a good part is produced we can state that there are 855 occasions out of 885 when the good part is from a correctly set up machine.

Further applications of probability

5. The rules described so far may be conveniently summarised and developed using examples.

Example 5

There are 100 students in a first year college intake. 36 are male and are studying accounting, 9 are male and not studying accounting, 42 are female and studying accounting, 13 are female and are not studying accounting.

Use these data to deduce probabilities concerning a student drawn at random.

Solution

The most effective way to handle these data is to draw up a table:

	Accounting A	Not Accounting $\bar{\text{A}}$	Total
Male M	36	9	45
Female F	42	13	55
Total	78	22	100

1. $P(M)$ $= \dfrac{45}{100} = \textbf{0.45}$

2. $P(F)$ $= \dfrac{55}{100} = \textbf{0.55}$

3. $P(A)$ $= \dfrac{78}{100} = \textbf{0.78}$

4. $P(\bar{A})$ $= \dfrac{22}{100} = \textbf{0.22}$

5. $P(M \text{ and } A) = \dfrac{36}{100} = \textbf{0.36} = P(A \text{ and } M)$

This is so because the same cells in the table are involved.

6. $P(M \text{ and } \bar{A}) = \dfrac{9}{100} = 0.09$

7. $P(F \text{ and } A) = \dfrac{42}{100} = 0.42$

8. $P(F \text{ and } \bar{A}) = \dfrac{13}{100} = 0.13$

The probabilities 1 to 4 may be expressed in terms of probabilities 5 to 8.

9. $P(M)$ $= P(M \text{ and } A) + P(M \text{ and } \bar{A})$
               $= 0.36 + 0.09$
               $= 0.45$

10. $P(F)$ $= P(F \text{ and } A) + P(F \text{ and } \bar{A})$
               $= 0.42 + 0.13$
               $= 0.55$

11. $P(A)$ $= P(A \text{ and } M) + P(A \text{ and } F)$
               $= 0.36 + 0.42$
               $= 0.78$

12. $P(\bar{A})$ $= P(\bar{A} \text{ and } M) + P(\bar{A} \text{ and } F)$
               $= 0.09 + 0.13$
               $= 0.22$

Now let the student be a male. This changes the problem because the female section of the table disappears.

Calculate the probability that the student is studying accounting. **P(A)** is now no longer appropriate, for the required probability is **P(A|M)**, which is the probability that the student is studying accounting given he is male.

$$P(A|M) = \frac{36}{45} = 0.80$$

$$= \frac{P(A \text{ and } M)}{P(M)}$$

$$\therefore P(A \text{ and } M) = P(M) \times P(A|M)$$
$$\therefore P(A \text{ and } M) = 0.45 \times 0.80$$
$$= 0.36$$

Now let the student be studying accounting. Calculate the probability that the student is male. The required probability is **P(M|A)**.

$$P(M|A) = \frac{36}{78} = 0.462$$

13

$$= \frac{P(A \text{ and } M)}{P(A)}$$

therefore $P(A \text{ and } M) = P(A) \times P(M|A)$

Note that $P(A|M) \neq P(M|A)$, but that there is a relation between the two through $P(A \text{ and } M)$ or $P(M \text{ and } A)$.

$$P(M) \times P(A|M) = P(A \text{ and } M) = P(A) \times P(M|A)$$

$$P(M) \times P(A|M) = P(A) \times P(M|A)$$

thus

$$P(A|M) \quad = \frac{P(A)(M|A)}{P(M)}$$

and

$$P(M|A) \quad = \frac{P(M) \times P(A|M)}{P(A)}$$

Example 6

A company has three production sections S_1, S_2, and S_3 which contribute 40%, 35% and 25%, respectively, to total output. The following percentages of faulty units have been observed:

S_1	2%	(0.02)
S_2	3%	(0.03)
S_3	4%	(0.04)

There is a final check before output is despatched. Calculate the probability that a unit found faulty at this check has come from Section 1 (S_1).

Solution

Let F represent a unit which has been found to be faulty.

Let $P(S_1)$ = probability that a unit chosen at random comes from S_1
Let $P(S_2)$ = probability that a unit chosen at random comes from S_2
Let $P(S_3)$ = probability that a unit chosen at random comes from S_3

$$P(S_1) = 0.40$$

$$P(S_2) = 0.35$$

$$P(S_3) = \frac{0.25}{1.00}$$

The percentages of faulty units are as follows:

$$P(F|S_1) = 0.02$$

$$P(F|S_2) = 0.03$$

$$P(F|S_3) = 0.04$$

The required probability may be expressed as

$$P(S_1|F)$$

The unknown probability is $P(F)$ to be slotted into the formula

$$P(S_1|F) = \frac{P(S_1) \times P(F|S_1)}{P(F)}$$

Note that

$$P(S_2|F) = \frac{P(S_2) \times P(F|S_2)}{P(F)}$$

$$P(S_3|F) = \frac{P(S_3) \times P(F|S_3)}{P(F)}$$

The faulty part can only have come from S_1 or S_2 or S_3 and so

$$P(S_1|F) + P(S_2|F) + P(S_3|F) = 1.0$$

Since $P(F)$ is a denominator and the sum equals unity then the expression

$$P(S_1) \times P(F|S_1) + P(S_2) \times P(F|S_2) + P(S_3) \times P(F|S_3)$$

must be equal to $P(F)$.

Thus
$$P(F) = (0.4 \times 0.02) + (0.35 \times 0.03) + (0.25 \times 0.04)$$

$$= 0.0080 + 0.0105 + 0.0100$$

$$= 0.0285$$

Substitution into $P(S_1|F) = \dfrac{P(S_1) \times P(F|S_1)}{P(F)}$

gives
$$P(S_1|F) = \frac{0.4 \times 0.02}{0.0285}$$

$$= 0.2807$$

Also
$$P(S_2|F) = \frac{0.35 \times 0.03}{0.0285}$$

$$= 0.3684$$

and
$$P(S_3|F) = \frac{0.25 \times 0.04}{0.0285}$$

$$= 0.3509$$

15

Note 0.2807 + 0.3684 + 0.3509 = 1.000

Thus if a faulty unit is chosen at random then the probability that it has come from S_1 is **0.2807**.

Bayes' rule

6. The principle behind the above result is known as *Bayes' Rule* and it is interesting to note that the process involves working backwards from effect to cause. In this case a faulty item or unit **was** found and it was possible to work out the probability that it **had** come from S_1.

Bayes' Rule is frequently used in the analysis of decisions using decision trees (see Chapter 3) where information is given in the form of conditional probabilities and the reverse of these probabilities must be found.

The general form of *Bayes' Rule* or *theorem* is:

$$P(A|B) = \frac{P(A) \times P(B|A)}{P(B)}$$

Permutations

7. A permutation is an ordered arrangement of items. Thus **AB** is a different permutation to **BA** even though the individual two items **A** and **B** are the same – they are in a different order. Frequently an analyst has to work out the number of ways that an event can occur in order to calculate a required probability. If each possibility has to be listed this can be time consuming and error prone. Formulae can be used to make the task quick and easy.

> **Example 7**
>
> A restaurant offers a choice of 3 starters, 4 main courses and 3 sweets. How many different meals are available?

The solution is:

$$3 \times 4 \times 3 = 36 \text{ different meals.}$$

This basic idea may be developed further.

> **Example 8**
>
> A transport manager has to plan routes for his drivers. There are 3 deliveries to be made to customers X, Y, and Z. How many routes can be followed?

Solution

In this case the routes can be listed:

Route	1	X	Y	Z
- - -	2	X	Z	Y
- - -	3	Y	X	Z
- - -	4	Y	Z	X
- - -	5	Z	X	Y
- - -	6	Z	Y	X

This covers all possibilities and so the number of routes is 6. Note if there were 6 customers the number of routes increases dramatically to 720. The question is how to work out the number of routes without listing them.

The first delivery may be chosen in 3 ways (X, Y, or Z). Let it be X. The second delivery may be made in two ways (Y or Z). Let it be Y. The third delivery can be made in only one way! The total number of routes is thus

$$3 \times 2 \times 1 = 6 \text{ ways.}$$

If there are 6 customers then the number of routes is:

$$6 \times 5 \times 4 \times 3 \times 2 \times 1 = 720$$

These arrangements are known as *permutations*.

Note: $3 \times 2 \times 1$ is more usually written as **3 !**, pronounced 'three factorial'.

In general, a permutation of **n** objects is written as:

$$n! = n(n-1)(n-2)(n-3) \ldots$$

Note that by definition **1 ! = 1** and **0 ! = 1**

Permutations of groups

8. The analyst may not wish to arrange all the items, but groups of items from within a list.

Example 9

A company has four training officers A, B, C, D, and it is required to assign one to each of two training sections. In how many different ways may the four officers be assigned to the two sections, X and Y?

Solution

The assignments are:

	X	Y
1	A	B
2	A	C
3	A	D
4	B	A
5	B	C
6	B	D

7	C	A
8	C	B
9	C	D
10	D	A
11	D	B
12	D	C

The first assignment may be made in 4 ways. For the second assignment there are only 3 officers left. There are no more assignments and so the number of permutations is:

$$4 \times 3^* = 12$$

Now let there be 6 officers and 3 sections

1st assignment = 6 ways; 2nd assignment = 5 ways; 3rd assignment = 4 ways

The number of permutations is:

$$6 \times 5 \times 4^* = 120$$

* *Note* the last figure in each case. Both are obtained in the same way.

Officers	Sections		
n	r		
4	−2	+1	= 3
6	−3	+1	= 4
(n	−r	+1)	

A convenient way to work out the number of permutations of n items r at a time is to use the following formula:

$$n(n-1)(n-2) \ldots (n-r+1)$$

Note that there is an alternative calculation:

$$\frac{n!}{(n-r)!}$$

For the case with 4 officers and 2 sections:

$$\frac{4!}{(4-2)!} = \frac{4 \times 3 \times 2 \times 1}{2 \times 1} = 12 \text{ ways}$$

Permutations with similar items

9. There will be occasions when the items to be arranged will not all be different. If this is the case then the number of permutations will be reduced.

Example 10

If there are 3 items x, y, z then there are

$3 \times 2 \times 1 = 6$ permutations, and these are

xyz   yxz   zxy

xzy   yzx   zyx

If however $x = y = p$ the arrangement becomes

ppz	ppz	zpp
ppz	ppz	zpp

The number of permutations is reduced to 3. The version of the permutation formula that will accommodate this case is

$$\frac{n!}{n_1! \times n_2! \times n_3!...n_x!}$$

where n_1 of the items are of one kind and n_2 of the items are of another type and so on up to x types. In the present case, $n_1 = 2$ and $n_2 = 1$, giving

$$\frac{3!}{2! \times 1!} = \frac{3 \times 2 \times 1}{2 \times 1 \times 1} = 3 \text{ ways}$$

Occasionally it is necessary to work out how many ways a group of people can be arranged around a table.

Example 11

A, B, C, D are to sit around a conference table: in how many ways may this be arranged?

Solution

Fix A in one place. This leaves three who may be placed using 3 ! as below:

```
      A                    A                    A
  B       D        B           C        C           D
      C                    D                    B

      A                    A                    A
  C       B        D           B        D           C
      D                    C                    B
```

There are thus 6 ways. Note that the 6 ways comes from 3 ! and 3 ! and 3 ! = (4 − 1)! and 4 is the number of people to arrange. Thus 7 people around a table can be arranged in 6! ways.

$$6! = 6 \times 5 \times 4 \times 3 \times 2 \times 1 = 720 \text{ ways}$$

Combinations

10. There will be occasions when selections will be made where the order does not matter meaning that the arrangement **a, b** will be same as **b, a**. This is known as a **combination**.

Example 12

6 apprentices A, B, C, D, E, F have to be paired into two's for an exercise. In how many ways may this be done? (A, B is the same as B, A, etc.)

Solution

$$
\begin{array}{lllll}
\text{AB} & \text{BC} & \text{CD} & \text{DE} & \text{EF} \\
\text{AC} & \text{BD} & \text{CE} & \text{DF} & \\
\text{AD} & \text{BE} & \text{CF} & & \\
\text{AE} & \text{BF} & & & \\
\text{AF} & & & & \\
\end{array}
$$

There are 15 ways.

It is not necessary to list all the ways. The following formula can be used for combinations:

$$\frac{n!}{r!(n-r)!}$$

where **n** is the total number of items and **r** is the number of items per arrangement. The above example works out as follows:

$$\frac{6!}{2!(6-2)!} = \frac{6 \times 5 \times 4 \times 3 \times 2 \times 1}{2 \times 1 \times 4 \times 3 \times 2 \times 1} = 15 \text{ ways}$$

This type of arrangement is known as a *combination* of **n** items **r** at a time and is denoted

$$^{n}C_{r}$$

$$\text{or} \quad \binom{n}{r}$$

This formula will be encountered again in Chapter 4 when dealing with the Binomial Distribution.

Note: A combination is a selection of items where the *order of sequence does not matter*, i.e.

$$\textbf{AB} = \textbf{BA} \text{ (because the same two items are selected)}.$$

A *permutation* is a selection where the order does matter, i.e.

AB ≠ BA

(although the same two items are present they are in a different sequence).

Expected value

11. The basic rules of probability so far outlined help to quantify the options open to management and thereby may help in coming to a decision. Where the options have values (so much profit, contribution, etc.) as well as probabilities, the concept of **expected value** is often used. The expected value of an event is its probability times the outcome or value of the event over a series of trials.

Example 13

Two projects are being considered and it is required to calculate the expected value of each project. The project data have been estimated as follows:

	Project A				Project B			
	£		p		EV	£		p		EV
Optimistic outcome	6,000	×	0.2	=	1,200	6,500	×	0.1	=	650
Most likely outcome	3,500	×	0.5	=	1,750	4,000	×	0.6	=	2,400
Pessimistic outcome	2,500	×	0.3	=	750	1,000	×	0.3	=	300
Project EV				**£3,700**					**£3,350**

EV = Expected Value

Solution

On the basis of Expected Value, Project A would be preferred as it has the higher value.

Notes

a) Although the EV of A is £3700, strictly this value would only be achieved in the long run over many similar decisions – extremely unlikely circumstances!

b) If the project was implemented, any of the three outcomes could occur, with the values stated.

Expected value advantages and disadvantages

12. Expected Value is a useful summarising technique, but suffers from similar advantages and disadvantages to all averaging methods.

Advantages:

a) Simple to understand and calculate

b) Represents whole distribution by a single figure

c) Arithmetically takes account of the expected variabilities of all outcomes.

Disadvantages:

a) By representing the whole distribution with a single figure it ignores the other characteristics of the distribution, e.g. the range and skewness.

b) Makes the assumption that the decision maker is risk neutral, i.e. he would rank equally the following two distributions:

	£	p	
Pessimistic Outcome	18,000	0.25	
Most likely Outcome	20,000	0.5	**EV = £20,000**
Optimistic Outcome	22,000	0.25	

and

	£	p	
Pessimistic Outcome	6,000	0.2	
Most likely Outcome	18,000	0.6	**EV = £20,000**
Optimistic Outcome	40,000	0.2	

It is of course unlikely that any decision maker would rank them equally due to his personal attitude to risk.

Although it appears to be widely used for the purpose, expected value is not particularly well suited to one-off decisions. Expected value can strictly only be interpreted as the value that would be obtained if a large number of similar decisions were taken with the same ranges of outcome and associated probabilities. Hardly a typical business situation!

Optimisation of levels of activity under conditions of uncertainty

13. Expected Value principles can be used to calculate the maximum stock or profit level when demand is subject to random variations over a period.

Example 14

A distributor buys perishable articles for £2 per item and sells them at £5. Demand per day is uncertain and items unsold at the end of the day represent a write off because of perishability. If he understocks he loses profit he could have made.

A 300-day record of past activity is as follows:

Daily demand (units)	No. of days	P
10	30	0.1
11	60	0.2
12	120	0.4
13	90	0.3
	300	1.0

What level of stock should be held from day to day to maximise profit?

Solution

It is necessary to calculate the Conditional Profit (**CP**) and Expected Profit (**EP**). **CP** = profit that could be made at any particular conjunction of stock and demand, e.g. if 13 articles were bought and demand was 10 then:

$$\textbf{CP} = (\textbf{10} \times \textbf{5}) - (\textbf{13} \times \textbf{2}) = \textbf{£24}$$

$$\textbf{EP} = \textbf{CP} \times \textbf{probability of the demand}$$

e.g. the **CP** above is £24 and p (demand = 10) = 0.1

$$\therefore \textbf{EP} = \textbf{£24} \times \textbf{0.1} = \textbf{£2.40}$$

Conditional and expected profit table. Table 1

Demand	p	Stock Options							
		10		11		12		13	
		CP £	EP £	CP £	EP £	CP £	EP £	CP £	EP £
10	0.1	30	3	28	2.8	26	2.6	24	2.4
11	0.2	30	6	33	6.6	31	6.2	29	5.8
12	0.4	30	12	33	13.2	36	14.4	34	13.6
13	0.3	30	9	33	9.9	36	10.8	39	11.7
	1.0		30		32.5		*34.0		33.5

* Optimum

The optimum stock position, given the pattern of demand, is to stock **12 units per day**.

Solution by marginal probability formula

14. An alternative approach to solving problems such as Example 14 above is to work out the marginal profit, **MP**, i.e. the profit from selling one more unit, and the marginal loss, **ML**, i.e. the loss from not selling the marginal unit, and to calculate the relationship between **MP** and **ML** in terms of probability.

Example 15

Using the same data as in Example 14, solve using marginal profitability and marginal loss relationships.

Solution

From Example 14

$$\textbf{MP} = \textbf{£3 and ML} = \textbf{£2}$$

It follows that additional items will be stocked whilst the following relationship holds:

p(making additonal sale) × **MP** > **p(not making additional sale)** × **ML**

23

If **P** denotes the probability of making the additional sale then **(1 − P) = probability of not making the sale**.

$$\text{at Break Even Point } \mathbf{P(MP)} = \mathbf{(1 - P)\ (ML)}$$

$$\mathbf{P(MP) + P(ML) = ML}$$

$$\mathbf{P(MP + ML) = ML}$$

$$\mathbf{P = \frac{ML}{MP + ML}}$$

Inserting the data from Example 14 we obtain

$$\mathbf{P = \frac{2}{3 + 2}}$$

$$= \mathbf{0.4} \text{ i.e. the probability at Break Even Point}$$

This probability is compared with the probability of demand at the various levels, the optimum position being the highest demand with a probability greater than **P**, i.e.

Demand	Probability	
Greater than or equal to 10 units	1.00	
Greater than or equal to 11 units	0.9	
Greater than or equal to 12 units	0.7	Break Even probability = **0.4**
Greater than or equal to 13 units	0.3	

12 items is the most profitable stock level.

Value of perfect information

15. Assume that the distributor in Example 14 could buy market research information which was perfect, i.e. it would enable him to forecast the exact demand on any day so that he could stock up accordingly. How much would the distributor be prepared to pay for such information? To solve this type of problem, the profit with perfect information is compared with the optimum **EP** from Table 1.

Profit with perfect information

				£
When demand is 10, stock 10	Profit	= (10 × £3) × 0.1	=	3.0
When demand is 11, stock 11	Profit	= (11 × £3) × 0.2	=	6.6
When demand is 12, stock 12	Profit	= (12 × £3) × 0.4	=	14.4
When demand is 13, stock 13	Profit	= (13 × £3) × 0.3	=	11.7
				35.7

As the **EP** from Table 1 was £34, the distributor could pay up to **£1.70** (£35.7 − 34) for the information.

The above illustrates the general rule for calculating the value of perfect information. First, calculate the value of the outcome using expected value and then the value assuming perfect knowledge and deduct the one from the other.

In practice, of course, information is not perfect, yet may still be valuable. Organisations pay large sums for information which is reasonably accurate and which will enable them to improve their decision making. Market research opinion surveys, test marketing and so on produce imperfect information, i.e. it is not perfectly accurate, yet it is still accurate enough to be relied upon. Under certain conditions it is possible to estimate the value of imperfect information.

Valuing imperfect information

16. Providing that there is some indication of how accurate the information is likely to be, it is possible to calculate the worth of imperfect information. For example, if a firm had used a market research agency on numerous occasions it would be able to make an assessment of the accuracy of their information.

The method used to value imperfect information uses posterior probabilities based on Bayes' Theorem and is demonstrated by the following example.

Example 16

Theta Limited are considering the launch of a new product, the Gamma.

Various prior estimates of outcomes have been made as follows and the expected values calculated.

Market state	Probability	Profit/(Loss)	Expected value
Good	0.2	60,000	12,000
Average	0.6	40,000	24,000
Bad	0.2	(40,000)	(8,000)
			EV = **£28,000**

In order to have more information on which to base their decision the management are considering whether to commission a market research survey at a cost of £1,000. The agency concerned produce reasonably accurate, but not perfect, information and Theta's Trade Association provided the following information about the Agency's performance.

Likely actual market state	Market research agency survey findings		
	Good	Average	Bad
Good	60%	30%	–
Average	40%	50%	10%
Bad	–	20%	90%

Solution

The first stage is to calculate the posterior probabilities, i.e. the probability of the market state being good, average or bad after the market research survey results have become available.

Prior probability	Market state	Market research results		Posterior probability
		State	Probability	
0.2	Good	Good	0.6	GG = 0.12*
		Average	0.4	GA = 0.08
0.6	Average	Good	0.3	AG = 0.18
		Average	0.5	AA = 0.30
		Bad	0.2	AB = 0.12
0.2	Bad	Average	0.1	BA = 0.02
		Bad	0.9	BB = 0.18
				1.00

* GG = 0.12 means that the probability of the market being good and the survey predicting a good state is $0.2 \times 0.6 = 0.12$.

From the table survey probabilities can be summarised thus:

- P (survey will show 'good')　　 = **GG** + **AG** = 0.12 + 0.18 = **0.3**
- P (survey will show 'average') = **GA** + **AA** + **BA** = 0.08 + 0.30 + 0.02 = **0.4**
- P (survey will show 'bad')　　 = **AB** + **BB** = 0.12 + 0.18 = **0.3**

Assuming that the product will be launched if the survey predicts a good or average market, and not launched if the bad state is predicted, the following table can be prepared:

Survey results	Decision	Actual market	Posterior probabilities	Profit (Loss) £	EV of Profits (Losses) £
Good	Launch	Good	0.12	60,000	7,200
		Average	0.18	40,000	7,200
Average	Launch	Good	0.08	60,000	4,800
		Average	0.30	40,000	12,000
		Bad	0.02	(40,000)	(800)
Bad	Do not launch		0.3	–	–
			1.00		£30,400

∴ value of imperfect information is

£30,400 − £28,000 = £2,400

As the survey cost is £1,000 it would appear to be worthwhile. However, there is only a relatively small gain (£2,400) from having more information and because of estimation problems it is possible that management would decide not to commission the survey.

Alternative decision rules

17.　The rule so far covered in this chapter is to choose the alternative which maximises the expected value. This is the most commonly encountered decision

2 Probability and decision making

rule and is the one which should be used unless there are clear instructions to the contrary. However, alternative rules do exist and these include:

- the *maximin* rule
- the *maximax* rule
- the *minimax regret* rule
- the *Hurwicz* rule or criterion

These rules are illustrated using the following payoff table showing potential profits and losses which are expected to arise from launching various products in three market conditions.

Payoff table in £'000s. Table 2

	Boom conditions	Steady state	Recession
Product A	+8	1	−10
Product B	−2	+6	+12
Product C	+16	0	−26

The probabilities are Boom 0.6, Steady State 0.3 and Recession 0.1 so that the expected values are

$$\text{Product A} = (0.6 \times 8) \quad + (0.3 \times 1) + (0.1 \times -10) = \textbf{4.1}$$

$$\text{Product B} = (0.6 \times -2) + (0.3 \times 6) + (0.1 \times 12) \quad = \textbf{1.8}$$

$$\text{Product C} = (0.6 \times 16) \quad + (0.3 \times 0) + (0.1 \times -26) = \textbf{7}$$

So using the expected value rule the ranking would be **C, A, B**.

What are the rankings using the alternatives?

- *Maximin* the 'best of the worst'

  This is a cautious decision rule based on maximising the minimum return that can occur. The worst losses are:

A	−10
B	−2
C	−26

  ∴ Ranking using the *Maximin* rule is **B, A, C**.

- *Maximax* the 'best of the best'

  This is an optimistic rule and maximises the maximum that can be gained. The maximum gains are:

A	+8
B	+12
C	+16

  ∴ Ranking using the *Maximax* rule is **C, B, A**.

- *Minimax regret*

  This decision rules seeks to 'minimise the maximum regret' that there would

2 Probability and decision making

be from choosing a particular strategy. To see this clearly it is necessary to construct a regret table based on the payoff table, Table 2. The regret is the opportunity loss from taking one decision given that a certain contingency occurs; in our Example whether there is boom, steady state, or recession.

Regret table in £'000s. Table 3

	Boom conditions	Steady state	Recession
Product A	8	5	22
Product B	18	0	0
Product C	0	6	38

Note: The above opportunity losses are calculated by setting the best position under any state to zero and than calculating the amount of shortfall there is by not being at that position. For example, if there is a recession Product B gains +12 but if Product A had been chosen there is a loss of -10 making a total shortfall, as compared with B, of 22, which is the opportunity loss.

The maximum regrets are:

$$\begin{array}{ll} \text{A} & 22 \\ \text{B} & 18 \\ \text{C} & 38 \end{array}$$

∴ the ranking using the *Minimax regret* rule is **B, A, C**.

- *Hurwicz criterion*

The Hurwicz criterion is a weighted average of the best and worst payoffs of each action. This may be considered more realistic than some of the earlier rules (Maximin, Maximax, etc.), which are based solely on extreme payoffs. It is calculated thus:

Weighted payoff $= \alpha \times$ worst payoff $+ (1 - \alpha) \times$ best payoff

α is a number between 0 and 1, sometimes called the pessimism-optimism index. The value of α chosen reflects the decision maker's attitude to the risk of poor payoffs and the chance of good payoffs. The option with the highest weighted payoff is chosen.

Example 17

Assume an α value of 0.7 and the values from the original payoff table

	Boom £'000s	Steady State £'000s	Recession £'000s	Worst £'000s	Best £'000s	Weighted Average £'000s
Product A	+8	1	−10	−10	+8	−4.6*
Product B	−2	+6	+12	−2	+12	+2.2
Product C	+16	0	−26	−26	+16	−13.4

Thus the ranking using the *Hurwicz criterion*, with an α value of 0.7, is **B, A, C**.

*Each weighted average is calculated the same way; for example,

$$0.7 \times -10 + 0.3 \times 8 = -4.6$$

If an α value over 0.5 is chosen, as above, this shows a risk avoidance attitude as it gives most weight to the worst result. If the decision maker was a risk seeker some value below 0.5 would be chosen, and this could result in a different ranking of the products.

For example, if an α value of 0.2 was chosen, the weighted average payoffs would be:

Product A	= +4.4
Product B	= +9.2
Product C	= +7.6

and the ranking B, C, A

Opportunity loss and expected value

18.   As a loss is the negative aspect of gain it is to be expected that opportunity loss and expected value are related. This is indeed so and the opportunity losses multiplied by the probabilities, i.e. the expected opportunity loss (EOL) can be used to arrive at the same ranking as the expected value (EV) rule except that where the maximum EV is chosen, the minimum EOL is required.

∴ *Minimising* **EOL** gives the same decision as *maximising* **EV**.

For example the EOLs of Table 3 are:

A	$(0.6 \times 8) + (0.3 \times 5) + (0.1 \times 22)$	= 8.5
B	$(0.6 \times 18) + (0.3 \times 0) + (0.1 \times 0)$	= 10.8
C	$(0.6 \times 0) + (0.3 \times 6) + (0.1 \times 38)$	= 5.6

∴ ranking in order of minimum EOL gives C, A, B which is identical to the ranking given by the expected value method.

Summary

19.   a)   Objective probabilities are rarely available in business decision making so that subjective probabilities, i.e. the quantification of judgement, are used.

b)   The two basic rules of probability are the multiplication rule (AND) and the addition rule (OR)

c)   Bayes' Rule finds a conditional probability **(A|B)** given its inverse **(B|A)**

d)   The general form of Bayes' Rule is:

$$\mathbf{P(A|B)} = \frac{\mathbf{P(A)} \times \mathbf{P(B|A)}}{\mathbf{P(B)}}$$

e)   A permutation is an ordered arrangement.

f)   The number of permutations is written as $^{n}\mathbf{P_r}$ and is calculated thus:

$$^nP_r = \frac{n!}{(n-r)!}$$

g) A combination is an arrangement without regard to order.

h) The number of combinations is written as ^nC_r and is calculated thus:

$$^nC_r = \frac{n!}{r!(n-r)!}$$

i) Expected Value **(EV)** has many uses in decision making and is calculated by multiplying the value of the outcome(s) by the probability. In general the option with the highest **EV** would be chosen.

j) Although useful, **EV** ignores the range and skewness of distributions, so that it needs to be used with judgement.

k) The value of perfect information is the difference between the profits obtainable from having perfect knowledge and the expected value of outcomes without perfect knowledge.

l) Alternative decision rules are *Maximin, Maximax, Minimax Regret* and the *Hurwicz criterion*.

m) Opportunity loss is the shortfall between the best possible value in the state and the actual result.

n) Using the minimum expected opportunity loss as the decision criterion gives the same ranking as expected value.

Points to note

20. a) The use of probability concepts and expected value is widespread in professional and college examinations so that this chapter should be studied carefully.

b) Questions involving probability may appear in any part of the syllabus (inventory control, investment appraisal, network analysis, etc.).

c) The nature of probability is essentially that of assuming repetitive trials or occurrences. Accordingly the use of probability in one-off decisions needs to be carefully considered.

d) The type of organisation, its objectives, style and ethos, are all involved in considering the appropriate decision criteria to use. A Public Sector organisation may be precluded from speculative investments and may have to fulfill statutory obligations at minimum cost. In such quite typical circumstances the use of the minimax criterion is likely to be appropriate whilst other types of organisations may tend to adopt a maximax rule.

Self review questions *Numbers in brackets refer to paragraph numbers*

1. Distinguish between objective and subjective probabilities. (3)
2. Define the Multiplication and Addition rules of Probability. (4)
3. What is the principle of Bayes' Rule? (6)

4. Distinguish between a permutation and a combination. (7–10)

5. How is Expected Value calculated? (11)

6. What are the advantages and disadvantages of using Expected Value as a decision criterion? (12)

7. Assuming that data on past demand and probabilities are available how might a value be imputed for receiving perfect information (14) or imperfect information? (16).

8. Define the minimax and maximax decision rules. (17)

9. How is the regret table calculated? (17)

10. How are opportunity loss and expected value related? (18)

Exercises with answers

1. A card is drawn from a shuffled pack of 52 cards. What is the probability of drawing a king or a heart?

2. Records of service requests at a garage and their probabilities are as follows:

Daily demand	Probability
5	0.3
6	0.7

Daily demand is independent.

What is probability that over a two day period the number of requests will be:

   a) 10 requests?

   b) 11 requests?

   c) 12 requests?

3. Analysis of questionnaire completed by holiday makers showed that 0.75 classified their holiday as good at Costa Lotta. The probability of hot weather in the resort is 0.6. If the probability of regarding the holiday as good given hot weather is 0.9, what is the probability that there was hot weather if a holiday maker considers his holiday good?

Answers to exercises

1.
$$P(\text{King}) = \frac{4}{52}$$

$$P(\text{heart}) = \frac{13}{52}$$

$$P(\text{King and heart}) = \frac{1}{52}$$

2 Probability and decision making

$$\therefore \mathbf{P}(\text{King or heart}) = \frac{4}{52} + \frac{13}{52} - \frac{1}{52} = \frac{16}{52} = \frac{4}{13}$$

Note: The probability of a king **and** a Heart (ie $^1/_{52}$) has to be deducted because we are trying to find the probability of drawing a king **or** a heart.

2.

10 Requests:	P(5 AND 5) = 0.3 × 0.3	= 0.09
11 Requests:	$\begin{cases} \text{P(5 AND 6)} = 0.3 \times 0.7 = 0.21\} \\ \text{P(6 AND 5)} = 0.7 \times 0.3 = 0.21 \end{cases}$	= 0.42
12 Requests:	P(6 AND 6) = 0.7 × 0.7	= 0.49
		= **1.00**

3. This exercise uses Bayes Theorem which has the general form

$$\mathbf{P(A|B)} = \frac{\mathbf{P(A)} \times \mathbf{P(B|A)}}{\mathbf{P(B)}}$$

Let $\mathbf{H}$ = hot weather

$\mathbf{G}$ = Good holiday

then $\mathbf{P(G)} = 0.75$ $\mathbf{P(H)} = 0.6$ and $\mathbf{P(G|H)} = 0.9$

$$\therefore \mathbf{P(H|G)} = \frac{\mathbf{P(H)} \times \mathbf{P(G|H)}}{\mathbf{P(G)}}$$

$$= \frac{0.6 \times 0.9}{0.75} = \underline{\underline{\mathbf{0.72}}}$$

3 Decision trees

Objectives

1. After studying this chapter you will
    - know what is meant by a Decision Tree;
    - be able to describe Decision Nodes and Outcome Nodes;
    - understand how to draw Decision Trees using the Forward Pass;
    - know that the Outcome Values are calculated using the Backward Pass;
    - be able to incorporate Bayes' Theorem into a Decision Tree.

Decision trees – definition

2. A pictorial method of showing a sequence of inter-related decisions and outcomes. All the possible choices are shown on the tree as branches and the possible outcomes as subsidiary branches. In summary, a decision tree shows: the decision points, the outcomes (usually dependent on probabilities) and the outcome values. Frequently, but not always, the tree is evaluated using Expected Values.

Structure of decision trees

3. The structure and typical components of a decision tree are shown in the following illustration:

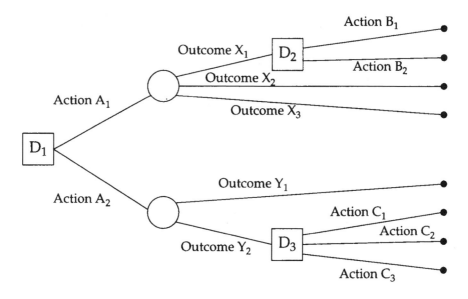

Figure 3/1

Notes:

a) It will be seen there are two types of nodes, Decision nodes depicted by squares and Outcome nodes depicted by circles.

b) The Decision nodes are points where a choice exists between alternatives and a managerial decision is made based on estimates and calculations of the returns expected.

c) The Outcome nodes are points where the events depend on probabilities, e.g. assume that Action A_1 in Figure 3/1 was – Build branch factory – then outcomes X_1, X_2, and X_3 could represent various possible sales; high, medium, and low, each with an estimated probability.

Drawing decision trees

4. The following example will be used to illustrate how to draw decision trees and evaluate the outcomes.

Example 1

A company is considering whether to launch a new product. The success of the idea depends on the ability of a competitor to bring out a competing product (estimated at 60%) and the relationship of the competitor's price to the firm's price.

Table A shows the profits for each price range that could be set by the company related to the possible competing prices.

Table A

Profits in £'000s

If Company's Price is	If Competitor's Price is			Profit if no Competitor
	Low	Medium	High	
Low	30	42	45	50
Medium	34	45	49	70
High	10	30	53	90

The Company must set its price first because its product will be on the market earlier so that the competitor will be able to react to the price. Estimates of the probability of a competitor's price are shown in Table B.

Table B

If Company's Prices are	Competitor's Price Expected to be		
	Low	Medium	High
Low	0.8	0.15	0.05
Medium	0.20	0.70	0.10
High	0.05	0.35	0.60

Tasks

a) Draw a decision tree and analyse the problem.

b) Recommend what the company should do.

Solution

The first stage is to draw the tree, starting from the left, showing the various decision points and outcome nodes. Concentrate first on the logic of the problem and don't bother with the values and probabilities to start with.

This results in Figure 3/2, the logic of which should be checked against the problem. (This is known as the *forward pass*).

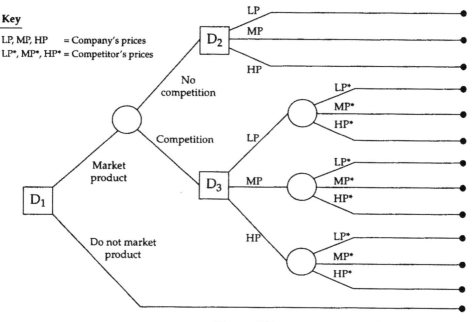

Key

LP, MP, HP = Company's prices
LP*, MP*, HP* = Competitor's prices

Figure 3/2

It will be seen that the tree is a mixture of decision points, D_1, D_2 and D_3 and outcome nodes which reflect whether or not there will be competition and the prices the competition will set.

The decision points can be summarised thus:

at D1: Decide whether to market the product or not.

at D2: Decide the Company's prices in the absence of competition.

at D3: Decide the Company's prices when there is competition.

The next step is to fill in the outcome values and probabilities from Tables A and B in the problem. See Figure 3/3.

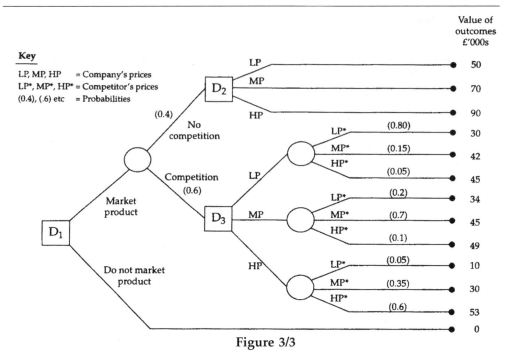

Figure 3/3

The final stage uses the probabilities and outcome values and calculates the Expected Values at various points so that the correct decisions are highlighted. This stage works from **right** to **left** and is known as the *backward pass*. See Figure 3/4.

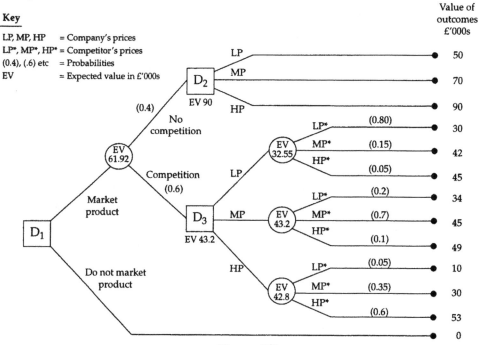

Figure 3/4

Figure 3/4 shows the expected values in the outcome nodes (i.e. probability × values) and at D_2 and D_3 the highest expected values from the subsequent part of the tree indicating the best decisions.

Interpretation and recommendation

Figure 3/4 shows that at $D-1$ the company should decide to market the product (EV 61.92 compared with EV of 0). If there is no competition, the D_2 decision should be to set a High Price (EV 90). If there is competition, at D_3, the company should set a Medium Price as this gives the best EV, i.e. 43.2.

Bayes' theorem and decision trees

5.   The ideas developed in Chapter 2 on Bayes' Theorem may be applied to the analysis of decision trees. Consider the following example.

Example 2

A company with an ageing product range is investigating the launch of a new range. Their business analysts have mapped out several possible scenarios which are given below.

Scenario 1

Continue with old range producing profits declining at 10% per annum on a compounding basis. Last year's profits were £60,000 from this range.

Scenario 2

Introduce a new range *without* any prior market research. If sales are *high*, annual profit is put at £90,000 with a probability which from past data is put at 0.7. If sales are *low*, annual profit is put at £30,000 with a probability of 0.3.

Scenario 3

Introduce a new range with prior market research costing £30,000. The market research will indicate whether future sales are likely to be '*good*' or '*bad*'. If the research indicates '*good*', then the management will spend £35,000 more on capital equipment and this will increase annual profits to £100,000 if sales are actually high. If however sales are actually low, annual profits will drop to £25,000. Should market research indicate '*good*' and should management *not* spend more on promotion then profit levels will be as for *Scenario 2*.

If the research indicates '*bad*' then the management will scale down their expectations to give annual profits of £50,000 when sales are actually low. However, if sales do turn out to be high, profits can only rise to £70,000 because of capacity constraints. Past history of the market research company indicates the following results.

		Actual sales	
		High	Low
Predicted sales levels	Good	*0.8	0.1
	Bad	0.2	0.9

* When Actual Sales were *high* the market research company *had* predicted *good* sales levels 80% of the time and so on.

Use a Time Horizon of 6 years to indicate to the management of the company which scenario they should adopt.

Ignore the time value of money.

Drawing the decision tree

6. As always the first step is to draw the decision tree working from *left* to *right*. The tree is shown in Figure 3/5.

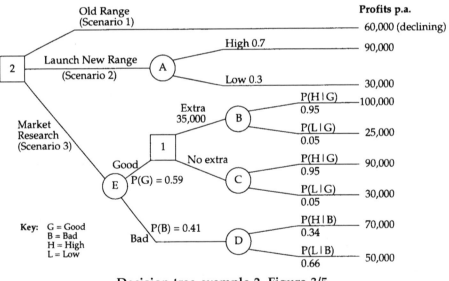

Decision tree example 2. Figure 3/5

The decision points can be summarised thus:

at D1: Decide whether or not to invest £35,000 on capital equipment given that the market research indicates that future sales are likely to be good.

at D2: Decide whether to continue with the old range or to launch the new range without market research or launch the new range with market research.

Note: The various probabilities for Scenario 3 are calculated below.

Calculating the probabilities

7. The decision tree reveals that the following probabilities must be calculated.

$$\left.\begin{matrix} P(G) \\ P(B) \end{matrix}\right\} \text{ for market research}$$

$$\left.\begin{matrix} P(H|G) \\ P(L|G) \\ P(H|B) \\ P(L|B) \end{matrix}\right\} \text{ for sales outcomes}$$

From the data supplied the following probabilities are available.

$$\left.\begin{matrix} P(G|H) & 0.8 \\ P(B|H) & 0.2 \\ P(G|L) & 0.1 \\ P(B|L) & 0.9 \end{matrix}\right\} \text{ These are } \textit{prior} \text{ probabilities}$$

The calculations are shown below:

	High 0.7	Low 0.3
Good	G & H = $P(H) \times P(G|H)$ $0.7 \times 0.8 = \mathbf{0.56}$	G & L = $P(L) \times (PG|L)$ $0.3 \times 0.1 = \mathbf{0.03}$
Bad	B & H = $P(H) \times P(B|H)$ $0.7 \times 0.2 = \mathbf{0.14}$	B & L = $P(L) \times P(B|L)$ $0.3 \times 0.9 = \mathbf{0.27}$

$$P(G) = P(H) \times P(G|H) + P(L) \times P(G|L)$$

$$= P(G\&H) + P(G\&L)$$

$$= 0.56 + 0.03$$

$$= \mathbf{0.59}$$

$$P(B) = P(H) \times P(B|H) + P(L) \times P(B|L)$$

$$= P(B\&H) + P(B\&L)$$

$$= 0.14 + 0.27$$

$$= \mathbf{0.41}$$

$$\textit{Note } P(G) + P(B) = 0.59 + 041 = \mathbf{1.00}$$

From Bayes' Rule,

$$P(H|G) = \frac{P(G|H) \times P(H)}{P(G)}$$

$$= \frac{0.56}{0.59} = 0.95$$

$$P(L|G) = \frac{P(G|L) \times P(L)}{P(G)}$$

$$= \frac{0.03}{0.59} = 0.05$$

$$P(H|B) = \frac{P(B|H) \times P(H)}{P(B)}$$

$$= \frac{0.14}{0.41} = 0.34$$

$$P(L|B) = \frac{P(B|L) \times P(L)}{P(B)}$$

$$= \frac{0.27}{0.41} = 0.66$$

These may be entered on the decision tree (see Figure 3/5).

Evaluating the financial outcome

8. It is now possible to evaluate the financial outcomes of the three scenarios using the information given in the question and the probabilities calculated above.

Scenario 1

Last year £60,000 profits.

Year 1	$60,000 \times 0.9$	=	54,000
2	$60,000 \times 0.9^2$	=	48,600
3	$60,000 \times 0.9^3$	=	43,740
4	$60,000 \times 0.9^4$	=	39,366
5	$60,000 \times 0.9^5$	=	35,429.5
6	$60,000 \times 0.9^6$	=	31,886.5
			£253022.0

Scenario 2

Expected value of Direct Launch.

Node (A): $0.7(90,000 \times 6) + 0.3(30,000 \times 6)$

$= £378,000 + 54,000 = $ **£432,000**

Scenario 3

Expected Value of Market Research

Node (B): $0.95(100,000 \times 6) + 0.05(25,000 \times 6)$

$= £570,000 + 7500 = $ **£577,500**

Deduct £35,000 for extensions

$= $ **£542,500**

Node (C):	$0.95(90,000 \times 6) + 0.05(30,000 \times 6)$
	$= £513,000 + 9,000 = £522,000$
Node 1:	Compare £542,500 and £522,000
	∴ It is worth spending more.
	Carrying £542,500 back to (E).
Node (D):	$0.34(70,000 \times 6) + 0.66(50,000 \times 6)$
	$£142,800 + 198,000 = £340,800$
Node (E):	$0.59 \times 542,500 + 0.41 \times 340,800$
	$£320,075 + 139,728 = £459,803$
	Deduct Market Research Expenditure
	$£459,803 - 30,000 = £429,803$
Node 2:	Final Decision Summary
	Scenario 1 EMV = £253,022
	Scenario 2 EMV = £432,000
	Scenario 3 EMV = £429,803

Therefore Choose Scenario 2 as this gives the greatest expected monetary value. However Scenario 3 is close. Had the cash flows been discounted, the situation may be different.

Points to note

9. a) Decision trees do not show any more information than could be shown in a tabular form, but they may enable a clearer overview to be obtained.

   b) Invariably decision tree problems comprise investments spread over a number of years.

   c) Because decision trees use expected values as the decision criterion they suffer from the same disadvantages as expected value which are given in Chapter 2.

Self review questions *Numbers in brackets refer to paragraph numbers*

1. What are Decision Trees? (2)
2. Distinguish between the types of nodes found in Decision Trees. (3)
3. What is the Forward Pass? (4)
4. What is the Backward Pass? (4)

Exercises with answers

1. A firm has developed a new product X. They can either test the market or abandon the project. The details are set out below.

   Test market cost £50,000; likely outcomes are favourable (P = 0.7) or failure (P = 0.3). If favourable, they could either abandon or produce it when demand is

anticipated to be

Low	P = 0.25	**Loss**	£100,000
Medium	P = 0.6	**Profit**	£150,000
High	P = 0.15	**Profit**	£450,000

If the test market indicates failure the project would be abandoned.

Abandonment at any stage results in a gain of £30,000 from the special machinery used. Draw the decision tree showing the nodes and probabilities.

2. Evaluate the decision tree in 1.

Answers to exercises

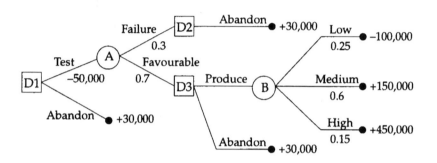

2. Expected value at Node B

$(0.25 \times -100,000) + (0.6 \times 150,000) + (0.15 \times 450,000) = \underline{\textbf{£132,500}}$

∴ comparison at D3 is + 30,000 and + 132,500

∴ produce with EV of £132,500

EV at Node A $(0.3 \times 30,000) + (0.7 \times 132,500) = \underline{\textbf{£101,750}}$

∴ comparison at D1 is

$101,750 - 50,000 = \underline{\textbf{51,750}}$ and 30,000

∴ Test market at cost of £50,000 and produce if there are favourable indications.

Assessment and revision Chapters 1 to 3

Examination questions with answers

Answers commence on page 519

A1. The production manager of Longfield Garden Supplies is currently considering the results to be expected on launching a new product. As part of this review he has estimated the probabilities that other competing companies will enter the market with products of their own. He has estimated that the probability that competitor A enters the market is 0.4, that competitor B enters is 0.7, and that competitor C enters the market is 0.2.

Required:

Calculate the probability that

i)   no competitors will enter the market;

ii) competitors A and B will enter the market, but C will not;

iii) there will be exactly one competitor.

ACCA Decision Making Techniques (part question)

A2. A pharmaceutical company has developed a new headache treatment which is being field tested on 1,000 volunteers. In a test some volunteers have received the treatment and some a placebo (a harmless neutral substance). The results of the test are as follows:

	Treatment received	Placebo received
Some improvement	600	125
No improvement	150	125

Requirement:

Calculate

a) the probability that a volunteer has shown some improvement;

b) the conditional probability that the volunteer has shown some improvement given that he or she has received the treatment;

c) the conditional probability that the volunteer has received the treatment given that no improvement has been observed;

d) the conditional probability that the volunteer has received the placebo given that some improvement has been observed.

e) Explain briefly what the evidence in the table suggests to you about the effectiveness of the treatment.

CIMA Cost Accounting and Quantitative Methods

A3. A construction company has a £1 million contract to complete a building by 31 March 1995, but is experiencing delays due to the complex design. The managers have to make a decision now whether to continue as at present, or to employ specialist enginerring consultants at a cost of £200,000.

If the company continues as at present, it estimates there is only a 30% chance of completing the building on time, and that the delay could be one, two or three months, with equal probability. If the building is late, there are penalties of £100,000 for each month's delay (or part of a month).

The managers believe that if they employ specialist engineering consultants, their chances of completing the contract on time will be trebled. But if the building is still late, it would be only one or two months late, with equal probability.

You are required

a) to draw a tree diagram to represent this decision problem, using squares for decision points, circles for random outcomes, and including probabilities, revenues and penalties;

b) to analyse the tree, using expected value techniques;

c) to write a short report for the managers, with reasons and comments, recommending which decision to make.

CIMA Quantitative Methods

A4. Taurus Press, a printing and publishing company, have recently purchased and installed a new photocopying machine. Spare parts for this machine are only available now, not at a later date, so Taurus Press have to make a decision on the number of spare parts to purchase for this machine. The machine has an estimated working life of five years and, based on past experience, the probability distribution for the number of failures over this time period is given in the table below.

Number of failures during machine lifetime	Probability
0	0.15
1	0.3
2	0.25
3	0.2
4	0.1

Each spare part for the machine costs £5,000, but if the machine failed and there was no spare part available, the cost to Taurus Press of mending the machine in present value terms is £20,000 (no discounting over time is necessary).

Required:

a) Calculate the expected number of failures during the machine lifetime.

b) Calculate the total cost for each failure number and order policy combination.

A.7 A bakery produces fresh cakes each day in anticipation of demand. The variable costs of production are £0.20 per cake and the retail price is £0.50 per cake. The daily demand for cakes over the last four months is shown in the table below. At the end of each day, any unsold cakes are sold to a local pig farmer for £0.05 per cake.

Daily demand (number of cakes)	1–39	40–79	80–119	120–159	160–199
Number of days	10	20	30	30	10

(For practical purposes, you may assume the midpoints of the class intervals to be 20, 60, 100, 140, 180.)

Required:

a) Prepare a 5×5 contribution table for each production/demand combination and calculate the expected contribution for each level of production.

b) Recommend, with reasons and comments, the most suitable production policy for the bakery.

CIMA Business Mathematics

Examination questions without answers

B1. A centralised kitchen provides food for various canteens throughout an organisation. Any food prepared but not required is used for pig food at a net value of 1p per portion.

A particular dish, D, is sold to the canteens for £1.00 and costs 20p to prepare.

Based on past demand, it is expected that during the 250-day working year the canteens will require the following quantities:

On 100 days	40
On 75 days	50
On 50 days	60
On 25 days	70

The kitchen prepares the dish in batches of 10 and has to decide how many it will make during the next year.

You are required to:

a) Calculate the highest expected profit for the year that would be earned if the same quantity of D were made each day;

b) calculate the maximum amount that could be paid for perfect information of the demand (either 40, 50, 60 or 70) if this meant that the exact quantity could be made each day.

CIMA Quantitative Techniques

B2. DCH Supplies Ltd needs to purchase a photocopying machine for its administration department. The model that seems to best fit their needs is available from the manufacturers at a retail price of £4,200. On the chance that the company could save some money, the departmental manager made enquiries regarding the availability of a second-hand model. One is available at a price of £2,700. After inspection the departmental manager believes that the basic body of the machine is fine but is suspicious that a main component, the central copying unit, would not be acceptable. From experience, he felt there was a 0.3 probability that the main component would not be acceptable. A new central copying unit would cost £2,400. After purchasing the second-hand photocopying machine, DCH might find that the main component was not acceptable, and a new one would have to be purchased at this price.

Required:

a) i) Represent the information in a decision tree. Use the method of expectation to make a suitable recommendation to the departmental manager.

     ii) What other features should the departmental manager bear in mind when making his decision?

b) Prior to making a decision, the second-hand photocopying machine could be sent for a thorough test to a local engineering firm. The charge for this test would be £200, but the company would then know for certain whether the main component was acceptable or not.

     Using the financial information only advise the departmental manager on this possibility.

c) DCH Supplies Ltd employs an engineer who could also test the photocopying machine at a cost of £100, but in this case the test would not be conclusive. From past experience the following table shows the probabilities relevant to the situation.

			Actual condition of main component	
			Acceptable	Faulty
Engineer's prediction for component	Satisfactory	0.8	0.8	0.2
	Unsatisfactory	0.2	0.3	0.7
			0.7	0.3

For example, the probability the engineer predicts the main copying component to be unsatisfactory is 0.2, and the probability the main copying component is actually acceptable despite the engineer predicting it to be unsatisfactory is 0.3.

Represent the information in a decision tree and advise the departmental manager appropriately.

ACCA Decision Making Techniques.

B3. As a result of an increase in demand for a town's car parking facilities, the owners of a car park are reviewing their business operations. A decision has to be made now to select one of the following three options for the next year:

Option 1: Make no change. Annual profit is £100,000. There is little likelihood that this will provoke new competition this year.

Option 2: Raise prices by 50%. If this occurs there is a 75% chance that an entrepreneur will set up in competition this year. The Board's estimate of its annual profit in this situation would be as follows:

2A. *With* new competitor		2B. *Without* new competitor	
Probability	Profit (£)	Probability	Profit (£)
0.3	150,000	0.5	200,000
0.4	120,000	0.3	150,000
0.3	80,000	0.2	100,000

Option 3: Expand the car park quickly at a cost of £50,000, keeping prices the same. The profits are then estimated to be like 2B, above, except that the probabilities would be 0.6, 0.3 and 0.1 respectively.

You are required

a) to draw a decision tree for this problem, including all relevant data;

b) using expected values, to analyse the decision tree and to recommend one option, with reasons;

c) to state any assumptions or reservations in your analysis.

CIMA Quantitative Methods

4 Statistics – introduction

Objectives

1.   After studying this chapter you will

- have refreshed your knowledge of Foundation Level Statistics;
- understand how to construct and use Frequency Distributions;
- be able to derive and use the Arithmetic Mean, the Median and the Mode;
- understand how to calculate the key measures of dispersion the Standard Deviation and the Variance;
- know what is meant by Skewness;
- be able to describe the Normal Distribution and use Normal Area Tables;
- know the characteristics of the Binomial and Poisson Distributions;
- understand when to use the Normal approximation to the Binomial.

Statistical analysis

2.   Statistical Analysis or, more simply, Statistics deals with quantitative data. Statistical Analysis is a scientific method of analysing masses of numerical data so as to summarise the essential features and relationships of the data in order to generalise from the analysis to determine patterns of behaviour, particular outcomes or future tendencies. Statistics can be applied in any field in which there is extensive numerical data, not only in accounting and business but in medicine, engineering, the sciences, public administration and many other areas.

Statistical theory is based on the mathematics of probability which provides the basis for determining not only the general characteristics of data but also the reliability of each generalisation. From a business viewpoint the basic steps in statistical analysis include:

a)   Collecting the data from records or other sources or from sample surveys.

b)   Analysing and interpreting the figures by means of statistical techniques.

c)   Using the calculated results together with probabilities, costs and revenues to make more rational decisions.

Frequency distributions

3.   Before data can be effectively used or analysed it is normal to group or arrange the raw data into a manageable form. An important form which is the basis of most of the applications described in this book is the *frequency distribution*.

This shows the variable values and the number of occurrences (i.e the frequency) of each value, or more commonly, of each class of values, in which case it is known as a *grouped frequency distribution*.

For example, the outputs of 50 operators were recorded during a shift as follows:

1162	1497	1391	1468	1406
1286	1329	1410	1340	1277
1419	1103	1383	1401	1343
1509	1292	1227	1288	1596
1322	1457	1184	1359	1461
1128	1339	1429	1540	1235
1384	1485	1498	1322	1520
1505	1572	1517	1417	1485
1401	1292	1255	1132	1336
1362	1305	1262	1396	1426

Output in units. Table 1

These data could be rearranged in ascending (or descending) order and would then be termed an *array*, or they could be listed in order but with each output listed only once and the frequency (usually denoted by f) written alongside. More simply the values would be grouped into classes and the frequency of the class entered.

Such an arrangement would be as follows.

Output			Frequency (f)
1100	to under	1200	5
1200	to under	1300	9
1300	to under	1400	14
1400	to under	1500	15
1500	to under	1600	7
			$50 = \Sigma f$

Grouped frequency distribution. Table 2

Where Σf is the total of the frequencies

Notes:

a) Such a table is a convenient and informative method of summarising the original raw data albeit with some loss of accuracy.

b) The above table uses equal class intervals. On occasions unequal or open ended class intervals are used.

c) It will be seen that $\Sigma f = 50$, i.e. the number of recordings in the original data.

Discrete or continuous data

4. Discrete data have distinct values with no intermediate points. Thus the number of employees in a department may be 0, 1, 2, 3, 4, etc., but not 2.7 or 4.35. Continuous data can have any values over a range either a whole number or any fraction. Thus, if weights were recorded they might be any value, e.g. 10.68 kg, 14.753 kg, 16 kg, 21.005 kg. If continuous data were formed into a frequency distribution it would be a continuous frequency distribution. Continuous frequency distributions are more amenable to statistical analysis and most of

the statistical tables are based on continuous values.

Because of this, where the statistical population is large, continuity of values is often assumed even though strictly the data are discrete.

For simplicity the same assumption is often made in examination questions.

Note: The term population is used in the statistical sense meaning *all* possible observations of a given quantity in which one is interested. An alternative term is *universe*.

Charting frequency distributions

5.   The most common method of charting a grouped frequency distribution is by a **histogram**. A histogram is a set of vertical bars or columns whose areas are proportional to the frequencies represented. Frequencies are plotted on the vertical axis and the class intervals of the values on the horizontal axis. Figure 4/1 shows the histogram of the distribution given in Table 2.

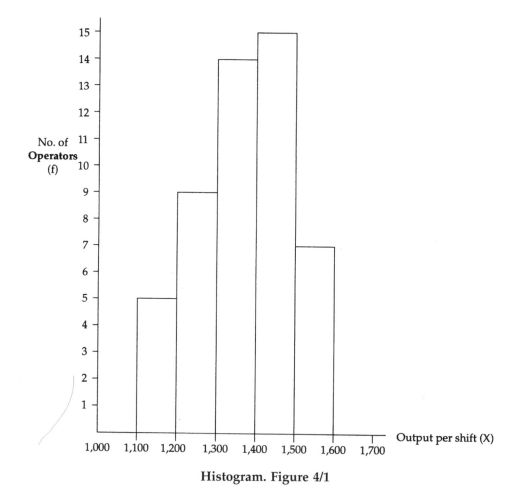

Histogram. Figure 4/1

Note: The histogram in Figure 4/1 has bar heights directly proportional to the frequencies the classes occurred. This is because the class intervals were equal. Where the class intervals are not equal the class frequency is represented by the **bar area** not the bar height. Thus if a class is *twice* the standard width the bar is drawn at *half* the class frequency; if a class is *half* the standard width the bar is drawn at *twice* the class frequency and so on.

An alternative to the histogram, based on the same scales, is the **frequency polygon** in which straight lines join the mid points of the class intervals. The area covered by the polygon is the same as the histogram. The data in Table 2 are shown as a frequency polygon in Figure 4/2.

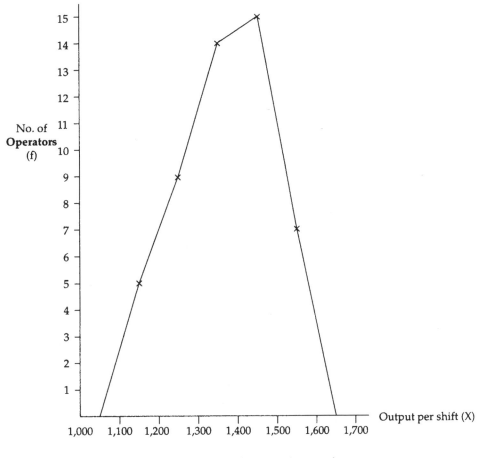

Frequency polygon. Figure 4/2

On occasions it is required to depict a frequency distribution by a *cumulative frequency curve* or *ogive*. These diagrams are formed by plotting the cumulative frequencies of the distribution against the upper limit of each class interval. The diagrams are an easy way of finding the median of a distribution.

As an example, the data in Table 2 are shown as an ogive and the output of the median operator illustrated. Table 2 is reproduced below with a cumulative frequency column and the ogive shown in Figure 4/3.

Output			Frequency	Cumulative Frequency
1100	to under	1200	5	5
1200	to under	1300	9	14
1300	to under	1400	14	28
1400	to under	1500	15	43
1500	to under	1600	7	50

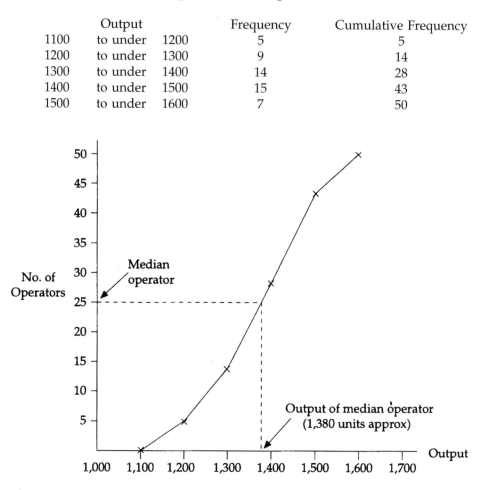

Cumulative frequency polygon of table 2 data. Figure 4/3

Frequency curves

6.   A smooth curve can be drawn to depict the frequency distributions of a population of continuous data. This is the limiting form of the histogram or frequency polygon as the number of classes becomes infinitely large and the class intervals become infinitely small. Some types of frequency curves are shown in Figure 4/4.

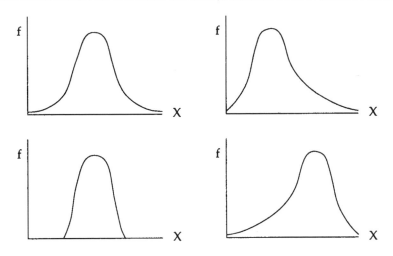

Examples of frequency curves. Figure 4/4

Characteristics of distributions

7. As previously explained, statistical analysis seeks to provide summary measures which describe the important characteristics of a distribution of values. There are four important characteristics of distributions:

    a) **Average**, i.e. the typical size. For our purposes the most important measures being the arithmetic mean, the median, and the mode.

    b) **Dispersion**, i.e. the variation, spread or scatter of the distribution for which the most important measures are the standard deviation and the variance. Other measures of dispersion are the range (the difference between largest and smallest values) and the semi-interquartile range (half the range of the middle 50% of items).

    c) **Skewness**, i.e. the lopsidedness or asymmetry of the distribution.

    d) **Kurtosis**, i.e. the peakedness of the distribution.

Average

8. By far the most important measure of central tendency is the **arithmetic mean**, usually denoted by $\bar{x}$. The arithmetic mean, or simply the **mean**, is calculated by the following formula.

$$\bar{x} = \frac{\sum x}{n}$$

where **n** is the number of values. The mean is always in the same units as the variable.

When the data are arranged in a grouped frequency distribution the formula is

$$\bar{x} = \frac{\sum fx}{\sum f}$$

where **f** is the number of values in an interval and **x** is the midpoint of the class interval.

The mean lends itself to subsequent analysis because it includes the values of all items. However it may not coincide with any actual value and may, in some circumstances, be unrepresentative because of the effect of extreme values.

The *median* of any set of data is the middle value in order of size if n is odd, or the mean of the two middle items if **n** is even. Where the data contain a few very large or small values the median value is often considered to be a more representative value than the mean although it cannot be used for subsequent calculations as is possible with the mean.

The *mode* is the value which occurs most often or the value around which there is the greatest degree of clustering. Ordinarily the mode is only meaningful if there is a marked clustering of values round a single point.

Note: In a symmetrical distribution the mean, median and mode have the same value.

Dispersion

9. Dispersion is the variation or scatter of a set of values. A measure of the degree of dispersion of data is needed for two basic purposes:

    a) To assess the reliability of the average of the data.

    b) To serve as a basis for control of the variability. For example, assessment of the degree of quality variation is an essential part of Quality Control procedures in factories, dealt with later in the book.

    The measure of dispersion which is the most important is the *standard deviation,* denoted by σ. The standard deviation is found by:

    a) Squaring the deviations of individual values from the arithmetic mean.

    b) Adding the squared deviations together.

    c) Dividing this sum by the number of items in the distribution.

    d) Taking the square root of the value in (c).

    The formula for ungrouped data is:

$$\sigma = \sqrt{\frac{\sum(x - \mu)^2}{N}}$$

where μ = population mean and **N** = population size.

Note: The above formula is the appropriate one where all the details are known of the whole population. Since in many circumstances the whole population is not known, and a sample is taken in order to estimate the population characteristics, the formula which provides the best estimate of the population standard deviation, based on the sample data, is:

$$s = \sqrt{\frac{\sum(x - \bar{x})^2}{n - 1}}$$

where **s** is the estimate of σ, the population standard deviation.

The standard deviation is in the same units as the variable measured. For example, if the mean and standard deviation of the ages of a group of people were calculated the answers might be as follows

$$\text{mean} = \bar{x} = 43.5 \text{ years}$$

$$\text{Standard deviation} = \sigma = 6.9 \text{ years}$$

The *variance* is the square of the standard deviation i.e.

$$\text{variance} = \sigma^2$$

Where the data are arranged in a grouped frequency distribution, the formulae are as follows:

For a population

$$\sigma = \sqrt{\frac{\sum f(x - \mu)^2}{\sum f}}$$

where $\sum f = N$ = population size

For a sample

$$s = \sqrt{\frac{\sum f(x - \bar{x})^2}{\sum f - 1}}$$

where $\sum f - 1 = n - 1$, and **n** is the sample size.

The above formulae are in a basic form and alternatives, which may reduce computation, are available. For example, the formula for the standard deviation of a population

$$\sigma = \sqrt{\frac{\sum(x - \bar{x})^2}{n}}$$

can be re-arranged to produce an alternative formulation thus:

$$\sigma = \sqrt{\frac{\sum x^2 - \frac{(\sum x)^2}{n}}{n}}$$

Another formulation which is convenient when a calculator is being used is as follows:

$$\sigma = \sqrt{\frac{\sum fx^2}{\sum f} - \left(\frac{\sum fx}{\sum f}\right)^2}$$

Unlike standard deviations, variances can be added together. This is a useful property when two distributions have to be combined, i.e. if two distributions A and B had variances of σ_A^2 and σ_B^2 then the variance of A + B = $\sigma_A^2 + \sigma_B^2$.

The following example demonstrates the measures covered in paragraph 8 and 9.

Example 1

Calculate the mean, standard deviation and variance of the following Sample data:

$$5, 8, 15, 29, 47, 47, 64, 71, 74$$

Solution

As there are 9 values, n = 9.

x	$x-\bar{x}$	$(x-\bar{x})^2$
5	−35	1,225
8	−32	1,024
15	−25	625
29	−11	121
47	+7	49
47	+7	49
64	+24	576
71	+31	961
74	+34	1,156
$\Sigma x = 360$		5786

$$\therefore \bar{x} = \frac{360}{9} = 40$$

$$\therefore \text{variance} = \frac{\sum(x - \bar{x})^2}{n - 1} = \frac{5,786}{9 - 1} = 723.25$$

and the standard deviation $= \sqrt{723.25} = 26.893$

Note: The median of the above data is the middle value, i.e. the 5th, which is 47. The mode is the value which occurs most frequently which is also 47 as it occurs twice.

Example 2 – Mean and standard deviation – Grouped data

Calculate the mean and standard deviation for the sample data in Table 2.

Class interval	mid-point x	f	fx	$f(x-\bar{x})^2$
1,100–1,200	1,150	5	5,750	242,000
1,200–1,300	1,250	9	11,250	129,600
1,300–1,400	1,350	14	18,900	5,600
1,400–1,500	1,450	15	21,750	96,000
1,500–1,600	1,550	7	10,850	226,800
		50	68,500	700,000

Table 2

Solution

$$\bar{x} = \frac{\sum fx}{\sum f} = \frac{68,500}{50} = \textbf{1,370 units}$$

Since Table 2 represents sample data the standard deviation is calculated as below:

$$s = \sqrt{\frac{\sum f(x - \bar{x})^2}{n - 1}}$$

$$= \sqrt{\frac{700,000}{49}}$$

$$= \underline{\textbf{119.52 units}}$$

Note: When the sample is large then s and σ are close in value. When n > 30 for practical purposes s and σ can be regarded as the same.

Relative and absolute dispersion

10. The standard deviation provides considerable information on the *absolute* dispersion of a given distribution, the higher the standard deviation the greater the amount of scatter.

On occasions it is useful to have a measure of the *relative dispersion* of a distribution, particularly if distributions are being compared. The measure which provides this relative view is the *coefficient of variation* or, as it is sometimes known, the *coefficient of dispersion*.

This is simply the ratio of the standard deviation of a distribution to the mean of the distribution, expressed as a percentage:

$$\text{Coefficient of variation} = \frac{\sigma}{\bar{x}} \times 100\%$$

For example, does distribution A vary relatively more than distribution B given that

$$\sigma_A = 5.5 \qquad\qquad \bar{x}_A = 45$$
$$\sigma_B = 2.6 \qquad\qquad \bar{x}_B = 18.5 \ ?$$

Coefficients of variation

$$A = \frac{5.5}{45} \times 100 = \textbf{12.2\%}$$

$$B = \frac{2.6}{18.5} \times 100 = \textbf{14.05\%}$$

$\therefore$ Distribution B is relatively more variable.

Skewness

11. Where there is a lack of symmetry in a distribution, skewness exists. The effect of this is that the mode, mean, and median have differing values. Figure 4/5 illustrates positive and negative skewness and the position of the mean, mode and median.

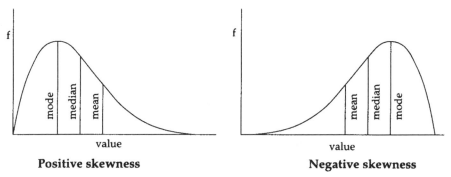

Positive skewness   Negative skewness

Figure 4/5

Calculating the accurate measure of skewness requires advanced techniques but an approximate measure of the amount and direction of skewness can be found by the *Pearson coefficient of skewness* thus

$$\textbf{Sk} = \frac{\textbf{3(Mean} - \textbf{Median)}}{\sigma}$$

The formula has limited practical use and skewness is usually treated in descriptive terms rather than summarised by a single measure.

Kurtosis

12. This relates to the peakedness of distributions. This can range from distributions which are *platykurtic* (flatter than normal) through *mesokurtic* (or normal peakedness) to *leptokurtic* (more peaked than normal).

    These are illustrated in Figure 4/6.

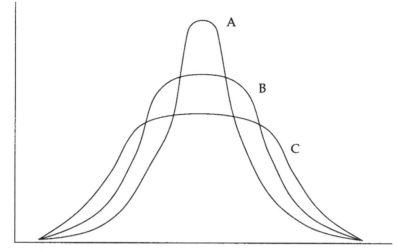

A = leptokurtic distribution

B = mesokurtic distribution

C = platykurtic distribution

Figure 4/6

The Normal Distribution

13. Having discussed the characteristics of distributions in general one particular distribution can now be examined in more detail. This is the *normal distribution or normal curve*. The normal distribution is the most important in statistics. It occurs frequently when describing natural occurrences and is of particular importance in sampling theory and statistical inference which are dealt with later in the book.

    The main features of the Normal Distribution are:

    a) It is continuous distribution.

    b) It is a perfectly symmetrical bell shaped curve.

    c) The 'tails' of the distribution continually approach, but never touch, the horizontal axis.

    d) The formula for the curve is

$$y = \frac{1}{\sigma\sqrt{2\pi}} e^{-\frac{1}{2}\left(\frac{x - \bar{x}}{\sigma}\right)^2}$$

(although the formula need not be learned, what is important to note is that all the properties of a normal distribution are described when the mean and standard deviation of the distribution are known).

e) The mean, mode the median pass through the peak of the curve and precisely bisect the area under the curve into two equal halves.

f) When the mean and standard deviation of a Normal Distribution are known, probabilities associated with the Distribution can be calculated using Normal Area Tables. The procedures are described in detail in the following paragraphs.

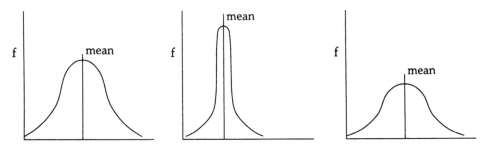

Normal curves. Figure 4/7

Areas under the normal curve

14. An important feature of the normal curve is the relationship of the area under the curve (i.e. the percentage of the population) and the standard deviation of the distribution.

For example, 68.26% of the total area of the curve lies between the mean plus and minus 1 standard deviation. It follows that as the normal curve is symmetrical the area between the mean and plus 1 standard deviation is 34.13% of the total and the mean minus 1 standard deviation is also 34.13%. This is shown in Figure 4/8, where μ (mu) is the population mean.

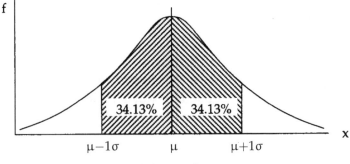

Figure 4/8

The area between any two perpendicular lines can be found from Table (**I**) 'Areas under the Normal Curve' but students will find that three particular values occur frequently in problems, so that these should be learned.

> **mean ± 1σ** includes **68.26%** of the area.
>
> **mean ± 2σ** includes **95.44%** of the area.
>
> **mean ± 3σ** includes **99.74%** of the area.

It will be seen that virtually all of the distribution lies within between the mean minus 3σ and the mean plus 3σ, i.e. a range of 6σ.

Note: A frequently used value in statistical testing relates to 95% of the area of a Normal Curve. Examination of Table 1 shows that 95% of the area is within the mean ± 1.96 σ.

Using the normal area table (Table I)

15. Normal area tables show the area under the normal curve between the mean and any given number of standard deviations (including fractional values). To use the tables it is necessary to calculate the ***standardised variate*** (often called the '**Z** score') which is simply the number of standard deviations that the required value is away from the mean.

    z is calculated thus:

$$z = \left| \frac{x - \mu}{\sigma} \right|$$

where   **Z** = number of standard deviations above or below the mean.

         x = the value of the variable being considered

         μ = the population mean.

         σ = the population standard deviation.

The parallel lines round the expression mean that we are normally only concerned with the number of standard deviations calculated, not whether the answer is negative or positive.

Example 3

An assembly line contains 2,000 of a component which has a limited life. Records show that the life of the components is normally distributed with a mean of 900 hours and a standard deviation of 80 hours.

     a)   What proportion of components will fail before 1,000 hours?

     b)   What proportion will fail before 750 hours?

     c)   What proportion of components fail between 850 and 880 hours?

     d)   Given that the standard deviation will remain at 80 hours what would the average life have to be to ensure that not more than 10% of components fail before 900 hours?

Solution

a)
$$\mathbf{z} = \left| \frac{1,000 - 900}{80} \right| = 1.25$$

(i.e. this means that the value being investigated, 1,000 hrs, is 1.25 standard deviations away from the mean of 900 hours).

If Table I is examined it will be seen that the value for a **z** score of 1.25 is 0.3944. As one half of the distribution is less than 900, the proportion which fail before 1,000 hours is 0.5 + 0.3944 = **89.44%.**

If required this could be expressed as the number of components which are expected to fail, thus

2,000 × .8944 = 1,788.8 which would be rounded to **1,789**

b)
$$\mathbf{z} = \left| \frac{750 - 900}{80} \right| = 1.875$$

and from the tables we obtain the value 0.4696. In this case as we require the proportion that will fail before 750 hours, the table value is deducted from 0.5.

∴ Proportion expected to fail before 750 hours:

$$= 0.5 - 0.4696 = \underline{0.0304} \text{ i.e. } \mathbf{3.04\%}$$

c) When it is required to find the proportion between two values, neither of which is the mean, it is necessary to use the tables, to find the proportion between the mean and one value and the proportion between the mean and the other value and then find the difference between the two proportions.

$$\mathbf{z} = \left| \frac{900 - 850}{80} \right| = 0.625 \quad \text{which gives a proportion of } \mathbf{0.2340}$$

$$\mathbf{z} = \left| \frac{900 - 880}{80} \right| = 0.25 \quad \text{which gives a proportion of } \mathbf{0.0987}$$

∴ Proportion between 850 and 880 is

$$0.2340 - 0.0987 = \underline{0.1353} \text{ i.e. } \mathbf{13.53\%}$$

Note: This part of the example illustrates the proportion between two values on the *same side* of the mean. If the two values are on opposite sides of the mean, the calculated proportions would be *added*.

d) This problem is the reverse of the earlier questions although based on the same principles. The earlier problems started with the mean and standard deviation, found the **z** score and thence the proportion from the tables. We now *start* with the proportion and work back, through the tables, to find a new mean value.

If not more than 10% should be under 900 it follows that 90% of the area of the curve must be greater than 900 i.e. as shown in Figure 4/9.

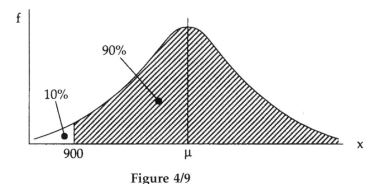

Figure 4/9

Bearing in mind that the tables only show values for half the distribution (because both halves are identical) we have to look in the tables for a value close to 0.4 (i.e. 0.9 – 0.5).

It will be see is a value in the body of the table of 0.3997, i.e. virtually 0.4. This value has a **z** score of 1.28.

$$\text{Thus } 1.28 = \left| \frac{\text{mean} - 900}{80} \right|$$

$$\therefore 102.4 = \text{mean} - 900$$

$$\therefore \text{mean} = 1002.4 \text{ hours.}$$

Thus if the mean life of the components is 1002.4 hours with a standard deviation of 80 hours, less than 10% of the components will fail before 900 hours.

See the appendix at the end of the chapter for a comment on normal distribution tables.

Other important distributions

16.   The binomial and Poisson distributions are probability distributions, i.e. they directly depict the probability of the required frequency of an event occurring. The binomial and Poisson distributions describe the behaviour of *attributes*, e.g. success or failure, good or bad, black or white, whilst the Normal Distribution describes the behaviour of variables or measurements, e.g. height, weight, length of life. The binomial and Poisson distributions are *discrete* distributions compared with the Normal Distribution which is *continuous*.

The binomial distribution

17.   The binomial distribution is useful for problems in which we are concerned with determining the number of times an event is likely to occur or not occur during a given number of trials and consequently the probability of the event occurring or not occurring.

Example 4

The probability that a salesman makes a sale on a visit to a prospect is 0.2.
What is the probability, in 2 visits, of:

making no sales?

making one sale?

making two sales?

Solution

Let p = probability of sale = 0.2

let q = probability of no sale = $1 - 0.2 = 0.8$

The various outcome possibilities are

Visit 1	Visit 2		Probabilities	
Sale	Sale	i.e. $p \times p$	$= p^2 = 0.2^2$	$= 0.04$
Sale	No sale	i.e. $p \times q$	$= 0.02 \times 0.8$	$= 0.16$
No sale	Sale	i.e. $q \times p$	$= 0.8 \times 0.2$	$= 0.16$
No sale	No sale	i.e. $q \times q$	$= q^2 = 0.8^2$	$= 0.64$
				$\underline{1.00}$

Thus P (no sales) = **0.64**

P (one sale) = **0.32** (i.e. 0.16 + 0.16)

P (two sales) = **0.04**

In this simple example it is easy to show the whole process but this becomes lengthy and cumbersome where the number of trials (visits, in the above example) becomes larger.

Fortunately there is a simpler approach which is by the expansion of the binomial expression. The general form of the binomial expression is:

$$(p + q)^n$$

where p = probability of an event occurring

q = probability of an event not occurring

and n = number of trials.

In the above example p = probability of a sale, i.e. 0.2, q = probability of no sale, i.e. 0.8 and n = number of visits, i.e. 2.

Use the binomial expansion to confirm the probabilities.

$$\therefore (p + q)^2 = p^2 + 2pq + q^2$$

$$= 0.2^2 + 2(0.2 \times 0.8) + 0.8^2$$

$$= 0.04 + 0.32 + 0.64, \text{ i.e. the values previously obtained.}$$

Where n becomes larger it is useful to be able to calculate the coefficients of each part of the expansion in a direct manner rather than writing out the whole expansion.

The binomial coefficients

18. Tables are not always available in examinations and it is therefore necessary to be able to solve a binomial problem from first principles.

Assume that the full expansion is required of the binomial formula $(\mathbf{p} + \mathbf{q})^5$.

The first step is to write down the successive terms within the brackets, i.e.

$$(p^5) + (p^4q) + (p^3q^2) + (p^2q^3) + (pq^4) + (q^5)$$

It will be noted that

- the number of terms (i.e. 6 in this case) is one more than the power of $\mathbf{n}$ (i.e. 5 in this case).
- the power of $\mathbf{p}$ starts with the $\mathbf{n}$th power (i.e. 5 in this case) and decreases from term to term whilst conversely the power of $\mathbf{q}$ increases from term to term.
- the powers of $\mathbf{p}$ and $\mathbf{q}$ in every expression total $\mathbf{n}$ (i.e. 5 in this case).

The next step is to calculate the coefficient of each term, i.e. the value in front of each bracket. This can be done by using either *Pascal's Triangle* or by the *combination formula*.

n	Pascal's Triangle
1	1　　1
2	1　　2　　1
3	1　　3　　3　　1
4	1　　4　　6　　4　　1
5	1　　5　　10　　10　　5　　1
6	1　　6　　15　　20　　15　　6　　1
7	1　　7　　21　　35　　35　　21　　7　　1

and so on for higher values of $\mathbf{n}$.

The values in any line are derived by adding adjacent figures in the line above as shown by the dotted triangle. It will be noted that the values are symmetrical on each line and that there is one more number on each line than the value of $\mathbf{n}$. The coefficients in the expansion can be read directly from the triangle against the value of $\mathbf{n}$. Where $\mathbf{n}$ is 5, as in our example, the coefficients are

$$1 \quad 5 \quad 10 \quad 10 \quad 5 \quad 1$$

Thus the expansion of $(\mathbf{p} + \mathbf{q})^5$ is

$$p^5 + 5(p^4q) + 10(p^3q^2) + 10(p^2q^3) + 5(pq^4) + q^5$$

Combination formula

19. Although useful for small values of **n**, Pascal's triangle becomes unwieldy for larger values but the combination formula can be used for any value of **n** with equal facility.

The general form of the combination formula dealt with in Chapter 2 enables us to find the number of combinations of **r** that are possible given that there is a total of **n**.

Using the combination formula to find the coefficients of our example, $(p + q)^5$, we obtain the following:

					Binomial coefficient
$\binom{5}{0}$ or $^5C_0$	=	$\dfrac{5!}{0! \times (5-0)!}$	$\dfrac{5.4.3.2.1}{5.4.3.2.1}$	=	1

(note: $0! = 1$)

$\binom{5}{1}$ or $^5C_1$	=	$\dfrac{5!}{1! \times (5-1)!}$	$\dfrac{5.4.3.2.1}{4.3.2.1}$	=	5
$\binom{5}{2}$ or $^5C_2$	=	$\dfrac{5!}{2! \times (5-2)!}$	$\dfrac{5.4.3.2.1}{2.1 \times 3.2.1}$	=	10
$\binom{5}{3}$ or $^5C_3$	=	$\dfrac{5!}{3! \times (5-3)!}$	$\dfrac{5.4.3.2.1}{3.2.1 \times 2.1}$	=	10
$\binom{5}{4}$ or $^5C_4$	=	$\dfrac{5!}{4! \times (5-4)!}$	$\dfrac{5.4.3.2.1}{4.3.2.1 \times 2.1}$	=	5
$\binom{5}{5}$ or $^5C_5$	=	$\dfrac{5!}{5! \times (5-5)!}$	$\dfrac{5.4.3.2.1}{5.4.3.2.1}$	=	1

The advantage of the binomial formula is that there is no need to calculate the whole expansion if only one value is required. For example, given that the probability of a good part is 0.8 and a bad part is 0.2 what is the probability of finding 10 good parts in a sample of 12?

This means calculating the probability of one particular term in the binomial expansion, i.e.

$$x(p^{10}q^2)$$

where x is the appropriate binomial coefficient and **p** is the probability of a good part and **q** the probability of a bad part.

To calculate the required coefficient:

$${}^{12}C_{10} = \frac{12!}{10! \times (12 - 10)!} = \frac{12 \times 11}{2} = 66$$

Thus the probability of finding 10 good parts in a sample of 12 is:

$$66(0.8^{10} \times 0.2^2) = \underline{\mathbf{0.28346}} \text{ i.e. } \underline{\mathbf{28.35\%}}$$

Mean and standard deviation of the binomial distribution

20. The mean of a binomial distribution is found by multiplying the probability of the event in which we are interested by n, the number of trials:

$$\textbf{Mean} = \textbf{np}$$

This value is the same as the expected value previously discussed in Chapter 2 and that used in Example 3 in this chapter.

The variance of a binomial distribution is calculated as:

$$\textbf{variance} = \textbf{npq}$$

so that the standard deviation is

$$\sigma = \sqrt{\mathrm{npq}}$$

Example 5

Assuming the same data as in Paragraph 19, i.e. sample of 12 with the probability of a good part being 0.8, what is the expected number of good parts in a sample of 12 and what is the standard deviation?

Solution

Expected number of good parts in a sample of 12 = $\textbf{np}$ = 12 × 0.8 = $\underline{\mathbf{9.6}}$

$$\text{Standard deviation} = \sqrt{\mathbf{npq}} = \sqrt{12 \times 0.8 \times 0.2} = \underline{\mathbf{1.386}}$$

Characteristics of the binomial distribution

21. a) It is a discrete distribution of the occurences of an event with two outcomes – success or failure, good or bad.

b) The trials must be independent of one another. This assumption implies sampling from an infinite population. Sampling with replacement fulfils this requirement, but where sampling without replacement is used, the binomial distribution is still useful provided that the sample size is less than 20.

c) As the number of trials grows and if $\mathbf{p} \approx 0.5$ then the binomial distributions approaches the Normal Distribution. For the Normal Distribution to be an appropriate approximation, $\mathbf{np}$ should be >5.

d) The main parameters of the binomial distribution are

$$\mu = np$$
$$\sigma = \sqrt{npq}$$

$$\text{Standardised skewness} = \frac{q - p}{\sqrt{npq}}$$

A practical example using the binomial distribution

22.

Example 6

Components are placed into bins containing 100. After inspection of a large number of bins the average number of defective parts was found to be 10 with a standard deviation of 3.

Assuming that the same production conditions continue, except that bins containing 300 were used:

a)   what would be the average number of defective components per larger bin?

b)   what would be the standard deviation of the number of defectives per larger bin?

c)   how many components must each bin hold so that the standard deviation of the number of defective components is equal to 1% of the total number of components in the bin?

Solution

Proportion defective $= \dfrac{10}{100} = \mathbf{0.1}$

Proportion good $= 1 - 0.1 = \mathbf{0.9}$

Mean $= \mathbf{np} = 100 \times 0.1 = 10$ components defective on average.

Standard deviation $= \sqrt{\mathbf{npq}} = \sqrt{100 \times 0.1 \times 0.9} = \mathbf{3}$

thus confirming the values given in the example.

a)   Larger bins of 300

$$\therefore n = \mathbf{300}$$

$\therefore$ Average number of defectives $= 300 \times 0.1 = \mathbf{30}$

b)   Standard deviation becomes

$$\sqrt{\mathbf{npq}} = \sqrt{300 \times 0.1 \times 0.9} = \mathbf{5.2}$$

c)   Standard deviation to be 1% of total

$$\therefore \sqrt{npq} = \frac{n}{100}$$

$$\therefore \sqrt{n \times 0.1 \times 0.9} = \frac{n}{100}$$

$$\therefore n \times 0.1 \times 0.9 = \frac{n^2}{10,000}$$

$$900n = n^2$$

$$900 = \frac{n^2}{n} = n$$

$\therefore$ Bins must hold **900** components.

Normal approximation to the binomial

23. As a broad generalisation it is safe to use the Normal approximation to the Binomial in the following circumstances:

if **np > 5** when **p < 0.5**
if **nq > 5** when **p > 0.5**

When *proportions* are being used instead of absolute numbers the formulae given previously do not apply and the following relationships hold:

$$\text{population mean} = \mu = p$$

$$\text{standard deviation} = \sigma = \sqrt{\frac{pq}{n}}$$

Example 7

Records show that 60% of students pass their examinations at the first attempt. Using the normal approximation to the binomial, calculate the probability that at least 65% of a group of 200 students will pass at the first attempt.

Solution

$$p = 0.6 \quad q = 1 - p = 1 - 0.6 = 0.4$$

$$\sigma = \sqrt{\frac{pq}{n}} = \sqrt{\frac{0.6 \times 0.4}{200}} = 0.035$$

$$z = \frac{0.65 - 0.60}{0.035} = 1.43 \text{ which gives } 0.4236 \ (42.36\%)$$

from Table I. This means that there is a $(0.5 - 0.4236) \ 0.0764 \ (7.64\%)$ chance of 65% or more passing at the first attempt. This is shown graphically below.

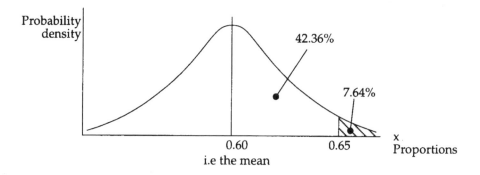

Poisson distribution

24. The Poisson distribution has similarities with the Binomial distribution and describes the number of events that occur within some given interval. It is applied where:

    a) **n** (the number of items or events) is very large and perhaps not readily known;

    b) events occur at random, ie. they are independent of one another;

    c) the probability of some event occurring (**p**) is very small relative to the probability of the event not occurring (**q**). Since **p** is very small, we are concerned with estimating the probability of a **rare** event.

    The Poisson distribution is a discrete distribution, i.e. the outcome can only be a whole number and is defined by its mean usually denoted by the letter **m**.

    For example, we may know that on a given stretch of motorway 6 accidents happen in a year. However, we are unable to express the number 6 as a probability because we do not know the value of **n**, the total number of accident possibilities. Furthermore the term **q**, i.e. the probability that an accident will not occur, has no meaning.

    All we can say is that 'n', the number of opportunities for accidents, is infinitely large and that the number of accidents that actually happen is relatively small, hence 'p', the probability of an accident occurring is small.

    In such circumstances i.e when **n** is infinitely large and **p** is small the Poisson distribution can be used. The Poisson distribution can be used without knowing the value of **n** or **p**, being based on the mean value of the distribution. Thus, if we know the average number of times than an event occurred over a period then the probability that an event will occur 0, 1, 2, 3, etc., times is given by the successive terms of the expansion:

    $$\frac{e^{-m}}{0!} + \frac{me^{-m}}{1!} + \frac{m^2e^{-m}}{2!} + \frac{m^3e^{-m}}{3!} \, ...$$

    where **e** is the constant 2.718 and m is the mean occurrences in the period.

    For example, in a transport fleet there is one breakdown a week on average which requires the vehicle to be recovered.

What is the expected pattern of recoveries in a 100 week period?

$$m = \text{mean breakdowns per week} = 1$$

Number of recoveries	Probability	Recovery Pattern for 100 weeks
0	$\dfrac{e^{-m}}{0!} = \dfrac{2.718^{-1}}{1} = \dfrac{1}{2.718} = 0.3679$	0.3679×100 = 37 weeks with no recoveries
1	$\dfrac{me^{-m}}{1!} = \dfrac{1 \times 2.718^{-1}}{1} = \dfrac{1}{2.718} = 0.3679$	0.3679×100 = 37 weeks with one recovery
2	$\dfrac{m^2e^{-m}}{2!} = \dfrac{1^2 \times 2.718^{-1}}{2 \times 1} = \dfrac{1}{5.436} = 0.1840$	0.184×00 = 18 weeks with two recoveries
3	$\dfrac{m^3e^{-m}}{3!} = \dfrac{1^3 \times 2.718^{-1}}{3 \times 2 \times 1} = 0.0613$	0.0613×100 = 6 weeks with three recoveries
4	$\dfrac{m^4e^{-m}}{4!} = \dfrac{1^4 \times 2.718^{-1}}{4 \times 3 \times 2 \times 1} = 0.0153$	0.0153×100 = 2 weeks with four recoveries
5	$\dfrac{m^5e^{-m}}{5!} = \dfrac{1^5 \times 2.718^{-1}}{5 \times 4 \times 3 \times 2 \times 1} = 0.0036$	0.0036×100 NIL
	Total Probability = 1.00	Total weeks = 100

It will be seen that the calculated probability of each successive term becomes smaller and smaller continually approaching, but never reaching, zero.

These calculations are tedious and tables are available showing Poisson probabilities for any value of **m**. The tables can show individual or cumulative probabilities and Table V shows an extract from each type of table.

Standard deviation of a Poisson distribution

25.　For once, this is simplicity itself.

The variance of a Poisson distribution equals the mean and the standard deviation is the square root of the variance, i.e.

$$\text{variance} = m$$

$$\text{Standard deviation} = \sqrt{m}$$

where **m** is the mean of the Poisson Distribution

Characteristics of Poisson distribution

26.　a)　It is a discrete distribution and is a limiting form of the binomial distribution when **n** is large and **p** or **q** is small.

　　b)　Mean and variance are equal.

c) Is usually definitely positively skewed but cannot be negatively skewed

d) The standardised skewness is found by $\dfrac{1}{\sqrt{m}}$ where **m** is the mean.

e) As **n** becomes very large the Poisson distribution approximates to the Normal Distribution.

f) The mean = **np**.

Applications of Poisson distribution

27. The Poisson distribution is similar to the Binomial but is used when **n**, the number of items or events, is large or unknown and **p**, the probability of an occurrence, is very small relative to **q**, the probability of non-occurrence. A rule of thumb is that the Poisson distribution may be used when n is greater than 50 and **np**, the mean, is less than 5. Some examples follow but it is important to realise that the Poisson distribution only applies when the events occur randomly i.e. they are independent of one another.

Example 8

Customers arrive randomly at a service point at an average rate of 30 per hour. Assuming a Poisson distribution calculate the probability that:

    a) no customer arrives in any particular minute;

    b) exactly one customer arrives in any particular minute;

    c) two or more customers arrive in any particular minute;

    d) three or fewer customers arrive in any particular minute.

Solution

The time interval to be used is one minute with a mean of $\dfrac{30}{60} = 0.5$

    a)   P(no customer)   = 0.6065 from Table V(a)
    b)   P(1 customer)   = 0.3033 from Table V(a)
    c)   P(2 or more)   = 1– 0.9098
                            = <u>0.0902</u>

The value of 0.9098 is the cumulative probability of 1 or fewer customers arriving in a particular minute from Table V(b). As the sum of the probability of every possible number of arrivals equals 1, the probability of 2 or more = 1 – P(1 or fewer).

    d)   P(3 or fewer)   = 0.9982 from Table V(b).

> **Example 9**
>
> A firm buys springs in very large quantities and from past records it is known that 0.2% are defective. The inspection department sample the springs in batches of 500. It is required to set a standard for the inspectors so that if more than the standard number of defectives is found in a batch the consignment can be rejected with at least 90% confidence that the supply is truly defective.
>
> How many defectives per batch should be set as the standard?

Solution

With 0.2% defective and a sample size of 500 $m = 500 \times 0.2\% = 1$.

To find the probability of 0, 1, 2, 3, etc., or more defectives the probabilities from Table V(b) are deducted from 1 as follows:

$$P(0 \text{ or more defectives}) = \text{certainty} = \mathbf{1}$$
$$P(1 \text{ or more defectives}) = 1 - 0.3679 = \mathbf{0.6321}$$
$$P(2 \text{ or more defectives}) = 1 - 0.7358 = \mathbf{0.2642}$$
$$P(3 \text{ or more defectives}) = 1 - 0.9197 = \mathbf{0.0803}$$
$$P(4 \text{ or more defectives}) = 1 - 0.9810 = \mathbf{0.0190}$$

These probabilities mean, for example, that there is a 26.42% chance that 2 or more defectives will occur at random in a batch of 500 with a 0.2% defect rate. If batches with 2 or more were rejected then there can be 73.58% $(1 - 0.2642)$ confidence that the supply is defective.

As the firm wishes to be at least 90% confident, the standard should be set at 3 or more defectives per batch. This level could only occur at random in 8.03% of occasions so that the firm can be 91.97% confident that the supply is truly defective.

Hints on what type of distribution applies

28. It is not always easy to recognise when a Normal, Binomial or Poisson distribution should be used. The following hints might help.

Binomial distribution

a) Outcomes have discrete values and do not have continuous ranges of possible values. For example, in dealing with people, there can only be discrete values (1, 2, 3, etc.); 2.35 people is not possible.

b) There are only two possible conditions: good/bad, black/white, male/female, acceptable/not acceptable and so on.

c) The probability of an item having one of the two possible conditions is $\mathbf{p}$, thus the probability of the item having the other condition is $\mathbf{(1 - p)}$. $\mathbf{(1 - p)}$

is usually referred to as **q**, so that **(p + q) = 1**.

d) When the number of items, **n**, is large and **p** is not close to 0 or 1 so that the distribution is approximately symmetric, the binomial probabilities can be approximated using a Normal Distribution with the same mean **(μ = np)** and standard deviation **(σ = $\sqrt{npq}$)**.

Poisson distribution

Similar to a Binomial Distribution, but used for rare events.

a) The number of items, **n**, is large; say greater than 50.

b) **p** is small in relation to **q** so that **np** (the mean of a Poisson Distribution) is less than, say, 5.

Normal distribution

The most commonly applied probability distribution.

a) It applies to variables with a continuous range of possible values. Examples include: time, weights, distances, sizes, growth rates, etc.

b) Where the quantities are large, the Normal Distribution can also be used for discrete variables (see *note* above under Binomial Distribution).

Summary

29. a) Statistical analysis is the process of analysing data to obtain useful information.

b) A frequency distribution shows the number of occurrences of each variable value.

c) Discrete data have distinct values with no intermediate points whereas continuous data may have any value.

d) The major summary measures of distributions are – Average – Dispersion – Skewness – Kurtosis.

e) The most useful average is the Arithmetic Mean and the most useful measure of dispersion is the Standard Deviation. The Coefficient of Variation links these two measures and gives a measure of relative dispersion.

f) The Normal Curve is a continuous distribution which is bell shaped and perfectly symmetrical. It is completely described when its mean and standard deviation are known.

g) Table I, Areas under the Normal Curve, gives the percentage of the area under the curve according to the number of standard deviations a point is away from the mean. The number of standard deviations away from the mean is the standardised variate or **z** score.

h) The Binomial and Poisson Distributions are discrete probability distributions which describe the behaviour of attributes, e.g. good or bad, black or white.

i) The general binomial formula is $(p + q)^n$ and the expansion of this formula, using Pascal's Triangle or the combination formula, provides the probability of any combination of **p** and **q**.

j) The mean and standard deviation of a Binomial Distribution are given by **np** and $\sqrt{\mathbf{npq}}$ respectively.

k) It is safe to use the Normal approximation to the Binomial if **np** > 5 when **p** < 0.5 or if **nq** > 5 when **p** > 0.5.

l) When **n** is infinitely large and **p** is small the Poisson distribution can be used. It can be used without knowing n or p, being based on the mean value.

m) The variance of a Poisson equals the mean, i.e. variance = **m** and standard deviation = $\sqrt{\mathbf{m}}$, where **m** is the mean of the Poisson Distribution. The Poisson distribution can be used as an approximation to the Binomial distribution if **n** is large, say >30, and **p** is small, say <0.1.

Points to note

30. a) The points which are **+1σ** and **−1σ** away from the mean of a Normal Curve are the *inflexion points* on the curve.

b) The grouping of data, as in a grouped frequency distribution, entails some loss of accuracy although obviously easing calculations.

c) Although the standard deviation is by far the most useful measure of dispersion, on occasions the *range* (difference between the highest and lowest values) and the *average deviation* are of some value.

Self review questions *Numbers in brackets refer to paragraph numbers*

1. What is statistical analysis? (2)
2. Describe a grouped frequency distribution. (3)
3. Distinguish between discrete and continuous data. (4)
4. What is a frequency polygon? (5)
5. What are the four important characteristics of distributions? (7)
6. What is the most important measure of central tendency? (8)
7. What is dispersion and what is the formula for the standard deviation? (9)
8. What is the measure of relative dispersion? (10)
9. Draw diagrams showing positive and negative skewness (11)
10. What is a platykurtic distribution? (12)
11. What are the main features of the Normal Distribution? (13)
12. What percentage of the area of a Normal Curve is between the mean ± 1σ? (14)
13. Define a probability distribution. (16)
14. What is the general form of the binomial expression? (17)
15. What are the alternative ways of establishing the binomial coefficients? (18)

16. What is the formula for the mean of a binomial distribution?, the standard deviation? (21)

17. When is it safe to use the Normal approximation to the binomial? (23)

18. Describe the Poisson distribution. (24)

19. What are the characteristics of the Poisson distribution? (26)

Appendix

The reader should note that there is more than one way to compile Normal Distribution Tables. Table I shows the area under the curve from 0 to **Z** as follows:

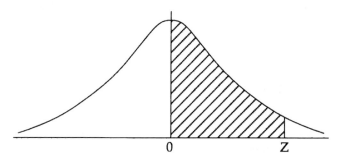

The alternative Table II shows the area from **Z** to ∞ thus:

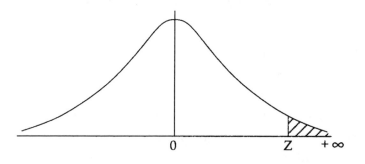

Clearly any answer to a problem will not differ because of this but the reader should be able to handle whichever form of the table is encountered.

For example in the diagram below the probability that **x** lies between **A** and **B** has to be calculated.

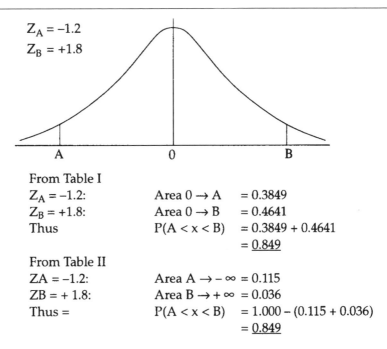

$Z_A = -1.2$
$Z_B = +1.8$

A   0   B

From Table I
$Z_A = -1.2$:   Area $0 \rightarrow A$ = 0.3849
$Z_B = +1.8$:   Area $0 \rightarrow B$ = 0.4641
Thus $P(A < x < B)$ = 0.3849 + 0.4641
                                = <u>0.849</u>

From Table II
$ZA = -1.2$:   Area $A \rightarrow -\infty = 0.115$
$ZB = +1.8$:   Area $B \rightarrow +\infty = 0.036$
Thus = $P(A < x < B)$ = 1.000 – (0.115 + 0.036)
                                = <u>0.849</u>

Exercises with answers

1. Calculate the mean and standard deviation of the following data:
   12, 15, 19, 34, 46, 65, 79, 94, 108.
2. A batch of 5,000 electric lamps have a mean life of 1,000 hours and a standard deviation of 75 hours. Assume a Normal Distribution.
   a) How many lamps will fail before 900 hours?
   b) How many lamps will fail between 950 and 1,000 hours?
   c) What proportion of lamps will fail before 925 hours?
   d) Given the same mean life, what would the standard deviation have to be to ensure that not more than 20% of lamps fail before 916 hours?
3. Assuming a Binomial Distribution what is the probability of a Salesman making 0, 1, 2, 3, 4, 5 or 6 sales in 6 visits if the probability of making a sale on a visit is 0.3?
   Do not use tables for this question.

Answers to exercises

1. $\bar{x} = 52.44$ $\sigma = 33.68$
2. a) The number of lamps that will fail *before* 900 hours is represented by the left-hand tail of the distribution below 900 hours. Accordingly, if the probability of the distribution above 900 is found and deducted from 0.5, the required number can be found, thus:

$$Z = \frac{1,000 - 900}{75} = 1.33 \text{ the probability of which is } 0.4082$$

∴ lamps failing before 900 hours = 5,000 (0.5 − 0.4082) = **459**

b)
$$Z = \frac{1,000 - 950}{75} = 0.67 \text{ the probability of which is } 0.2486$$

∴ lamps failing between 1,000 and 950 hours = 5,000 (0.2486) = **1,243**

c)
$$Z = \frac{1,000 - 925}{75} = 1 \text{ the probability of which is } 0.3413$$

∴ Proportion failing before 925 hours = 0.5 − 0.3413 = **15.87%**

d) A probability of 0.3 (0.5 − 0.2) is found in the tables with a Z score of 0.84.

$$\text{Thus } 0.84 = \frac{1,000 - 916}{\text{s.d.}} \quad \therefore \text{ s.d.} = \frac{84}{0.84} = \mathbf{100}$$

3.   Binomal expansion of $(0.3 + 0.7)^6$

0.3^6	=	0.0007	=	P(6 sales)
$6(0.3^5 \times 0.7)$	=	0.0102	=	P(5 sales)
$15(0.3^4 \times 0.7^2)$	=	0.0595	=	P(4 sales)
$20(0.3^3 \times 0.7^3)$	=	0.1852	=	P(3 sales)
$15(0.3^2 \times 0.7^4)$	=	0.3241	=	P(2 sales)
$6(0.3 \times 0.7^5)$	=	0.3025	=	P(1 sale)
0.7^6	=	0.1176	=	P(0 sales)

5 Statistics – statistical inference

Objectives

1. After studying this chapter you will
    - understand what is meant by statistical inference;
    - know the main characteristic of various sampling methods, including random, systematic, stratified, quota;
    - be able to calculate and use the standard error of the mean;
    - understand the principles of confidence limits;
    - be able to use the Finite Population Correction Factor;
    - know how to estimate population proportions;
    - understand when and how to use Students' t distribution.

Statistical inference defined

2. Statistical inference can be defined as the process by which conclusions are drawn about some measure or attribute of a population (e.g. the mean or standard deviation) based upon analysis of *sample data*. This is a large and important area of the application of statistical theory and is the basis of many examination questions.

    There are numerous reasons why samples are taken and analysed in order to draw conclusions about the whole population. For example, if it was desired to calculate the average weight of the adult population of Britain it would be impractical (except at enormous cost) to weigh each individual and calculate the average weight. More simply a sample would be taken and the population average deduced from the sample measurements. Alternatively, the measurement or testing process may be destructive so that sampling is the only feasible method. There would, for example, be little point in an electric bulb manufacturer testing all bulbs to destruction to discover the average life of his bulbs.

    Sampling is cheaper, quicker and is often the only feasible method of finding information about the population. The type of sampling with which we are concerned is *random sampling* (sometimes known as *probability sampling*). In this type of sampling each item in the population has an equal probability of being chosen.

    Random does not mean haphazard. Random selection is determined objectively, usually by means of random number tables or by computer generated random numbers. Although simple random sampling is rarely used alone in business applications it is important because it is the basis of more sophisticated sampling systems and because it illustrates the fundamental principles of sampling.

    An outline of the more important sampling systems is given below.

Sampling

3.   Sampling is the process of examining a representative set of items (people or things) out of the whole population or universe. The purpose is to gain an understanding about some feature or attribute of the whole population, based on the characteristics of the sample. We may wish to make an estimate of the mean or standard deviation of the population or to discover the type of distribution (e.g. Poisson, Normal and so on) which the population most resembles. A sample only provides an estimate of a population characteristic and the accuracy of the estimate will depend on:

a)   the size of the sample. In general the larger the sample the greater the probability that the sample is truly representative of the population.

b)   how the sample is selected

c)   the extent of variability in the population.

In practice the design and operation of sampling investigations is a complex and technical matter but the overall objective is to select as representative sample as possible, without bias, at a realistic cost.

Pure random sampling, i.e. where every item in the population has an equal chance of being chosen, is the best theoretical sampling method but it does require a *sampling frame*. A sampling frame is a list of every item or member of the population. Thus if it was required to select a random sample of rate payers in a district, the sampling frame would be the Register of Rate Payers maintained by the Local Authority. In practice a complete sampling frame is difficult, if not impossible, to obtain. This is because:

a)   Even where lists are available they are often inaccurate or out of date.

b)   There are many circumstances where a full sampling frame is not feasible. For example, a brewer wishes to sample lager drinkers. How could a sampling frame be constructed in these circumstances? The answer is that it could not be done.

c)   The population may be continually growing and changing.

d)   The costs of constructing a sample frame may be prohibitive and greater than the benefits of increased confidence in the accuracy of sample estimates.

Care is necessary where partial sample frames exist. For example, if it was wished to sample motorists' opinions and AA and RAC membership lists were used as a sampling frame then this would not be a random sample. This is because many motorists are not members of either organisation and would thus be excluded from the sample.

Because of the many practical problems various quasi-random sampling methods have been designed and are widely used.

Quasi-random sampling

4.   The various methods to be described in this paragraph all seek to provide representative samples without undue bias having regard to cost and practicability. The methods are:

Systematic sampling
Stratified sampling
Multi-stage sampling
Cluster sampling
Quota sampling

Systematic sampling

This method is frequently used in production and quality control sampling. After a randomly selected start point(s) a sample item would be selected every nth item. Assume that it was decided to sample every 100th item and a start point of 67 was chosen randomly, the sample would be the following items:

67th, 167th, 267th, 367th and so on

The gap between selections is known as the sampling interval and is itself often randomly selected.

Care is needed with this system to ensure that there is no regular pattern which by coinciding, or not coinciding, with the sampling interval, causes bias.

Stratified sampling

This is a commonly used method where the population can be divided into groups or strata.

Random samples are taken from within each group in the proportions that each group bears to the population as a whole. For example, assume it was required to sample 100 staff at a group of hospitals whose total staff could be stratified as follows:

		Proportion
Doctors	200	10%
Nurses	600	30%
Ancillary workers	800	40%
Administrators	400	20%
	2,000	100%

With a sample of 100 these proportions would result in the following numbers being selected:

Doctors	10
Nurses	30
Ancillary workers	40
Administrators	20
	100

Multi-stage sampling

This is a practical system widely used to reduce the travelling time for interviewers and the subsequent costs. Multi-stage sampling is similar to stratified sampling except that the groups and sub-groups are selected on a geographical/location basis rather than some social characteristic. A country could be divided into areas (say counties) and a random choice of, say, 5 areas made. From each of these selected areas would be drawn a random sample of, say, 6 towns from each of which would be drawn a random sample of people to interview. The process of stratification might continue below the level of a town and would depend on the homogeneity of the population.

Cluster sampling

This is a useful system for reducing sampling costs and dealing with the lack of a satisfactory sampling frame. A few geographical areas (perhaps a village or a street in a town) are selected at random and every single household is interviewed in the selected area.

Quota sampling

This type of sample is not pre-selected but is chosen by the interviewer on the spot up to the level of a quota. To avoid undue bias the quota is sub-divided into various categories, e.g. male/female, old/young, working/unemployed and so on. The interviewer is given quotas for each category and uses discretion to select the interviewees. If done badly, substantial bias can be introduced with this system so close control and checks are employed. The method is cheap and reasonably effective and in consequence is widely used.

The above notes provide a brief outline of common sampling designs. In most examinations where some statistical analysis has to be carried out on the results of a sample, the statement is made that the results were obtained from a random sample. In such circumstances it can be assumed that the sample is truly representative and is without bias.

Types of inference

5. Statistical inference can conveniently be divided into two types – *estimation* and *hypothesis testing*. Estimation deals with the estimation of population characteristics (such as the population mean and standard deviation) from sample characteristics (such as the sample mean and standard deviation). The population characteristics are known as *population parameters* whilst the sample characteristics are known as *sample statistics*. Estimation is dealt with in this chapter whilst hypothesis testing – i.e. the process of setting up a theory or hypothesis about some characteristic of the population and then sampling to see if the hypothesis is supported or not – is dealt with in the next chapter.

Initially it will be assumed that all the samples with which we are dealing are *large* samples (above, say, 30) drawn from an infinitely large population or that the sampling is carried out with replacement, which amounts to the same thing. Small samples and finite populations require somewhat different treatments which are dealt with subsequently.

Properties of good estimators

6. There are four properties of a good estimator.

   a) **Unbiased**. An estimator is said to be unbiased if the mean of the sample means $\bar{x}$ of all possible random samples of size **n**, drawn from a population of size **N**, equals the population parameter μ. Thus the mean of the distribution of sample means would equal the population mean.

   b) **Consistency**. An estimator is said to be consistent if, as the sample size increases, the precision of the estimate of the population parameter also increases.

   c) **Efficiency**. An estimator is said to be more efficient than another if, in repeated sampling, its variance is smaller.

   d) **Sufficiency**. An estimator is said to be sufficient it if uses all the information in the sample in estimating the required population parameter.

   Rarely, in practical situations, do we have all the population values so that we are able to calculate the population parameters. Invariably, we sample in order to estimate the population parameters. Because of this, it is necessary to distinguish between the symbols used for sample statistics and population parameters as follows:

	Sample Statistic	Population Parameter
Arithmetic Mean	$\bar{x}$	μ
Standard deviation	s	σ
Number of items	n	N

Estimation of the population mean

7. The use of the sample mean $\bar{x}$, to make inferences about the population mean is a common procedure in statistics.

   If a series of samples of size **n** (**n** $\geq$ 30) is taken from a population it will be found that:

   a) Each of the sample means is approximately equal to the population mean. (It would be pure coincidence if a sample mean was exactly equal to the population mean.)

   b) The sample means cluster much more closely around the population mean than the original individual values.

   c) The larger the samples the more closely would their means cluster round the population mean.

   d) The distribution of sample means follows a normal curve.

   The latter point is one aspect of what is called the *central limit theorem*. This has been called the most important single theorem in statistics. This, in essence, states that the means of samples (and the median and the standard deviation) tend to be normally distributed as **n** increases in size almost regardless of the shape of the original population.

The central limit theorem gives the normal distribution its key place in the theory of sampling.

The sample mean $\bar{x}$ is a random variable and because the distribution of the sample means is normal, or nearly so, it can be completely described by its mean and its standard deviation. The distribution of all possible sample means is known as the **sampling distribution of the mean** and its standard deviation is given the special name of *standard error*. This is convenient because whenever 'standard error' is used we know at once that a sampling distribution is referred to.

$$\text{Standard error of the mean} = s_{\bar{x}} = \frac{s}{\sqrt{n}}$$

Note: this formula is satisfactory when we have a large sample and a large population, i.e. $n > 30$ and $n < 5\%$ of N.

The smaller the standard error the greater the precision of the sample value. The word 'error' is used in place of 'deviation' to emphasize that variation among sample means is due to sampling errors. In this context 'error' does not mean mistake. It simply means that samples are unlikely to be truly representative of the population.

An example of the mean and standard error of the mean

8.

> **Example 1**
>
> A firm purchases a very large quantity of metal offcuts and wishes to know the average weight of an offcut. A random sample of 625 offcuts is weighed and it is found that the mean sample weight is 150 grams with a sample standard deviation of 30 grams. What is the estimate of the population mean and what is the standard error of the mean? What would be the standard error if the sample size was 1,225?

Solution

The sample mean is 150 grams so that the estimate of the population mean is 150 grams.

$$\text{i.e. } \bar{x} = 150 \text{ grams} = \hat{\mu} = 150 \text{ grams}$$

where ˆ means 'best estimate of'.

$$\text{When } n = 625$$

Standard error of the mean

$$= \frac{s}{\sqrt{n}} = \frac{30}{\sqrt{625}} = 1.2 \text{ grams}$$

$$\text{When } n = 1{,}225$$

$$s_{\bar{x}} = \frac{30}{\sqrt{1,225}} = 0.857 \text{ grams}$$

It will be seen that increasing the sample size reduces the standard error. This accords with common sense; we would expect a larger sample to be better than a smaller one.

The estimate obtained of the population mean, 150 grams, is what is known as a *point estimate*.

This estimate was easily made but so far we have ignored the all important question – how good an estimate is it? This is established by using what are known as *confidence limits*.

Confidence limits

9. Confidence limits are the outer limits to a confidence zone or interval. This is a zone of values within which we may be confident that the true population mean (or the parameter being considered) does lie. Confidence limits, for population means, are based on the sample mean, the standard error of the mean and on the known characteristics of Normal Distribution.

It is known that a Normal Distribution has the following characteristics.

mean $\pm 1.96\sigma$ includes 95% of the population.

mean $\pm 2.58\sigma$ includes 99% of the population.

These characteristics can be used to calculate confidence limits for the population mean when we have established the sample mean and the standard error.

For example, what are the 95% confidence limits for the estimate of the population mean given the data from the example in Paragraph 8, i.e.

$$\bar{x} = 150 \text{ grams and } s_{\bar{x}} = 1.2 \text{ grams}$$

$\therefore$ At the 95% confidence level μ is between the two values

$$\bar{x} \pm 1.96 \times s_{\bar{x}}$$
$$150 \pm 1.96(1.2)\text{grams}$$
$$\mathbf{150 \pm 2.35 \text{ grams}}$$

This means that we are 95% confident that the population mean lies within the confidence zone, i.e. somewhere between 147.65 grams and 152.35 grams.

At the 99% confidence level μ is between the two values:

$$\mathbf{150 \pm 2.58(1.2)\text{grams}}$$

i.e. a range from **146.90 to 153.10 grams**

Notes:

a) Raising the confidence factor from 95% to 99% increases the assurance that the confidence zone contains the population mean but it makes the estimate less precise.

b) Any confidence level could be chosen and the appropriate number of standard errors found from Normal Area Tables. However, the 95% and 99% values of 1.96σ and 2.58σ are widely used and should be remembered.

The principles involved in setting confidence limits can be used to determine what sample size should be taken, if we wish to achieve a given level of precision.

For example what sample size would be required in the example in Paragraph 8 if we wished to be 95% confident that the population mean is within 2 grams of the sample mean?

Because the confidence level is given, the only way that the precision can be increased is by reducing the **standard error** which can be done by *increasing the sample size*.

To obtain 95% limits of 2 grams would require a standard error of 1.02 grams, i.e.

$$\frac{2 \text{ grams}}{1.96}.$$

$$\therefore \text{ as } s_{\bar{x}} = \frac{s}{\sqrt{n}} \text{ then } 1.02 = \frac{30}{\sqrt{n}}$$

$$\therefore \sqrt{n} = \frac{30}{1.02}$$

$$n = \underline{\underline{865}}$$

The principles of confidence limits have been illustrated in connection with means but their use is not limited solely to the estimation of population means from a sample. The principles can be applied across a broad range of statistical applications.

Finite population correction factor

10. Large samples drawn from infinitely large populations increase the accuracy of the sample standard deviation as an estimate of the population standard deviation, reduce the standard error and therefore increase the reliability of the sample mean as an estimate of the population mean.

Obviously not all populations are of infinite size or effectively so and a given population may be relatively small. In these circumstances the proportion of the population included in the sample be relatively high. Where the sample contains in excess of 5% of the population, the standard error should be adjusted by multiplying it by the finite population correction factor, i.e.

$$\text{Finite Population Correction Factor} = \sqrt{\frac{N - n}{N - 1}}$$

where **N** is the population size and **n** is the sample size.

With this correction the standard error becomes:

Standard error (when sample size is > 5% of the population)

$$s_{\bar{x}} = \frac{\sigma}{\sqrt{n}} \times \sqrt{\frac{N - n}{N - 1}}$$

Alternatively where the population, although small, is greater than 100 an approximation of the correction factor can be used:

$$s_{\bar{x}} = \frac{\sigma}{\sqrt{n}} \times \sqrt{1 - \frac{n}{N}}$$

The use of the finite population correction factor always reduces the standard error.

Example 2

A sample of 80 is drawn at random from a population of 800. The sample standard deviation was found to be 6 grams.

What is the finite population correction factor?

What is the approximation of the correction factor?

What is the standard error of the mean?

Solution

$$\text{Correction factor} = \sqrt{\frac{N - n}{N - 1}} = \sqrt{\frac{800 - 80}{800 - 1}} = \underline{\underline{0.9493}}$$

$$\text{Approximation to correction factor} = \sqrt{1 - \frac{n}{N}} = \sqrt{1 - \frac{80}{800}} = \underline{\underline{0.9486}}$$

It will be seen that to 3 significant figures, accurate enough for all practical purposes, the two formulae give the same result, i.e. 0.949.

$$\text{Standard Error of the means } \mathbf{S}_{\bar{x}} = \frac{\sigma}{\sqrt{n}} \times \sqrt{1 - \frac{n}{N}}$$

$$= \frac{6}{\sqrt{80}} \times 0.949$$

$$= \mathbf{0.637 \text{ grams}}$$

Note: The standard error without correction is $\frac{6}{\sqrt{80}} = \mathbf{0.671 \text{ grams}}$

Thus the precision of the sample estimate, measured by the standard error, is determined not only be the absolute size of the sample but also to some extent by the proportion of the population sampled.

Estimation of population proportions

11. So far the process of statistical inference has been applied to the arithmetic mean. A similar process can be used for other types of statistical measures – medians, standard deviations, proportions and so on. In each case the analysis includes three essential elements:

    a) The required measure as found from the sample.

    b) The standard error of the measure involved.

    c) The sampling distribution of the measure.

    As explained in the previous chapter, a proportion represents an *attribute* of a population rather than the value of a variable. The proportion may represent the ratio of defective to good production, the proportion of consumers who plan to buy a given product or some similar piece of information.

    Estimating population proportions from sample statistics follows the same pattern as outlined for means. The major difference being that the Binomial Distribution is involved. Statistical inference based on the Binomial Distribution involves complex technical difficulties caused by the discreteness of the distribution and the asymmetry of confidence intervals.

    Fortunately when 'n' is large and **np** and **nq** are over 5 then the Binomial Distribution can be approximated by the Normal Distribution. This greatly simplifies the analysis and the concepts outlined for the mean can be applied directly to the proportion.

    The relevant formula for the standard error of a sample proportion is

    $$s_{ps} = \sqrt{\frac{pq}{n}}$$

    Using this value we are able to set confidence limits for the estimate of the population proportion based on the sample proportion in exactly the same manner as outlined previously for the mean.

Example 3

A random sample of 400 rail passengers is taken and 55% are in favour of proposed new timetables.

With 95% confidence what proportion of all rail passengers are in favour of the timetables?

Solution

$$n = 400 \quad p = 0.55 \quad \text{and} \quad q = 1 - 0.55 = 0.45$$

as **np** = 220, i.e. well over 5, the normal approximation can be used.

$$s_{ps} = \sqrt{\frac{pq}{n}} = \sqrt{\frac{0.55 \times 0.45}{400}} = 0.025$$

∴ We are 95% confident that the population proportion is between the two values

$$p_s \pm 1.96s_{ps}$$
$$= 0.55 \pm 1.96(0.025) = 0.55 \pm 0.049$$
$$= 0.501 \text{ to } 0.599$$

Estimation from small samples

12. So far it has been assumed that all the samples have been large ($n > 30$). In such circumstances s, the sample standard deviation, is used as an estimate of σ, the population standard deviation. Further, it is known that the distribution of sample means is approximately normal so that the properties of the Normal Distribution can be used to calculate confidence limits using the standard error of the mean.

However, the properties and relationships summarised above are not true if the sample size is small ($n < 30$), because the arithmetic means of small samples are not normally distributed. In such circumstances *Student's t distribution* must be used.

To use this distribution all that is necessary is to substitute a given value of 't' (from Table III) for the value of z obtained from Table I, Areas under the Normal curve, thus:

Large sample

μ is between the two values $\bar{x} \pm z \; s_{\bar{x}}$

At the 95% confidence level $z = 1.96$

At the 99% confidence level $z = 2.58$

(the values of z from Normal Area Tables)

Small sample

μ is between the two values $\bar{x} \pm t \; s_{\bar{x}}$

Where the value of t is obtained from Table III for the required confidence level. It will be seen that the processes are very similar with the exception that the Students distribution is used when the sample is small.

The 't' distribution

13. This distribution is used when there are small samples and when it is necessary to make an estimate of σ, the population standard deviation, based on s, the sample standard deviation.

The statistic, **t**, is as follows:

$$t = \frac{\bar{x} - \mu}{\frac{s}{\sqrt{n}}} \quad \text{where } s \text{ is an estimate of } \sigma$$

$$s = \sqrt{\frac{\sum(x - \bar{x})^2}{n - 1}} \quad \text{as previously defined.}$$

The characteristics of the distribution are:

a) It is an exact distribution which is unimodal and symmetric about 0.

b) It is flatter than the Normal Distribution, i.e. the areas near the tails are greater than the Normal Distribution.

c) As the sample size becomes larger the **t** distribution approaches the Normal Distribution.

To use the **t** distribution tables (Table III) it is necessary to find the *degrees of freedom* usually denoted by **v**, thus

$$v = n - 1$$

where **n** is the number of items in the sample. For example, find the **t** value, for a 95% confidence level, with a sample size of 10.

$v = 10 - 1 = 9$ and from the column in Table III headed '.05' we find the value **2.262**.

Example 4

A random sample of 10 items is taken and is found to have a mean weight of 60 grams and a standard deviation of 12 grams.

What is the mean weight of the population.

    a) With 95% confidence?

    b) With 99% confidence?

Solution

$$\bar{x} = 60 \text{ grams } s = 12 \text{ } v = 10 - 1 = 9 \text{ degrees of freedom}$$

μ is between the two values $\quad \bar{x} \pm t\dfrac{s}{\sqrt{n}}$

At 95% confidence

μ is between the two values $\quad 60 \text{ grams} \pm 2.262\left(\dfrac{12}{\sqrt{10}}\right)$

$$= 60 \pm 8.58 \text{ grams}$$

Therefore we can state with 95% confidence that the population mean is between 51.42 and 68.58 grams.

At 99% confidence

$$\mu \text{ is between the two values } 60 \text{ grams} \pm 2.262 \left(\frac{12}{\sqrt{10}} \right)$$

$$= 60 \pm 12.33 \text{ grams}$$

Therefore we can state with 99% confidence that the population mean is between 47.67 and 72.33 grams.

Summary

14. a) Statistical inference is the process of drawing conclusions about the population from samples.

  b) Random sampling means that every item in the population has an equal chance of being chosen and that there is no bias.

  c) Estimation is concerned with estimating population parameters from sample statistics.

  d) Where the sample size is relatively large the sampling distribution of means is a normal distribution.

  e) The Central Limit Theorem states that the means of samples (and the medians and standard deviations) tend to be normally distributed almost regardless of the shape of the original population.

  f) The standard deviation of the distribution of means is called the standard error of the mean, i.e.

$$s_{\bar{x}} = \frac{s}{\sqrt{n}}$$

  g) Confidence limits provide the limits to a confidence interval or zone in which we may be confident, at a given level, that the true population parameter lies.

  h) The most commonly used confidence limits are the 95% level ($\pm 1.96\sigma$) and the 99% level ($\pm 2.58\sigma$).

  i) Where the sample contains in excess of 5% of the population, the Finite Population Correction Factor

$$\left(\sqrt{\frac{N - n}{N - 1}} \right) \text{ needs to be applied.}$$

  j) Finding population proportions from sample information follows the usual estimation process provided that **np** and **nq** are over 5 so that the Normal approximation can be used.

  k) Where **n** < 30 the sample is small, the Students t distribution must be used instead of the Normal Distribution.

Points to note

15. a) The standard error of the mean, $\sigma_{\bar{x}}$, is always smaller than σ, the standard deviation of the values.

   b) An alternative view of, say a 95% confidence level, is that there is a 5% or 1 in 20 chance of the population parameter being outside the confidence interval.

   c) The Central Limit Theorem is valid for sampling from Normal and non Normal Distributions as in all cases the distribution of the means of samples tends towards a Normal Distribution.

   d) The **t** distribution used for small samples also has important applications in hypothesis testing dealt with in the next chapter.

Self review questions Numbers in brackets refer to paragraph numbers

1. What is statistical inference? (2)
2. What is the main feature of random sampling? (2)
3. What is the purpose of estimation? (5)
4. What are the properties of good estimators? (6)
5. What is the main feature of the Central Limit Theorem and why is it so important? (7)
6. What is the standard error of the mean? (7)
7. What are confidence limits? (9)
8. When is the Finite Population Correction Factor used? What is the formula? (10)
9. How are population proportions estimated? (11)
10. What adjustments are required when small samples are used? (12)
11. What are the characteristics of the t distribution? (13)
12. How are the number of degrees of freedom calculated? (13)

Exercises with answers

1. From a random sample of 529 televisions off the production line it was found that each set had 8 faults on average with a standard deviation of 3.45 faults.

   What are the confidence limits for the production as a whole: at the 99% level; at the 95% level?

2. An inspection department is trying to determine the appropriate sample size to use. They wish to be within 1% of the true proportion with 99% confidence. Past records indicate that the proportion defective is 3 in a 100.

   What sample size should they use?

3. A random sample of 10 packets was taken and found to have a mean weight of 50 grams and a standard deviation of 9 grams.

   What is the mean weight of the population.

a) with 95% confidence.

b) with 99% confidence?

Answers to exercises

$$\text{Standard Error} = \frac{s}{\sqrt{n}} = \frac{3.45}{\sqrt{529}} = \underline{\underline{0.15}}$$

At 99% level the average faults are in the range $8 \pm 2.58\,(0.15) = 7.613$ to 8.387

At the 95% level the average faults are in the range $8 \pm 1.96\,(0.15) = 7.706$ to 8.294

2. $\qquad\qquad\qquad\qquad\mathbf{p} = 0.03 \qquad\qquad \mathbf{q} = 1 - \mathbf{p} = 0.97$

$\mathbf{d}$ = degree of accuracy required and

$$s = \sqrt{\frac{pq}{n}}$$

$\therefore$ at 99% level $\qquad\qquad\qquad\qquad\qquad d = 2.58\sqrt{\frac{pq}{n}}$

$$\therefore d^2 = \frac{2.58^2 pq}{n}$$

$$\therefore n = \frac{2.58^2 pq}{d^2}$$

and substituting when $\mathbf{d} = 0.01 \qquad\qquad \therefore n = \dfrac{2.58^2\,(0.03)(0.97)}{0.01^2}$

$$= \underline{\underline{1937}}$$

3. $\qquad\qquad\qquad\qquad\qquad\qquad \bar{x} = 50 \quad s = 9 \quad v = 10 - 1 = 9$

μ is between the two values $\qquad\qquad\qquad \bar{x} \pm t\dfrac{s}{\sqrt{n}}$

at 95% confidence level $\qquad\quad \mu$ is between the two values $\quad 50 \pm 2.262\left(\dfrac{9}{\sqrt{12}}\right)$

$$= \underline{\underline{50 \pm 5.877 \text{ grams}}}$$

at 99% confidence level $\qquad\quad \mu$ is between the two values $\quad 50 \pm 3.25\left(\dfrac{9}{\sqrt{12}}\right)$

$$= \underline{\underline{50 \pm 8.444 \text{ grams}}}$$

6 Statistics – hypothesis testing

Objectives

1. After studying this chapter you will
    - be able to define hypothesis (or significance) testing;
    - understand Type I and Type II errors;
    - know how to calculate significance levels;
    - be able to distinguish between one tail and two tail tests;
    - understand significance tests of means, proportions and the difference between means and proportions;
    - know when and how to use Students t distribution;
    - understand the chi-square (χ^2) distribution;
    - be able to apply Yates correction;
    - understand the objectives of Statistical Process Control;
    - be able to prepare basic Control Charts.

Hypothesis testing defined

2. Hypothesis testing, alternatively called significance testing, is in many ways similar to the processes of estimation dealt with in the last chapter. Random sampling is involved and the properties of the distribution of sample means and proportions are still used.

    A hypothesis is some testable belief or opinion, and hypothesis testing is the process by which the belief is tested by statistical means.

    For example, from a large batch of components a random sample may be taken to test the belief (i.e. the hypothesis) 'That the mean diameter of the population of components is 50 mm'. Based on the results obtained from the sample the hypothesis would either be accepted or rejected. The hypothesis to be tested is termed the **Null Hypothesis** designated H_0

Results of hypothesis testing

3. There are only four possible results when we test the hypothesis, H_0:
    a) We *accept a true* hypothesis – a *correct* decision.
    b) We *reject a false* hypothesis – a *correct* decision
    c) We *reject a true* hypothesis – an *incorrect* decision. This is known as a *Type I* error.
    d) We *accept a false* hypothesis – an *incorrect* decision. This is known as a *Type II* error.

    As an example of the four possible results consider the following:

Components made by a machine are sampled and measured to assess whether the process is in control or not. If the process is deemed to be out of control, production will be stopped. Naturally, there will always be some inherent variability, but the object of the inspection is to decide whether the process is operating within safe limits, i.e. is it in control? It is possible that the process is out of control even though the measurements are within tolerances and, conversely, it is possible that the process is satisfactory even though the inspected output sample is outside normal tolerances. The four possible results are summarised below:

Decision	Position	
	Process in Control	Process out of Control
Continue production	Correct	Error (Type II)
Stop production	Error (Type I)	Correct

Type I and type II errors

4. Obviously, efforts are made to avoid any type of error but as it is not possible to make a correct decision with 100% certainty when a hypothesis is tested by sampling, there is always a possibility of either a Type I or Type II error.

The errors are mutually exclusive; an error can be Type I or Type II but not both. The errors are split into two types because there are situations where it is much more important to avoid one type of error rather than the other.

For example, if we have the hypothesis, 'The water is shallow', the consequences of a Type II error could be fatal for a non-swimmer.

The risk associated with the two types of error are denoted by α and β, thus:

$$P\ (\text{Type I error}) = \alpha$$

$$P\ (\text{Type II error}) = \beta$$

The α risk is the level of significance chosen for the hypothesis test, most commonly 1% or 5%.

Significance levels

5. When a sample is taken to test some hypothesis it is likely that the information gleaned from the sample data (e.g. the sample mean or standard deviation) does not completely support the hypothesis. The difference could be due to either the original hypothesis being wrong *or* the sample being slightly unrepresentative, which virtually all samples will be to a greater or lesser extent.

It is clearly of importance to be able to test which of the two possibilities is the more likely and this is the main objective of hypothesis testing. The tests will show whether any differences can be attributed to ordinary random factors or not. If the difference is probably *not* due to chance factors the difference is said to be *statistically significant*.

Because we are dealing with samples and random factors, we cannot say with 100% certainty that a difference is significant. Accordingly, various levels of significance are chosen, most commonly 5% or 1%, and thus the result of a particular test might be expressed as follows:

'The difference between the sample mean and the hypothetical population mean is significant at the 5% level'.

Significance levels and confidence limits, used in the last chapter for Estimation, are complementary and are based on the similar principles.

Notes:

a)   As with confidence limits, any level of significance could be used but 5% and 1% are the most common.

b)   The phrase used above, 'The difference between the sample mean and the population mean is significant at the 5% level', could alternatively be stated as:

'There is 95% confidence that the difference between the sample mean and the population mean is not due to chance factors'.

c)   It is because the significance level must be chosen before the test is carried out and because it is a critical factor in deciding whether to accept or reject a hypothesis that the term *'significance testing'* is commonly used instead of hypothesis testing.

An example of hypothesis testing

6.

Example 1

A machine fills packets with spice which are supposed to have a mean weight of 40 grams. A random sample of 36 packets is taken and the mean weight is found to be 42.4 grams with a standard deviation of 6 grams. It is required to conduct a significance test at the 5% level.

To do this we set up the following hypothesis:

'The population mean weight, μ, is 40 grams'.

This hypothesis is known as the *Null Hypothesis*, designated H_0, and means that we are assuming that there is no contradiction between the supposed mean and the sample mean and that any difference can be ascribed solely to random factors.

The *alternative hypothesis*, designated H_1, is that 'the population mean does not equal 40 grams'.

More simply these hypotheses can be shown thus

$$H_0 : \mu = 40 \text{ grams}$$

$$H_1 : \mu \neq 40 \text{ grams}$$

The null hypothesis is the one which is tested. If H_0 is accepted, H_1 is rejected whilst if H_0 is found to be false, H_1 is accepted.

Based on similar principles to those covered in the previous chapter, we know that for 95% confidence limits (i.e. equivalent to the 5% level of significance) the sample mean must be within ±1.96 standard errors so for H_0 to be accepted, 95% of the means of all samples must be within the range

$$\mu - 1.96\ s_{\bar{x}} \text{ to } \mu + 1.96\ s_{\bar{x}}$$

The standard error in this example is

$$s_{\bar{x}} = \frac{6}{\sqrt{36}} = 1$$

so this gives the range

$$40 \pm 1.96\ (1) \text{ grams, i.e.}$$

38.04 to 41.96 grams

This is shown diagrammatically in Figure 6/1.

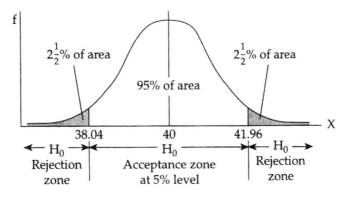

Distribution of sample means. Figure 6/1

As the sample mean, 42.4 grams, is outside the acceptance zone we reject the H_0 hypothesis and accept the H_1 hypothesis, i.e. $\mu \neq 40$ grams, i.e. the difference between $\bar{x}$ and μ is *significant at the 5% level.*

As we have been concerned with *both* tails of the distribution this is known as a *two tail* test of significance. Some significance tests are *one tail* tests, i.e. we are only concerned about differences in one direction. Examples of these are given later in the chapter.

Notes:

1. Because H_0 was rejected in Example 1 there is a possibility of a Type I error.

2. In this example and for all hypothesis testing the calculations are based on the assumption that H_0 is true.

A simpler approach to hypothesis testing

7. The preceding paragraph showed a particular method of solution mainly to illustrate the rationale of significance testing and the idea of a two-tail test. There is, however, a simpler approach to the same conclusion and this is the approach recommended, provided that the underlying principles of testing a hypothesis are understood.

Example 1 will be reworked using the recommended approach, i.e.

$$H_0 : \mu = 40 \text{grams}$$
$$H_1 : \mu \neq 40 \text{ grams (a two-tail test)}$$
$$\bar{x} = 42.4 \text{ grams (given)}$$
$$s = 6 \text{ grams (given)}$$
$$n = 36 \text{ grams (given)}$$

$$\text{Standard error} = s_{\bar{x}} = \frac{6}{\sqrt{36}} = 1$$

$$\therefore Z = \frac{\bar{x} - \mu}{s_{\bar{x}}} = \frac{42.4 - 40}{1} = \underline{\underline{2.24}}$$

At the 5% significance level the Z value must be within ±1.96.

∴ As the calculated score (2.24) is greater than 1.96, H_0 is rejected.

This then is the usual method of testing a hypothesis: calculate the Z score and compare the calculated value with the appropriate value for the required level of significance, e.g. 1.96 for 5%, 2.58 for 1% (for two tail tests).

This approach is used for the next example.

Example 2

Rework Example 1 using a 1% level of significance.

$$H_0 : \mu = 40 \text{ grams}$$
$$H_1 : \mu \neq 40 \text{ grams (a two-tail test)}$$
$$\bar{x} = 42.4 \text{ grams (given)}$$
$$s = 6 \text{ grams(given)}$$
$$n = 36 \text{ grams (given)}$$

$$\therefore s_{\bar{x}} = \frac{6}{\sqrt{36}} = 1$$

$$\therefore Z = \frac{42.4 - 40}{1} = \underline{\underline{2.24}}$$

At the 1% level Z must be lower than 2.58.

$\therefore$ as the calculated score is lower than this value, H_0 can be accepted *at the 1% level*.

i.e. the difference between the sample mean and the hypothetical population mean is *not significant at the 1% level*.

Notes:

a) It is is important to realise that if the null hypothesis is accepted, this is not proof that the assumed population mean is correct. Testing a hypothesis may only show that an assumed value is probably false.

b) It will be apparent from the two examples that the 1% level of significance is a more severe test than the 5% level. The greater the level of significance, the greater the probability of making a Type I error.

One-tailed tests

8. These are tests in which the alternative hypothesis is concerned with only one of the tails of the distributions. For example, if the breaking strain of lifting cables was being studied there would only be concern if they were too weak. If the cables were designed to have a mean breaking point of 50 tonnes we would not be interested in the hypothesis that the true population mean was 50 tonnes *or more* but only if it was *less*.

The procedures for one-tailed tests are very similar to those previously described for two-tailed tests except for one important difference.

This is, that the number of standard errors for a given level of significance is different for one-tailed compared to two-tailed tests. The values for the most common levels, 1% and 5% are shown below:

	Number of standard errors	
	Two-tailed tests	One-tailed tests
5% level of significance	1.96	1.65
1% level of significance	2.58	2.33

The reason for this is that when we are concerned with both tails of the distribution, the 5% (or 1% as the case may be) is shared between both tails as depicted in Figure 6/1, whereas in a one tailed test the 5% or 1% is only in the single tail. Values for any other required significance level can be obtained from Table I, Areas under the Normal Curve.

An example of a one-tailed test follows.

Example 3

Assume the same data as in Example 1 except that we require to test whether the population mean is greater than the assumed value of 40 grams, using a 5% level of significance.

---

i.e.:
$$H_0 : \mu = 40 \text{ grams}$$
$$H_1 : \mu > 40 \text{ grams}$$

This is a one-tailed test as we are only interested in the right-hand tail of the distribution.

As previously, the Z score is 2.24 and for a one-tailed test the Z score should be lower than 1.65.

$\therefore$ as the calculated Z score is greater than the 5% value, we reject the null hypothesis. We can state:

'At the 5% level, H_1 is accepted; the population mean is greater than 40 grams'.

Note: It is vital to ascertain whether a one- or two-tailed test is required so that the correct number of standard errors for the chosen significance level is used.

Hypothesis testing of proportions

9.   This follows a broadly similar approach to that outlined for means except that the standard error used is the *standard error of a proportion*, i.e.

$$s_p = \sqrt{\frac{pq}{n}}$$

The Z score is calculated as follows

$$Z = \frac{p - \pi}{s_p}$$

where

$$p = \text{proportion found in the sample}$$
$$\pi = \text{the hypothetical proportion.}$$

Example 4

It is required to test the hypothesis that 50% of households have a freezer. A random sample of 400 households found that 54% of the sample had freezers. The significance level is 5%.

Note: This is a two-tailed test because we wish to test the hypothesis as it is and not against a specific alternative hypothesis that the real proportion is either larger or smaller.

i.e.:
$$H_0 : \pi = 50\% \text{ of all households}$$
$$H_1 : \pi \neq 50\% \text{ of all households}$$

$$s_p = \sqrt{\frac{pq}{n}} = \sqrt{\frac{0.5 \times 0.5}{400}} = \underline{\underline{0.025}}$$

$$Z = \frac{0.54 - 0.50}{0.025} = \underline{\underline{1.6}}$$

At the 5% level of significance for a two-tailed test the appropriate value is 1.96.

$\therefore$ as the calculated Z score is 1.6 we can say that the difference is *not significant* and that H_0 should not be rejected.

Hypothesis testing of the difference between two means

10. Where two random samples are taken, frequently it is required to know if there is significant difference between the two means. This hypothesis test follows the general pattern except that, once again, the standard error calculation differs.

The distribution of sample mean differences is normally distributed and remains normally distributed whatever the distribution of the population from which the samples are drawn. Where $n > 30$, i.e. large samples, the Normal Area tables are used, and when $n < 30$ the t distribution applies.

$$\text{The standard errors of the difference of means} = s_{(\bar{x}_A - \bar{x}_B)} = \sqrt{\frac{s^2_A}{n_A} + \frac{s^2_B}{n_B}}$$

where s_A = standard deviation of sample A, size n_A

$\quad\quad s_B$ = standard deviation of sample B, size n_B

and the Z score is calculated thus

$$Z = \frac{\bar{x}_A - \bar{x}_B}{s_{(\bar{x}_A - \bar{x}_B)}}$$

Example 5

Machine A and machine B produce indentical components and it is required to test if the mean diameter of the components is the same. A random sample of 144 from machine A had a mean of 36.40 mm and a standard deviation of 3.6 mm, whilst a random sample of 225 from machine B had a mean of 36.90 mm and a standard deviation of 2.9 mm.

Are the means significantly different at the 5% level?

Solution

$$H_0 : \text{mean of A} = \text{mean of B}$$
$$H_1 : \text{mean of A} \neq \text{mean of B}$$

i.e. as we are not concerned with the direction of the variation this is a *two-tail test*.

$$S_{(\bar{x}_A - \bar{x}_B)} = \sqrt{\frac{3.6^2}{144} \quad \frac{2.9^2}{225}} = \underline{\underline{0.357}}$$

$$Z = \left| \frac{36.40 - 36.90}{0.357} \right| = \underline{\underline{1.4}}$$

The score for a two tailed test at the 5% level is 1.96.

Therefore as the calculated Z score of 1.4 is lower than this value there is nothing to suggest that there is any difference between the two means, i.e. H_0 would be accepted.

Hypothesis testing of the difference between proportions

11. In a similar manner it may be required to test the differences between the proportions of a given attribute found in two random samples.

    The following notation will be used:

	Sample 1	Sample 2
Sample size	n_1	n_2
Sample proportion of successes	p_1	p_2
Population proportion of successes	π_1	π_2

    The Null Hypothesis will be $\pi_1 = \pi_2 (= \pi$, say), i.e. that the two samples are from the same population. This being so the best estimate of the standard error of the difference of p_1 and p_2 is given by pooling the samples and finding the pooled sample proportion thus

    $$p = \frac{p_1 n_1 + p_2 n_2}{n_1 + n_2}$$

and the Standard error is:

$$s(p_1 - p_2) = \sqrt{\frac{pq}{n_1} + \frac{pq}{n_2}}$$

and

$$Z = \frac{(p_1 - p_2) - (\pi_1 - \pi_2)}{s_{(\pi_1 - \pi_2)}}$$

but where the Null Hypothesis is $\pi_1 = \pi_2$ the second part of the numerator disappears.

Example 6 – A one-tail test

A market research agency take a sample of 1,000 people and finds that 200 know of Brand X. After an advertising campaign a further sample of 1,091 people is taken and it is found that 240 know of Brand X.

It is required to know if there has been an increase in the number of people having an awareness of Brand X at the 5% level.

Solution

$$H_0 : \pi_2 = \pi_1$$
$$H_1 : \pi_2 > \pi_1 \text{ (one-tail test)}$$

$$p_1 = \frac{200}{1,000} = 0.2$$

$$p_2 = \frac{240}{1,091} = 0.22$$

pooled sample proportion

$$p = \frac{200 + 240}{1,000 + 1,091} = 0.21$$

and

$$q = 1 - p = 1 - 0.21 = 0.79$$

$$\therefore s\,(p_1 - p_2) = \sqrt{\frac{0.21 \times 0.79}{1,000} + \frac{0.21 \times 0.79}{1,091}} = 0.0178$$

$$\therefore Z = \left|\frac{0.20 - 0.22}{0.0178}\right| = 1.12$$

The critical value for a one-tailed test at the 5% level is 1.64 so that as the calculated value is lower than this value we conclude there is insufficient evidence to reject the null hypothesis.

Example 7 – A two-tail test

Surveys were conducted in Birmingham and London to ascertain viewers habits regarding Channel 4 television. In Birmingham 1,000 people were interviewed and 680 said they viewed Channel 4. In London 600 people were interviewed and 444 said they viewed Channel 4.

Is there a significant difference between the viewing habits in Birmingham and London at the 5% level? at the 1% level?

Solution

In Birmingham

$$p = \frac{680}{1,000} = 0.68 \text{ and } q = 0.32$$

In London

$$p = \frac{444}{600} = 0.74 \text{ and } q = 0.26$$

Pooled sample proportion

$$= \frac{680 + 444}{1,000 + 600} = \mathbf{0.7025}$$

$$\therefore \mathbf{q} = 1 - \mathbf{p} = 1 - 0.7025 = \mathbf{0.2975}$$

$$\text{Standard Error} = \sqrt{\frac{0.7025 \times 0.2975}{1,000} + \frac{0.7025 \times 0.2975}{600}} = \mathbf{0.0236}$$

$$\mathbf{Z} = \frac{0.74 - 0.68}{0.0236} = \mathbf{2.54 \ standard \ errors}$$

At the 5% level for a two-tail test $\mathbf{Z} = \pm 1.96$ and the calculated value is greater.

∴ there is a significant difference at 5%.

At the 1% level $\mathbf{Z} = \pm 2.58$ and the calculated value is lower.

∴ there is not a significant difference in the viewing habits at the 1% level.

Distributions other than normal

12. So far the discussion of statistical inference and hypothesis testing has been based on large samples resulting in a sampling distribution for the sample mean or proportion that is approximately normal. However, many sampling situations are not covered by this assumption. For example – it may be possible to use only a small sample – we may have an attribute that is classified into more than two categories – there may be samples from two or more populations to evaluate simultaneously – these and other special circumstances require distributions other than the Normal.

Accordingly the previous material is extended to cover two more distributions; the **t** distribution and the χ^2 (chi-square) distribution.

Hypothesis testing using the 't' distribution

13. There are many occasions where it is only feasible to use small samples, which is usually taken to be when **n** < 30. The means of small samples are distributed around the population mean in a manner similar to, but not exactly, a normal distribution. The probability distribution of small sample means is known as a **t distribution** or **Students t distribution**. When the **t** distribution applies, the value from the Normal Area Tables used so far in this chapter (e.g. 1.96 at the 95% level of confidence) is replaced by the **t** value from Table III. This value varies with **v**, the degrees of freedom.

For a single sample this is $\mathbf{n} - 1$, where two samples are being compared the degrees of freedom are $\mathbf{n_1} + \mathbf{n_1} - 2$

Apart from these differences the general procedures for hypothesis testing using the **t** distribution follow the pattern previously described.

Example 8

A firm ordered sacks of chemicals with a nominal weight of 50 kg. A random sample of 8 sacks was taken and it was found that the sample mean was 49.2 kg with a standard deviation of 1.6 kg. The firm wish to test whether the mean weight of the sample of sacks is significantly less than the nominal weight, using a 5% level of significance.

Solution

This is a one-tail test as we are only interested in whether the mean weight is less than the nominal.

$$H_0 : \textbf{mean} = \textbf{50 kg}$$
$$H_1 : \textbf{mean} < \textbf{50 kg}$$

Degrees of freedom $= \textbf{n} - 1 = 8 - 1 = 7$

$$\text{Standard error of the mean} = \frac{s}{\sqrt{n}} = \frac{1.6}{\sqrt{8}} = \underline{\underline{0.566}}$$

$$\text{Therefore } \mathbf{t} = \frac{\bar{x} - \mu}{s_{\bar{x}}} = \left| \frac{49.2 - 50}{0.566} \right| = \underline{\underline{1.413}}$$

In order to find a one-tailed 5% probability point we look up the .10 (two-tailed) point in Table III for 7 degrees of freedom. This value is 1.895.

Thus as the calculated **t** score, 1.413, is less than 1.895 we can accept the null hypothesis and reject the alternative hypothesis at the 5% level.

Note: The 10% probability point in Table III is spread between both tails, i.e. 5% in each tail.

t distribution tests of the difference between means

Previously this test was carried out using large samples. If the samples are small, the t distribution can be used to test for differences in population means. In such circumstances two additional assumptions are necessary:

a) That the two sampled populations are normally or near normally distributed.

b) That the two standard deviations are equal or at any rate not significantly different. The best estimate of the population standard deviation is obtained by pooling the two sample standard deviations as shown below.

Given that two samples are taken and that n_1, x_1, s_1 and n_2, x_2, s_2 are the sample sizes, means and standard deviations respectively then the common or pooled estimate of the standard deviation for both populations (i.e. s_p) can be found as follows:

$$s_p = \sqrt{\frac{(n_1 - 1)s_1^2 + (n_2 - 1)s_2^2}{n_1 + n_2 - 2}}$$

s_p is then the best estimate of the standard deviation in each population. The standard error for each sample mean can be calculated thus:

$$s_{\bar{x}_1} = \frac{s_p}{\sqrt{n_1}} \text{ and } s_{\bar{x}_2} = \frac{s_p}{\sqrt{n_2}}$$

and the sampling error of the distribution of differences in sample means is

$$s_{(\bar{x}_1 - \bar{x}_2)} = \sqrt{s_{\bar{x}_1}^2 + s_{\bar{x}_1}^2}$$

The value can be used to find the **t** score in the normal manner:

$$t = \frac{\bar{x}_1 - \bar{x}_2}{s_{(\bar{x}_1 - \bar{x}_2)}} \text{ with } n_1 + n_2 - 2 \text{ degrees of freedom.}$$

Example 9

The monthly bonuses of two groups of salesmen are being investigated to see if there is a difference in the average bonus received. Random samples of 12 and 9 are taken from the two groups and it can be assumed that the bonuses in both groups are approximately normally distributed and that the standard deviations are about the same. A 5% level of significance is to be used.

The sample results were

$n_1 = 12$　　　　　　　　　　$n_2 = 9$

$\bar{x}_1 = £1,060$　　　　　　　$\bar{x}_2 = £970$

$s_1 = £63$　　　　　　　　　$s_2 = £76$

Solution

$$H_0 : \mu_1 - \mu_2 = 0$$

$$H_1 : \mu_1 - \mu_2 \neq 0$$

It will be seen that this is a *two-tailed* test.

The common standard deviation is calculated first.

$$s_p = \sqrt{\frac{(n_1 - 1)s_1^2 + (n_2 - 1)s_2^2}{n_1 + n_2 - 2}}$$

$$= \sqrt{\frac{(12-1)63^2 + (9-1)76^2}{12+9-2}}$$

$$= \underline{68.77}$$

$$\therefore s_{\bar{x}_1} = \frac{s_p}{\sqrt{n_1}} = \frac{68.77}{\sqrt{12}} \quad = \underline{\underline{19.85}}$$

$$s_{\bar{x}_2} = \frac{s_p}{\sqrt{n_2}} = \frac{68.77}{\sqrt{9}} \quad = \underline{\underline{22.92}}$$

$$\therefore s_{(\bar{x}_1 - \bar{x}_2)} = \sqrt{s_{\bar{x}_1}^2 + s_{\bar{x}_2}^2} \quad = \underline{\underline{30.32}}$$

and, finally, the **t** score can be calculated

$$t = \frac{\bar{x}_1 - \bar{x}_2}{s_{(\bar{x}_1 - \bar{x}_2)}} = \frac{1{,}060 - 970}{30.32} = \underline{\underline{\mathbf{2.97}}}$$

From Table III the 5% value, with $(n_1 + n_2 - 2) = (12 + 9 - 2) = 19$ degrees of freedom is **2.093**.

Since the calculated value is greater than this, the null hypothesis can be rejected at the 5% level, i.e. we conclude that *there is a significant difference* between the mean monthly incomes.

Non-parametric tests

15. The significance tests covered so far depend, to a greater or lesser extent, on the assumption, or presence, of the normal distribution.

They are also concerned with the *parameters* of the distribution, e.g. mean, proportions, etc. Hence, they have been given the name of *parametric tests*.

However, on occasions, the data are non-normal or contain extreme values or not enough is known to be able to make any assumption about the type of distribution. In such circumstances *non-parametric* or *distribution-free* tests may be used. In addition, non-parametric tests can be used on data ranked in some order as, for example, when a consumer in a market test is asked to rank their preferences for a group of products.

Advantages of non-parametric tests:

a) No assumptions need be made about the underlying distributions.

b) They can be used on data ranked in some order.

c) The mathematic concepts are simpler than for parametric tests.

Disadvantages of non-parametric tests:

a) They are less discriminating than parametric tests, i.e. they are more prone to error and less powerful.

b) Although simple, the arithmetic may take a long time.

An important example of a test which does not make any assumption about the distribution from which the sample is taken and hence is often classified as non-parametric, is the Chi-square x^2 test.

The chi-square (x^2) distribution

16. The x^2 test is an important extension of hypothesis testing and is used when it is wished to compare an actual, observed distribution with a hypothesised, or expected distribution. It is often referred to as a 'goodness of fit' test.

The formula for the calculation of x^2 is as follows:

$$x^2 = \sum \frac{(O - E)^2}{E}$$

where

O = the observed frequency of any value.

E = the expected frequency of any value.

The x^2 value obtained from the formula is compared with the value from Table IV of x^2 for a given significance level and the number of degrees of freedom, i.e. the usual hypothesis testing procedures.

Table IV, the Distribution of x^2, shows for the various degrees of freedom (v) the cut off values at the two most common significance level, 5% and 1%.

The use of x^2 will be explained using the following simple example.

A random sample of 400 householders is classified by two characteristics:

Whether they own a colour television and by what type of householder (i.e. owner-occupier, private tenant, council tenant). The results of this investigation are given below:

	Actual frequencies			
	Owner occupier	Council tenant	Private tenant	Total
Colour TV	150	60	20	230
No colour TV	45	68	57	170
	195	128	77	400

Table 1

It is required to test at the 5% level, the following hypotheses:

H_0: The two classifications are independent (i.e. no relation between classes of householder and colour TV ownership).

H_1: The classifications are *not* independent.

Solution

The first stage in the solution is to calculate the *expected frequencies* for each category which will then be compared with the *actual frequencies* shown above.

To find the expected frequencies

These are found by making what is essentially the null hypothesis, i.e. assume there is no difference in the proportion of TV owners in each of the groups. The expected frequency in each cell in the table is found by apportioning the total of the type of householder in the ratio of colour TV: No colour TV, i.e. 230 : 170.

Thus, the 195 owner occupiers are split in the 230 : 170 proportion, i.e. 112 : 83. The tenants are split in a similar fashion resulting in the following tables.

	Expected frequencies			
	Owner occupier	Council tenant	Private tenant	Total
Colour TV	112	74	44	230
No colour TV	83	54	33	170
	195	128	77	400

Table 2

The χ^2 calculation can now be made.

Observed frequencies (O)	Expected frequencies (E)	(O – E)	$(O - E)^2$	$\dfrac{(O - E)^2}{E}$
150	112	+38	1,444	12.89
45	83	−38	1,444	17.40
60	74	−14	196	2.65
68	54	14	196	3.63
20	44	−24	576	13.09
57	33	+24	576	17.45
			χ^2	= **67.11**

It is now necessary to find the appropriate χ^2 value from Table IV. This is done by establishing **v**, the degrees of freedom. This is found by multiplying the number of rows in the table less one, by the number of columns less one, i.e.

$$\mathbf{v} = (\text{Rows} - 1)(\text{Columns} - 1)$$

in this case

$$\mathbf{v} = (2 - 1)(3 - 1)$$

$$= \underline{2 \text{ degrees of freedom}}$$

The value of the cut-off point of χ^2 for 2 degrees of freedom at the 5% level from Table IV is **5.991**.

As the calculated value (67.11) is greater than the table value we reject the null hypothesis and accept that there is a connection between the type of householder and colour TV ownership.

Notes:

a) Tables 1 and 2 illustrated in the example for the observed and expected frequencies are known as contingency tables or two way classification tables.

b) In calculating the number of degrees of freedom by the (Rows − 1) (Columns − 1) formula the row and column total cells are ignored.

c) The larger the value of χ^2 the bigger the differences between actual and expected. If actual results were exactly as expected then (O − E) would equal zero and χ^2 would be zero.

Additive qualities of χ^2

17. χ^2 has the property (like variances) of being additive to other χ^2 values. If the results of a number of samples yield χ^2 values of χ_1^2, χ_2^2 χ_3^2, etc., with v_1, v_2, v_3, etc., degrees of freedom then the result of all the samples can be summarised with a single χ^2 value given by $\chi_1^2 + \chi_2^2 + \chi_3^2 + \ldots$ with $v_1 + v_2 + v_3 + \ldots$ degrees of freedom.

'Goodness of fit' tests

18. In addition to the use of χ^2 for comparing observed and expected frequencies in contingency tables as already explained, the χ^2 test can be used to determine how well empirical distributions, i.e. those obtained from sample data, fit theoretical distributions such as the Normal, Poisson and Binomial. The procedure is similar to that previously described where the observed frequencies are compared with the expected frequencies. The expected frequencies are calculated in accordance with the assumed distribution characteristics based on the appropriate tables (Normal, Poisson, etc.) and χ^2 calculated in the manner previously described.

One slight difference is the way that the number of degrees of freedom are calculated.

For a Poisson distribution where $\bar{x}$ is the sample mean, the degrees of freedom, $v = n − 2$, where n is the number of classes or frequency values in the sample.

For a Binomial distribution with a predicted value of p and q the degrees of freedom, $v = n − 2$, where n is the number of classes or frequency values in the sample.

The use of χ^2 for goodness of fit tests is illustrated by the following examples.

Example 10

Torch bulbs are packed in boxes of 5 and 100 boxes are selected randomly to test for the number of defectives.

Number of defectives	Number of boxes	Total defectives
0	40	0
1	37	37
2	17	34
3	5	15
4	1	4
5	0	0
	100	90

The probability of any individual bulb being a reject is $\dfrac{90}{100} \times 5 = 0.18$

and it is required to test at the 5% level whether the frequency of rejects conforms to a binomial distribution.

Solution

We are testing whether the observed number of defects fits a binomial distribution, thus:

H_0: the observed number of defects conforms to a binomial distribution of the form $(p + q)^5$ where $p = 0.18$.

H_1: that the observations do not conform.

The observed frequencies are already given so it is only necessary to calculate the frequencies expected from a binomial distribution to the power 5, i.e. $(p + q)^5$, where $p = 0.18$ and $q = 1 - 0.18 = 0.82$.

The probabilities of the various values of p and q can be found from binomial probability tables if available. Alternatively they can be calculated from the binomial expansion as explained in Chapter 4 i.e.

$$p^5 + 5(p^4q) + 10(p^3q^2) + 10(p^2q^3) + 5(pq^4) + q^5$$

This shows the probabilities for 5, 4, 3, 2, 1 and 0 defectives and when $p = 0.18$ (the probability of a bulb being defective) and $q = 0.82$ (probability of not being defective) the probabilities range from 0.0002 (i.e. 0.18^5) for five defectives to 0.3711 (i.e. 0.82^5) for no defectives. When the probabilities are known they are multiplied by 100 boxes to find the expected frequencies which are used in the normal χ^2 procedures.

The table below summarises the calculations:

Defectives	No. of boxes observed	Binomial probabilities	Expected frequency	$(O - E)^2$	$\dfrac{(O - E)^2}{E}$
0	40	0.3711	37.11	8.35	0.22
1	37	0.4069	40.69	13.62	0.33
2	17	0.1786	17.86	0.74	0.04
3	5	0.0392	3.92		
4	1	0.0040	0.4	2.76	0.64
5	0	0.0002	0.02		
				χ^2 =	1.23

Note: Because of the very small values of the expected frequencies for 3, 4 and 5 defectives they have been combined into one but it makes little difference to the result if they are not combined.

The calculated χ^2 value is compared with an χ^2 value for the appropriate degrees of freedom. Because the last three classes have been combined there are four classes remaining, i.e. for 0, 1, 2, and the combined class for 3 – 5 rejects, thus v = n – 2 = 4 – 2 = 2.

The χ^2 value for 2 degrees of freedom at the 5% level from Table IV is 5.991 and, as the calculated value of 1.23 is well below this, we accept the Null hypothesis and conclude that the observed values fit a binomial distribution to the power 5 where **p** = 0.18.

Note: If the last three classes had not been combined, the calculated χ^2 score would have been 1.8 and there would have been 4 degrees of freedom. At 4 degrees of freedom the score from Table IV is 9.488, so the conclusion would be the same, i.e. we accept the Null hypothesis.

Example 11

An area of a city is divided into 600 squares and the frequency of burglaries noted in each square. The data were as follows:

Number of burglaries	Number of squares	Total burglaries
0	279	0
1	200	200
2	90	180
3	25	75
4	5	20
5	1	5
	600	480

$$\text{Average number of burglaries} = \frac{480}{600} = 0.8$$

Test the fit of the observed distribution to a Poisson distribution with a mean of 0.8, at the 5% level.

Solution

This test is to see whether the observed number of burglaries fit a Poisson distribution, thus:

H_0: The observed number of burglaries conforms to a Poisson distribution with a mean of 0.8.

H_1: That the observations do not conform.

This follows a similar pattern to the previous example except that Poisson probabilities are used (Table V)

The summary table is thus.

Number of burglaries	Observed no. of squares	Poisson probabilities	Expected frequency ($P \times 600$)	$(O - E)^2$	$\dfrac{(O - E)^2}{E}$
0	279	0.4493	269.58	88.73	0.33
1	200	0.3595	215.7	246.49	1.14
2	90	0.1438	86.28	13.83	0.16
3	25	0.0383	22.98	4.08	0.18
4	5 $\rbrace$	0.0077	4.62 $\rbrace$		
5	1	0.0012	0.72	0.44	0.08
				χ^2	= **1.89**

There are 5 (O − E) classes so the degrees of freedom are 5 − 2 = 3 and the χ^2 value for 3 degrees of freedom is 7.815 (Table V). As the calculated value is less than this we conclude that the observed values fit a Poisson distribution well.

Note: A similar procedure would be used for testing the goodness of fit of an observed distribution to a Normal Distribution except, of course, the probabilities and frequencies appropriate to a Normal Distribution would be used). Also, we would use up three degrees of freedom fitting a Normal distribution to an observed distribution; as the distributions must have the same mean, total and standard deviation. This means that the χ^2 value would be for **n − 3** degrees of freedom, where n is the number of classes of expected frequencies.

Yates correction

19. When the chi-square test is used on a 2 × 2 contingency table (i. when there is only 1 degree of freedom) serious errors can arise. Yates showed that the answer can be made less approximate if 0.5 is deducted from the absolute value of (O − E) for each item. This is known as Yates correction and should be used whenever there is a 2 ×2 chi-square test.

Example 12 – Using Yates correction

A survey was carried out in a firm of the smoking habits of men and women employees with the following results:

	Men	Women
Smokers	48	27
Non-smokers	58	57

It is required to test whether, at the 5% level of significance, the survey reveals any difference in the smoking habits of men and women.

Solution

H_0, the null hypothesis, is that there is no difference in the smoking habits of men and women.

There is 1 degree of freedom since $(2 - 1) \times (2 - 1) = 1$. χ^2 at the 5% level with 1 degree of freedom is 3.84 so our decision rules are:

$$\text{Accept } H_0 \text{ and reject } H_1 \text{ if } \chi^2 < 3.84$$

$$\text{Reject } H_0 \text{ and accept } H_1 \text{ if } \chi^2 > 3.84$$

The contingency table is:

	Total	Men Observed	Men Expected	Women Observed	Women Expected
Smokers	75	48	41.8	27	33.2
Non-smokers	115	58	64.2	57	50.8
	190	106	106	84	84

Yates correction is now applied, and we calculate not $(O - E)$ but $(|O - E| - 0.5)$

Observed	Expected	$(O - E)$	$(|O - E| - 0.5)$	$(|O - E| - 0.5)^2$	$\dfrac{(|O - E| - 0.5)^2}{E}$
48	41.8	+6.2	5.7	32.49	0.77
58	64.2	−6.2	5.7	32.49	0.51
27	33.2	−6.2	5.7	32.49	0.98
57	50.8	+6.2	5.7	32.49	0.64
					2.90

Thus as χ^2 is 2.9 and the critical value is 3.84, we accept the null hypothesis and conclude, at the 5% level of significance, that there is no difference in the smoking habits of the male and female employees.

Statistical process control

20. Statistical process control, sometimes called statistical quality control, was first developed in the 1920s in the USA and has become one of the most important applications of statistics in industry. In outline, samples are taken from production, tested, weighed or measured as the case may be, and statistical tests applied to see whether the process is performing satisfactorily or not. It is a form of hypothesis testing.

All production methods and processes are subject to variability. For example, a machine filling 5 litre containers of oil will be set to work to an average of 5 litres, but inevitably some containers will contain slightly more than 5 litres, others slightly less. The variability can arise from two causes:

- **variability due to inherent causes**

   These are the unavoidable causes of variability inherent in the nature of the process, the machines used, the material, etc. These causes produce a distribution of values (sizes, weights, volumes, etc.) which is approximately Normal and which is predictable. If all the variability of the samples was due to inherent causes, the process would be said to be *in control*.

- **variability due to special causes**

   These are variations caused by specific changes in the process or environment, e.g. the wrong material is used, the machine is incorrectly adjusted, a component fails, etc. These causes produce instability in the distributions, which are no longer predictable and the process is *out of control*.

   The general purpose of process or quality control is to indicate if the process remains in control or whether it is out of control or likely to go out of control if present trends continue. The statistical methods used cannot, of course, identify the specific cause of any variation. What they can do is to act as an early warning system which tells management that something odd has probably happened.

It will be seen that the procedure outlined above is much the same as Hypothesis testing where:

Null Hypothesis : H_0 = **Process in control and operating within limits**.

Alternative Hypothesis : H_1 = **Process out of control**.

In practice, statistical process control is usually carried out by means of *control charts*.

Control charts

21. These are charts on which the values from the samples can be directly plotted. The values might be, for example, the mean diameter of the sample, the range of the sample, the proportion defective and so on. The charts would already contain control limits (previously calculated using statistical principles) and the sample values would be checked against the control limits. There are normally two levels

117

of control limit: a Warning Limit and an Action Limit. If a sample plotting is outside the control limits action would be taken to trace the specific cause of the variation.

The charts thus provide a visual means of distinguishing between the variability due to inherent causes (process in control) and the variability due to special causes (process out of control). They can be used and understood by people with little or no statistical background. There are various types of control charts dealing with means, ranges, proportions and so on. The general principles of construction and use are shown in the following illustration of a mean control chart.

Mean control chart illustration

22.

Example 13

A machine produces 100 mm diameter bolts and is known to work to a standard deviation of 0.15 mm. Random samples of 6 bolts are taken every 20 minutes and each bolt measured. The results for 5 consecutive samples are given below:

Sample No.	1	2	3	4	5
Diameters (mm)	100.2	99.8	100.2	100.4	100.1
	100.1	99.9	100.1	99.8	100.0
	99.7	100.3	99.8	100.2	99.8
	100.0	100.1	100.1	99.8	100.2
	100.2	99.8	100.2	100.2	100.1
	100.1	99.8	99.7	100.3	100.1
Sample mean	100.05	100.02	99.95	100.12	100.07

a) Calculate the upper and lower warning and action control lines based on 95% confidence limits.

b) Plot the control lines and sample results on a mean control chart.

c) Interpret the chart.

Solution

a) Calculation of control lines.

Based on sampling theory, we know that the mean ±1.96 standard errors contains 95% of the population. (Alternatively we could say that there is only a 5% chance of getting a sample mean outside these values). At the 99% level, it would be the mean ±2.58 standard errors.

Thus in this example:

Warning limits(95% level) = mean ± 1.96 standard errors

$$= 100 \pm 1.96 \frac{0.15}{\sqrt{6}}$$

$$= \textbf{100} \pm \textbf{0.12}$$

(If a sample mean is outside these limits, this could only happen by chance in 5% of cases, so this could be taken as a warning that something may well be wrong.)

Action limits (99% level) = mean ± 2.58 standard errors

$$= 100 \pm 2.58 \frac{0.15}{\sqrt{6}}$$

$$= \textbf{100} \pm \textbf{0.158}$$

(If a sample mean is outside these limits, this could only happen by chance in 1% of cases, so such a sample mean is a sign that almost certainly something is wrong, and action should be taken.)

b) See Figure 6/2.

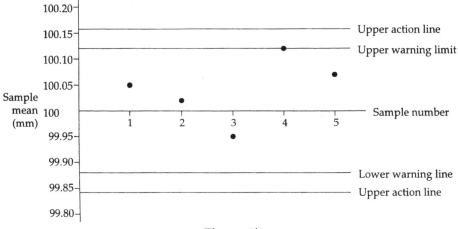

Figure 6/2

c) The mean for sample 4 is just on the upper warning line, but the mean of the next sample is well within the control lines, so we assume that there is no problem. It is normal practice to take a further sample immediately a sample is found to be on or beyond the control lines.

Note: In this illustration probabilities of 95% and 99% were used, resulting in values of 1.96 and 2.58. Of course, other limits may be chosen and the number of standard errors adjusted accordingly.

Other types of control charts

23. A process may give a satisfactory average, but the variation about the mean may cause concern. In these circumstances, the **range** of values in each sample

(largest–smallest) could be monitored using a range control chart. With these charts, it is usual to define only the **upper** warning and action lines since, generally, only large ranges lead to problems.

Alternatively control charts could also be prepared for *attributes* such as a chart showing the proportion that is defective in some way. The central line would be set at the average proportion defective expected and the actual proportion found to be defective would be plotted on the chart.

All the charts use the same general principles covered earlier which are based on the properties of normal distributions and sampling theory. Statistical process control is, in effect, continuous hypothesis testing.

Summary

24. a) Hypothesis or significant testing is testing a belief or opinion by statistical methods.

   b) Based on the Null Hypothesis, H_0, a Type I error is rejecting a true hypothesis and a Type II error is accepting a false hypothesis.

   c) Significance levels are complementary concepts to confidence limits. 1% and 5% are the most usual levels.

   d) The Null hypothesis (H_0) usually assumes there is no difference between the observed and believed values.

   e) A one-tail test is concerned with only one tail of the distribution, i.e. a difference in one direction only.

   f) Hypothesis testing is commonly used for testing sample means and proportions.

   g) The difference between the means and proportions of two samples are also used in significance testing.

   The standard error of the distribution of means is:

   $$s_{(\bar{x}_A - \bar{x}_B)} = \sqrt{\frac{s^2_A}{n_A} + \frac{s^2_B}{n_B}}$$

   The standard error of the difference between proportions is:

   $$s_{(p_1 - p_2)} = \sqrt{\frac{p_1 q_1}{n_1} + \frac{p_2 q_2}{n_2}}$$

   h) The 't' distribution is used for significance testing when the sample size is less than 30. The number of degrees of freedom, v, is $n - 1$ or, for two samples $n_1 + n_2 - 2$.

   i) The chi-square (χ^2) distribution is used for comparing an actual distribution with an expected distribution. The formula is

$$\chi^2 = \sum \frac{(O - E)^2}{E}$$

j) χ^2 values, like variances, can be added.

Points to note

26. a) A significant difference is one which is unlikely to have arisen by chance.

b) A shorthand way of expressing the level of significance for a test is $\chi^2_{0.05}$, i.e. a χ^2 test using a 5% level of significance, $t_{0.01}$, i.e. a t test using a 1% level of significance and so on.

c) The statistical tests described generally require assumptions regarding the underlying distribution from which the sample is taken, most assuming Normality. There are tests which do not require such assumptions; they are called *non parametric* or *distribution free* tests.

Self review questions *Numbers in brackets refer to paragraph numbers*

1.   What is hypothesis testing? (2)

2.   Define Type I and Type II errors. (4)

3.   What is a significance level? (5)

4.   What is a difference that is statistically significant? (5)

5.   What is the Null Hypothesis? – the Alternative Hypothesis? (6)

6.   What is a two-tail test? (6)

7.   What are the appropriate number of Standard Errors to use in a one-tailed test at the 5% level?, at the 1% level? (8)

8.   What is the standard error of a proportion? (9)

9.   How is the standard error of the distribution of means calculated? (10)

10. When is the t distribution used? (12)

11. What is the best estimate of the population standard deviation when two samples are taken? (14)

12. What is the χ^2 formula? (15)

13. How are the 'expected frequencies' calculated for use in the formula? (16)

15. What is a 'goodness of fit' test? (18)

16. What is Yates correction? (19)

17. What is statistical process control? (20)

18. How are the control lines on control charts calculated? (21)

19. What is a Range Control Chart? (23)

Exercises with answers

1.   The output of two workers was compared over a number of days with the following results:

	Average output per day	Standard deviation	Number of days Observed
Man A	30	6	50
Man B	32	5	60

Is there a significant difference in output at the 95% level?

2. Express Packets guarantee 95% of their deliveries are on time. In a recent week 80 deliveries were made and 6 were late and the Managing Director says that, at the 95% level, there has been a significant improvement in deliveries.

   a) Can the MD's statement be supported?

   b) If not, at what level of confidence could it be supported?

3. A batch of weighing machines has been purchased and one machine selected at random for testing. Ten weighing tests have been conducted and the errors found noted as follows:

Test	Errors (gms)
1	4.6
2	8.2
3	2.1
4	6.3
5	5.0
6	3.6
7	1.4
8	4.1
9	7.0
10	4.5

The Purchasing manager has previously accepted machines with a mean error of 3.8 gms and asserts that these tests are below standard.

Test the assertion at the 5% level.

4. A company makes a micro-chip in batches of 6. In a sample of 100 batches the following number of rejects were found:

Number of rejects found in batch	Number of batches
0	17
1	32
2	21
3	18
4	9
5	2
6	1
	100

Test at the 5% level to see whether the frequency of rejects in a batch conforms to a Binomial Distribution.

Answers to exercises

1.

$$\text{S.e. Man A} = \frac{6}{\sqrt{50}} = 0.849$$

$$\text{S.e. Man B} = \frac{5}{\sqrt{60}} = 0.645$$

$$\text{S.e. of difference of means} = \sqrt{0.849^2 + 0.645^2} = 1.066$$

$$Z = \frac{30 - 32}{1.066} = 1.876$$

Z score for two-tailed test at the 5% level is 1.96 $\therefore$ difference not significant.

2. a)

$$\text{S.e.} = \sqrt{\frac{pq}{n}} = \sqrt{\frac{0.1 \times 0.9}{80}} = 0.033$$

where **p** = proportion late

actual $\mathbf{p} = \frac{6}{80} = 0.075$

$$\therefore \frac{0.1 - 0.075}{0.033} = 0.75 \text{ standard errors.}$$

This is less than the standard errors at the 5% level $\pm$ 1.96 so we conclude there is no significant improvement in deliveries.

b) 0.75 standard errors from the mean would cover $\pm$ 0.2734, i.e. 54.68%, say 55% of the population, so the MD's claim could be accepted at any level of confidence below 55%, which is a very low level of confidence

3. Calculated mean of sample = 4.68 gms

Calculated sample s.d. = 1.968

$$\therefore \text{ best estimate of population s.d.} = 1.986\sqrt{\frac{n}{n-1}} = 1.986\sqrt{\frac{10}{9}} = 2.093$$

$$\therefore \text{ std. error of mean} = \frac{2.093}{\sqrt{10}} = 0.66$$

There are $10 - 1 = 9$ d.f and it is a one-tailed test. From Table III the 5% value for a one tail test is 1.833.

$\therefore$ The sample mean should be within 1.833×0.66 gms = 1.21 gms.

The actual difference is $4.68 - 3.8$ gms = 0.88 gms so the figures do not support the Purchasing Manager's assertion.

4.

Number of Rejects in batch	Number of Batches	
x	f	fx
0	17	0
1	32	32
2	21	42
3	18	54
4	9	36
5	2	10
6	1	6
	100	180

$$\text{Probability of an individual unit being a reject} = \frac{180}{6 \times 100} = 0.3$$

H_0: It is assumed that there is no difference between the observed frequency and a binomial distribution of the form $(p + q)^6$ where $p = 0.3$.

The binomial probabilities for $p = 0.3$ and $n = 6$ are as follows:

$$p^6 + 6(p^5q) + 15(p^4q^2) + 20(p^3q^3) + 15(p^2q^4) + 6(pq^5) + q^6$$

$$= 0.3^6 + 6(0.3^5 \times 0.7) + 15(0.3^4 \times 0.7^2) + 20(0.3^3 \times 0.7^3) + 15(0.3^2 \times 0.7^4)$$

$$+ 6(0.3 \times 0.7^5) + 0.7^6$$

$$= 0.001 + 0.01 + 0.059 + 0.19 + 0.32 + 0.30 + 0.12$$

These probabilities are used in the following table:

Rejects per batch	Observed Frequency (O)	Calculated Probability	Excepted Frequency (E)	$(O - E)^2$	$\dfrac{(O - E)^2}{E}$
0	17	0.12	12	25	2.08
1	32	0.30	30	4	0.13
2	21	0.32	32	121	3.78
3	18	0.19	19	1	0.05
4	9	0.059	6	9	1.5
5	2	0.01	1	1	1.0
6	1	0.001	0	1	0
	100	1.00			8.54

There are seven possible values for the number of rejects so the degrees of freedom are $7 - 2 = 5$.

From Table IV at the 5% level of significance the χ^2 value is 11.07, and as the calculated value is below this we accept the Null Hypothesis that the number of rejects in a batch conforms to a binomial distribution where $p = 0.3$ and $n = 6$.

7 Linear correlation and regression

Objectives

1. After studying this chapter you will
   - be able to describe the relationship between variables;
   - understand the types of correlation between variables;
   - know how to calculate and interpret the Product Moment Coefficient of Correlation (r);
   - understand that y = a + bx is the general equation for any straight line;
   - be able to calculate the constants a and b using Least Squares;
   - know how to calculate the Coefficient of Determination;
   - understand how to calculate the Spearman Rank Coefficient of Correlation.

Relationships between variables

2. There are many occasions in business when changes in one factor appear to be related in some way to movements in one or several other factors. For example, a marketing manager may observe that sales increase when there has been a change in advertising expenditure. The transport manager may notice that as vans and lorries cover more miles then the need for maintenance becomes more frequent.

Certain questions may arise in the mind of the manager or analyst. These may be summarised as follows:

a) Are the movements in the same or in opposite directions?

b) Could changes in one phenomenon or variable be causing or be caused by movements in the other variable? (This is an important relationship known as a *causal relationship*).

c) Could apparently related movements come about purely by chance?

d) Could movements in one factor or variable be as a result of combined movements in several other factors or variables?

e) Could movements in two factors be related, not directly, but through movements in a third variable hitherto unnoticed?

f) What is the use of this knowledge anyway?

Very frequently, the manager or analyst is interested in prediction of some kind. For example, the quality control manager may want to know what might be the effect on the number of faulty parts discovered if the amount of expenditure on inspection were increased. The Sales Manager may wish to predict sales levels if advertising were increased by, say, 20%. Here there is clearly some kind of causal model in the minds of the two managers.

Methodology

3.   Suppose that a manager has sensed that two variables or phenomena are behaving in some 'related' way, how might that manager proceed to investigate the matter further? A possible methodology might be as follows.

   a)   Observe and note what is happening in a systematic way.

   b)   Form some kind of theory about the observed facts.

   c)   Draw a graph or diagram of what is being observed.

   d)   Measure what is happening.

   e)   Use the results.

   This methodology is developed throughout this Chapter and the various stages are illustrated using the problem shown in Example 1.

Example 1

The managers of a company with ten operating plants of similar size producing small components have observed the following pattern of expenditure on inspection and defective parts delivered to the customer:

Observation number	Inspection expenditure per 1,000 units (pence)	Defective parts per 1000 units delivered
1	25	50
2	30	35
3	15	60
4	75	15
5	40	46
6	65	20
7	45	28
8	24	45
9	35	42
10	70	22

They are wondering how strong the relationship is between inspection expenditure and the number of faulty items delivered and to what extent they may predict the number of faulty parts delivered from a knowledge of expenditure on inspection.

Drawing a diagram for Example 1

4.   Clearly in this problem the managers have already noted and recorded what is happening in a systematic manner. They would also reasonably deduce that there is likely to be a causal relationship between the expenditure on inspection

and the number of defectives parts delivered to the customer; the higher the expenditure, the fewer defective parts are delivered. Based on this assumption – which is a form of hypothesis – the data can be graphed using the accepted convention that the horizontal or x axis is used for the independent or causing variable in this case, expenditure. The y or vertical axis is used for the dependent variable, in this case, defective parts delivered. This type of diagram is known as a *scatter diagram*.

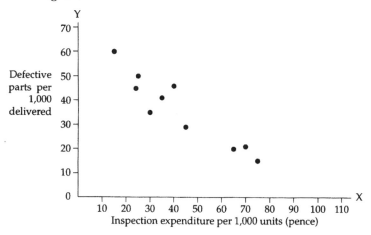

Scatter diagram (based on Example 1). Figure 7/1

Figure 7/1 shows a clear drift downwards in defectives delivered as inspection expenditure increases. This is known as a *negative slope* or *negative relationship*.

Different relationships between variables

5.   Figure 7/1 was based on the data from Example 1. Sometimes other possibilities exist ranging from a perfect negative or perfect positive relationship to no discernible relationship at. A perfect relationship is one where a single straight line can be drawn through all the points, for example 2.1 and 2.2 in Figure 7/2.

It will be seen that the points plotted in Figure 7/1 are similar to 2.4 in Figure 7/2 so we can conclude there is a high negative relationship between the data in Example 1, but not a perfect relationship.

Correlation

6.   When the value of one variable is related to the value of another, they are said to be *correlated*. Thus, correlation means an inter-relationship or association. For example, there is likely to be some correlation between the heights and weights of people.

Variables may be:

a)   perfectly correlated (move in perfect unison)

b)   partly correlated (some inter-relationship but not exact)

c)   uncorrelated (no relationship between their movements).

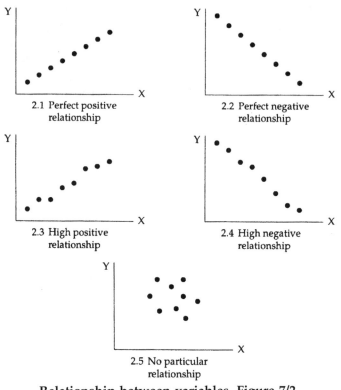

2.1 Perfect positive
relationship

2.2 Perfect negative
relationship

2.3 High positive
relationship

2.4 High negative
relationship

2.5 No particular
relationship

Relationship between variables. Figure 7/2

Movements in one variable may cause movements *in the same direction* in the other variable. This is known as *positive* correlation; an example being height and weight. Alternatively, movement in one variable could cause change *in the opposite direction* in the other variable. This is known as *negative* correlation. For example, if the price of an item is increased, then fewer will be bought.

The degree of correlation between two variables can be measured and there are two measures of correlation, denoted by **r** and **R**.

a) The **Product Moment Coefficient of Correlation**, denoted by **r**.

    This provides a measure of the strength of association between two variables; one the dependent variable, the other the independent variable. **r** can range from +1, i.e. perfect positive correlation where the variables change value in the same direction as each other, to −1, i.e. perfect negative correlation where **y** decreases linearly as **x** increases.

b) The **Rank Correlation Coefficient**, denoted by **R** (dealt with in Para 26).

    This provides a measure of the association between two sets of ranked or ordered data. **R** can also vary from +1, perfect positive rank correlation to −1, perfect negative rank correlation.

Whichever type of coefficient is being used it follows that a coefficient of zero or near zero generally indicates no correlation.

Product moment coefficient of correlation (r)

7. This coefficient gives an indication of the strength of the linear relationship between two variables. There are several possible formulae but a practical one is:

$$r = \frac{n\sum xy - \sum x \sum y}{\sqrt{n\sum x^2 - (\sum x)^2} \times \sqrt{n\sum y^2 - (\sum y)^2}}$$

This formula is used to find **r** from the data in Example 1.

The data have to be re-arranged in ascending order of Inspection Expenditure.

X	Y	X²	Y²	XY
15	60	225	3,600	900
24	45	576	2,025	1,080
25	50	625	2,500	1,250
30	35	900	1,225	1,050
35	42	1,225	1,764	1,470
40	46	1,600	2,116	1,840
45	28	2,025	784	1,260
65	20	4,225	400	1,300
70	22	4,900	484	1,540
75	15	5,625	225	1,125
424	363	21,926	15,123	12,815
$\sum X$	$\sum Y$	$\sum X^2$	$\sum Y^2$	$\sum XY$

Table 1

Using the formula above

$$r = \frac{10 \times 12,815 - 424 \times 363}{\sqrt{(10 \times 21,926 - 424^2)} \times \sqrt{(10 \times 15,123 - 363^2)}}$$

$$= \frac{128,150 - 153,912}{\sqrt{(219,260 - 179,776)} \times \sqrt{(151,230 - 131,769)}}$$

$$= \frac{-25,762}{\sqrt{(39,484)} \times \sqrt{(19,461)}} = -0.93$$

Thus the correlation coefficient is −0.93 which indicates a strong negative linear association between expenditure on inspection and defective parts delivered. It will be seen that the formula automatically produces the correct sign for the coefficient.

Note: A strong correlation between two variables would produce an r value in excess of +0.9 or −0.9. If the value was less than, say 0.5 there would only be a very weak relationship between the variables.

Interpretation of the value of r

8. Care is needed in the interpretation of the calculated value of **r**. A high value (above +0.9 or −0.9) only shows a strong association between the two variables if there is a *causal relationship*, i.e. if a change in one variable *causes* changes in the other. It is possible to find two variables which produce a high calculated **r** value yet which have no causal relationship. This is known as *spurious* or *nonsense correlation*.

    An example might be the wheat harvest in America and the number of deaths by drowning in Britain. There *might* be a high apparent correlation between these two variables but there clearly is no causal relationship.

    A low correlation coefficient, somewhere near zero, does not always mean that there is no relationship between the variables. All it says is that there is no *linear* relationship between the variables – there may be a strong relationship but of a *non-linear* kind.

    A further problem in interpretation arises from the fact that the Product Moment Coefficient of Correlation measures the relationship between a *single* independent variable and dependent variable, whereas a particular variable may be dependent on *several* independent variables in which case *multiple correlation* should have been calculated rather than the simple two-variable coefficient.

    Note that in Figure 7/2,

Diagram 2.1	$r = +1$
Diagram 2.2	$r = -1$
Diagram 2.3	r is close to $+1$
Diagram 2.4	r is close to -1
Diagram 2.5	r is close to *zero*

The significance of r

9. Frequently the set of **X** and **Y** observations is based upon a sample. Had a different sample been drawn then the value of **r** would be different, although the degree of correlation in the parent population would remain the same. In the same way that a knowledge of $\bar{x}$ enables an estimate to be made of μ then the knowledge of **r** enables the analyst to make an estimate of p, the population co-efficient of correlation.

    Generally in examination questions the sample size is limited to some figure that can be dealt with in the time allowed. It is questionable whether the sample size given in examinations gives enough data for a credible judgement to be formed about a possible relationship between the **X** and **Y** values. If **r** is high, does this mean that there *is* really a close relationship between the **X** and **Y** values or is it just that the particular samples gives this impression?

    Conversely, if **r** is low does it really imply a lack of a relationship? There may indeed be a close relationship but the data has not revealed it. Further, the relationship may exist, but it may not be linear or it may not be direct.

It is possible to test whether the value of is sufficiently different from zero for the analyst to decide whether the X and Y values are correlated. The test may be stated in summary

$$H_0 : \rho = 0$$
$$H_1 : \rho \neq 0$$

It is a **t** test for which the test statistic is given by:

$$|\,t\,| = \left| \frac{r - \rho}{\sqrt{1 - r^2}} \times \sqrt{n - 2} \right|$$

Using the values from Example 1, i.e. $r = -0.93$ and $n = 10$ we obtain:

$$|\,t\,| = \left| \frac{-0.93 - 0}{\sqrt{1 - 0.93^2}} \times \sqrt{10 - 2} \right|$$
$$= 2.53 \times 2.83$$
$$= \mathbf{7.16}$$

The tabulated value for $n - 2$ for 8 degrees of freedom using a 5% level of significance is 2.306. Since 7.16 is greater than 2.306 the numerical evidence is strong enough to reject the null hypothesis and conclude that the value of p is not zero.

Using the results for prediction

10. Given that the manager is satisfied with the value of r (−0.93 based on Example 1 data), it is possible that he wishes to predict likely levels of faulty parts delivered for levels of inspection expenditure not yet recorded. For example, how many defective parts might be expected if 50p per 1,000 units was spent on inspection?

Inspection of Figure 7/1 and the sample data shows that there is a general downwards drift of the scatter points. It will be seen that making predictions from, say, two values of **x** would be inaccurate because they may not be representative of the general relationship between **x** and **y**. What is required is to be able to predict an expected or mean value of **y** (i.e. defective parts) for a given value of **x** (i.e. expenditure) using the whole, known relationship between **x** and **y**. The process by which this is calculated is known as *regression analysis* and is developed below using the properties of a straight line graph.

Straight line graph

11. Straight line, or linear, relationships are commonly encountered in business and personal life. For example, many public utilities charge their customers on the basis of a fixed charge plus a charge per unit for gas, electricity, or telephone time units consumed. Suppose the charge for using the telephone is £15 per quarter

plus 5p for each unit of time that the user is connected on a call. If the subscriber uses 50 units then the telephone bill for the quarter may be predicted thus:

$$Telephone\ bill = £15.00 + 50 \times 0.05$$

$$= £15.00 + 2.50$$

$$= £17.50$$

A table of charges may be built up as follows

Units	Rental £	Call charge £	Total £
50	15.00	2.50	17.50
100	15.00	5.00	20.00
150	15.00	7.50	22.50
200	15.00	10.00	25.00

This table may be shown on a graph which will be seen to be similar to a scatter diagram.

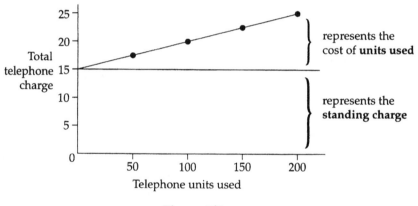

Figure 7/3

It is possible for a subscriber to read off from Figure 7/3 the total telephone charge. If say 130 units of telephone time were used the charge would be £21.50. The important point to note here is that it is possible to make a unique prediction of Y from a value of X.

The equation of a straight line

12. The total telephone charge graph may be made more general. Let the following symbols be used for actual values as follows:

$$£15 : a$$

$$£0.05 : b$$

The total charge may now be expressed as:

$$y = a + bx$$

Where **x** is the number of units of telephone time used, '**a**' is the constant factor and '**b**' is the rate at which the charge rises per unit. *This is the general form of the equation for any straight line.*

'**a**' represents the fixed element

'**b**' represents the slope of the line equivalent to the variable element.

Figure 7/4 shows the graph of the two commonly encountered straight lines. One has a positive value of **b** and hence a positive slope (like the telephone charge example) and one has a negative value of **b** and hence a negative slope (like Example 1).

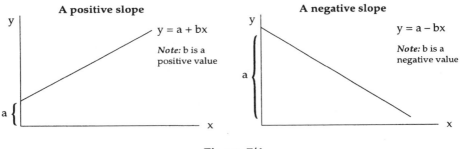

Figure 7/4

Regression analysis or curve fitting

13. This is a statistical technique which can be used for medium term forecasting which seeks to establish the line of 'best fit' to the observed data. The data can be shown as a scatter diagram with various freehand attempts at fitting a line, but naturally the lines drawn in this way would vary according to the judgement of individuals, for example: Figure 7/5 is a reproduction of Figure 7/1 with two of the many possible lines drawn. Although the freehand method is quick, no claims can be made for its accuracy and it is more normal to calculate the line of best fit mathematically using the method of least squares.

Least squares

14. To find the line of best fit mathematically it is necessary to calculate a line which minimises the total of the squared deviations of the actual observations from the calculated line. This is known as the method of *least squares* or the *least squares method of linear regression*.

Least squares regression analysis gives equal importance to all the items in the time series, the older and the more recent. Consequently if the data in the recent past were obtained from conditions significantly different from long past conditions then it is unlikely that good forecasts will be achieved using least squares regression analysis. It is because of this that forecasts based on regression analysis should only be made for the near to medium term future.

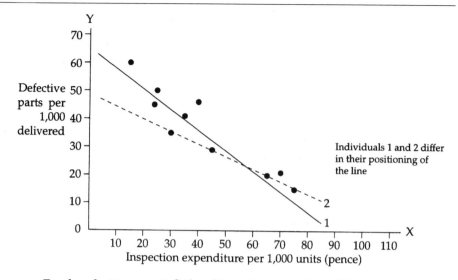

Freehand attempts at fitting lines for Example 1. Figure 7/5

Note: There are many non-linear methods of regression analysis, some of which are dealt with in Chapter 8.

Calculating the values of a and b

15. In the general form of the equation for a straight line

$$y = a + bx$$

where **a** and **b** are constants and **a** represents the fixed element and **b** and the slope of the line, i.e. the ratio of the vertical increase in y to horizontal increase in x.

To find **a** and **b** it is necessary to solve two simultaneous equations known as the *Normal Equations* which are

$$an + b \Sigma x = \Sigma y \qquad \text{.... Equation 1}$$
$$a\Sigma x + b\Sigma x^2 = \Sigma xy \qquad \text{.... Equation 2}$$

where **n** = number of pairs of figures

Note: The slope of the line, **b**, is sometimes called the regression coefficient.

The use of these equations will be demonstrated using the Example 1 data contained in Table 1, para 7.

The equations become

$$10\,a + 424\,b = 363 \qquad \text{.... Equation I}$$
$$424\,a + 21{,}926\,b = 12{,}815 \qquad \text{.... Equation II}$$

Simultaneous equations are solved algebraically by eliminating one of the variables and solving for the other. The value calculated is then substituted in one of the equations to find the one originally eliminated.

In this example Equation I is multiplied by 42.4 and then deducted from Equation II to eliminate **a** and thus find the value of **b**:

$$424a + 21{,}926b = 12815 \quad .... \text{ Equation II}$$

$$424a + 17977.6b = 15391.2 \quad ....\text{I} \times 42.4$$

$$3948.4b = -2576.2$$

$$\therefore \textbf{b} = \textbf{-0.65} \text{ to 2 decimal places}$$

This value is substituted in one of the original equations to find the value of a thus:

$$10a + (424 \times -0.65) = 363$$

$$10a = 363 + 275.6$$

$$\therefore \textbf{a} = \textbf{63.86}$$

Therefore, the regression line for Example 1 is

$$\underline{\textbf{y} = \textbf{63.86} - \textbf{0.65x}}$$

Note: the Normal Equations automatically produce the correction sign (+ or −) for the regression coefficient **b**; in this case, minus.

The calculated values can be used to draw the mathematically correct line of best fit on a graph. This is usually done by plottings based on three values of **x**. The lowest, highest and mean.

Based on Example 1 the three values of **x** are

$$15, \ 42.4 \text{ and } 75$$

Each of these values are substituted into the calculated regression line and the result values plotted on the graph.

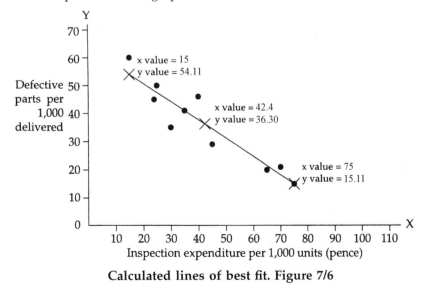

Calculated lines of best fit. Figure 7/6

Note: The values of a and b have been calculated in the example above by substituting in the Normal Equations. An alternative is to transpose the Normal Equations so as to be able find **a** and **b** directively. The formulae are as follows:

$$a = \frac{\Sigma y - b\Sigma x}{n}$$

$$b = \frac{n\Sigma xy - \Sigma x\Sigma y}{n\Sigma x^2 - (\Sigma x)^2}$$

It is often more convenient to use these alternative forms especially when using a calculator. Values for **a** and **b** are re-calculated using the transposed formulae and the Table 1 data.

$$b = \frac{10 \times 12,815 - 424 \times 363}{10 \times 21,926 - (424)^2}$$

$$= -0.65$$

$$a = \frac{363 - (-0.65 \times 424)}{10}$$

$$= 63.86$$

Using the results of the regression analysis

16.   When the values have been calculated for **a** and **b**, predictions or forecasts can be made for values of x which have not yet occurred. The predictions can be read from the graph on which the line of best fit has been plotted, Figure 7/6, or the values inserted into the straight line formula.

Reverting to Example 1 it will be recalled that the manager wished to know the likely number of defects if 50p per 1000 was spent on inspection.

From Figure 7/6 it will be seen that the number of defects would be 31 per 1000. The formula can also be used, thus:

$$y = 63.86 - 0.65x$$
$$\text{so when } x \text{ is } 50$$
$$y = 63.86 - 0.65(50) = 31$$

Thus the manager would conclude that, on average, 31 defects per 1000 would be found if 50p per 1000 was spent on inspection.

Great care must be taken with any form of prediction otherwise absurd results may be obtained. Whilst any value of x can be used mechanically to make a prediction this does not necessarily make a practical forecast. Using Example 1 data, a forecast based on values of x outside the range 15 to 75 (the highest and lowest recorded values) implicitly assumes that the calculated relationships will continue to apply, which may or may not be true. Taking this to the extreme if x = 98p per 1000 then the formula and graph would 'predict' zero defects which is extremely unlikely.

It is important to realise that any prediction, for example the 31 defects calculated above, is no more than a single or point estimate and, like the use of sample means as an estimate of the population mean, can be qualified by the use of a confidence interval.

Note: Predictions *outside* the observed values (15 to 75 in Example 1) are known as *extrapolations* Predictions *within* the observed range are known as *interpolations*.

Accuracy of the regression line

17. However wide the scatter of the data, a line of best fit can be calculated using least squares. Although such a line can always be calculated, it does not follow that the 'best fit' line is likely to be much use for predictive purposes, unless it is an accurate representation of any trend in the data. To find out how good the line of best fit really is, a measure called the *coefficient of determination* is calculated.

Coefficient of determination

18. This measure denoted by r^2 (because it is the square of the correlation coefficient, **r**) calculates what proportion of the variation in the actual values of may be predicted by changes in the values of **x**

$$\text{thus } r^2 \text{ is the ratio } \frac{\textbf{Explained variation}}{\textbf{Total variation}} = \frac{\Sigma(YE - \overline{Y})^2}{\Sigma(y - \overline{Y})^2}$$

where **YE** = estimate of **y** given by the regression equation for each value of **x**

$\overline{Y}$ = mean of actual values of **y**

y = individual actual values of **y**

r^2 will be calculated for the data in Example 1 for which it will be recalled that the regression line was

$$y = 63.86 - 0.65x$$

$$\overline{y} = \frac{363}{10}$$

$$= 36.3$$

X	Y	YE	YE $-\bar{Y}$	$(YE-\bar{Y})^2$	$Y - \bar{Y}$	$(Y - \bar{Y})^2$
15	60	54.18	17.88	319.61	23.7	561.69
24	45	48.31	12.01	144.13	8.7	75.69
25	50	47.65	11.35	128.89	13.7	187.69
30	35	44.39	8.09	65.46	−1.3	1.69
35	42	41.13	4.83	23.31	5.7	32.49
40	46	37.87	1.57	2.45	9.7	94.09
45	28	34.60	−1.70	2.88	−8.3	68.89
65	20	21.55	−14.75	217.44	−16.3	265.69
70	22	18.29	−18.01	324.29	14.3	204.49
75	15	15.03	−21.27	452.43	21.3	453.69
	363			1680.89		1946.10

$$r^2 = \frac{1680.89}{1946.10}$$

$$= 0.8637$$

which, expressed as a percentage, is **86.37%**

This result may be interpreted that in the problem 86.37% of the variation in actual faulty parts delivered may be predicted by change in the actual value of x amount spent on inspection. Factors other than changes in the value of x account for 13.63% of the variation in y.

An alternative formula for r^2 is:

$$r^2 = \frac{(n \sum xy - \sum x \sum y)^2}{(n \sum x^2 - (\sum x)^2) \times (n \sum y^2 - (\sum y)^2)}$$

Notes:

a) If the values obtained for **a** and **b** are based on a sample, then they are only estimates of the true values of the population regression coefficients usually denoted by α (alpha) and β (beta). This is only an extension of the principle covered in the previous chapters on statistics whereby the sample mean (or standard deviation or proportion) is used as an estimate of the true population value.

b) To keep the example small only 10 pairs of values were used. In practice many more are likely to be involved.

c) The regression analysis above is the regression line of **y** on **x**. The vertical axis represents **y**, the dependent variable and the horizontal axis represents **x**, the independent variable.

d) The 'y on x' regression analysis given is the most commonly used, but students should be aware that there exists the regression line of **x** on **y**, where **x** is the dependent variable. Always check carefully that **x** is the independent variable and **y** is the dependent variable whose value we wish to forecast for particular values of **x**. This is to ensure that the most common 'y on x' regression line can be used. In a time series, **x** will always represent the time periods.

e) If r^2 is low then the analyst should look for a non-linear relationship between **x** and **y**, or some other causal factors.

Standard error of regression

19. As with means and proportions, dealt with in earlier chapters, we have to consider the sampling errors associated with the estimates **a** and **b**. The inferences about these estimates can be made, as previously, either as significance tests or more usefully in the case of regression analysis as confidence limits. In either case the standard error of regression must be calculated. There are several possible formulae but the one given below is a useful, practical example.

$$\text{Standard error of regression} = \mathbf{S_e} = \sqrt{\frac{\sum y^2 - a\sum y - b\sum xy}{n-2}}$$

This formula does not provide an exact standard error because it involves the values of a and b which are themselves estimates.

This standard error, is also known as the *residual standard deviation*.

The use of the formula is shown below again based on the data in Table 1.

$$= \sqrt{\frac{15,123 - 63.86 \times 363 - (-0.65) \times 12,815}{10-2}}$$

$$= 5.38 \text{ defective parts}$$

This value can be used below in setting confidence limits for the calculated regression line in a similar manner to that shown in Chapter 6.

Standard errors of the intercept (a) and the gradient (b)

20. If **a** and **b** are calculated from a sample, then they may be looked upon as estimates of the population intercept and gradient α and β. In a manner similar to the distribution of sample means a distribution of values of **a** and a distribution for values of **b** emerge from repeated sampling.

The mean value for **a** is expressed as α and is the population value for the intercept.

The mean value for **b** is expressed as β and is the population value for the gradient.

Both these distributions have standard deviations known as standard errors which are shown below:

$$\text{The intercept } \mathbf{S_a} = \mathbf{S_e}\sqrt{\frac{\sum x^2}{n\sum x^2 - (\sum x)^2}}$$

$$\text{The gradient } \mathbf{S_b} = \frac{S_e}{\sqrt{\sum x^2 - \frac{(\sum x)^2}{n}}}$$

where $\mathbf{S_e}$ is the standard error of regression.

The confidence intervals for α and β may be established as follows:

For the intercept

$$\alpha = a \pm t \times s_a$$

For the gradient

$$\beta = b \pm t \times s_b$$

The value of t is based upon $n - 2$ degrees of freedom, and the chosen confidence level. In addition, it is possible to construct a test of significance for α and β.

For the Intercept

$$H_0 : \alpha = \text{some chosen value}$$

$$H_1 : \alpha \neq \text{some chosen value}$$

The test statistic is

$$t = \frac{a - \alpha}{S_a}$$

For the Gradient

$$H_0 : \beta = 0$$

$$H_1 : \beta \neq 0$$

The test statistic is

$$t = \frac{b - \beta}{S_b}$$

In both cases, the calculated value of t is compared with the tabulated value for $n - 2$ degrees of freedom at the chosen level of significance.

In the case of the gradient, $\beta = 0$ is generally used because if β is found not to be significantly different from 0 then $Y = a + bx$ collapses into $Y = a$ and since the line of best fit passes through X and Y it will be horizontal at the value of Y. Thus for all values of X the forecast of Y will be $\overline{Y}$. The significance test for β is probably the more important of the two for practical purposes.

Using the formulae

21. The above formulae may be illustrated using Example 1 data.

$$n = 10 \qquad\qquad a = 63.86$$
$$X^2 = 21926 \qquad\qquad b = -0.65$$
$$\Sigma X = 424 \qquad\qquad t = 2.306$$
$$S_e = 5.76$$

The standard error of the intercept

$$S_a = 5.76\sqrt{\frac{21,926}{39,484}}$$

$$= \underline{4.29}$$

The 95% confidence interval of the intercept is

$$\alpha = 63.86 \pm 2.306 \times 4.29$$

$$= 63.86 \pm 9.89$$

which gives an upper limit of 73.75 and a lower limit of 53.97.

Significance test for the intercept

$$\mathbf{H_0 : \alpha = 0}$$

$$\mathbf{H_1 : \alpha \neq 0}$$

$$\mathbf{t = \frac{a - \alpha}{S_a}}$$

$$= \frac{63.86 - 0}{4.29} = \underline{14.89}$$

Since 14.89 is much greater than 2.306 (the value from **t** tables) $\mathbf{H_0}$ can be rejected.

Standard error of the slope

$$S_b = \frac{5.76}{\sqrt{21,926 - \frac{424^2}{10}}}$$

$$= \underline{0.092}$$

The 95% confidence interval for the slope is

$$\beta = -0.65 \pm 2.306 \times 0.092$$

$$= -0.65 \pm 0.212$$

giving an upper limit of -0.438 and a lower limit of -0.862.

Significance test for slope

$$\mathbf{H_0 : \beta = 0}$$

$$\mathbf{H_1 : \beta \neq 0}$$

$$\mathbf{t = \frac{b - \beta}{S_b}}$$

$$= \frac{0.65 - 0}{0.092} = \underline{7.07}$$

Since $7.07 > 2.306$, $\mathbf{H_0}$ can be rejected.

On the basis of this evidence the regression equation $\mathbf{y = 63.86 - 0.65x}$ could be used as a basis of prediction for Example 1.

Characteristics of linear regression

22. a) Useful means of forecasting when the data has a generally linear relationship. Over operational ranges linearity (or near linearity) is often assumed for such items as costs, contributions and sales.

b) A measure of the accuracy of fit (r^2) can be easily calculated for any linear regression line.

c) To have confidence in the regression relationship calculated it is preferable to have a large number of observations.

d) With further analysis confidence limits can be calculated for forecasts produced by the regression formula.

e) Any form of extrapolation, including that based on regression analysis, must be done with great care. Once outside the observed values relationships and conditions may change drastically.

f) Regression is not an adaptive forecasting system, i.e. it is not suitable for incorporation in, say a stock control system where the requirements would be for a forecasting system automatically producing forecasts which adapt to current conditions. These types of systems are covered in Chapter 9.

g) In many circumstances it is not sufficiently accurate to assume that **y** depends only on one independent variable as discussed above in simple linear regression. Frequently, a particular value depends on two or more factors in which case multiple regression analysis is employed.

For example, an analysis of a firm might produce the following multiple regression equation:

$$\text{Overheads} = £10,500 + 6.8x + 9.2y + 2.7z$$

where x = labour hours worked
y = machine hours
z = tonnage produced

This concept is dealt with in Chapter 8.

The rank correlation coefficient (R)

23. This coefficient is also known as the *Spearman Rank Correlation Coefficient*. Its purpose is to establish whether there is any form of association between two variables when the variables are arranged in a *ranked* form.

The formula is as follows:

$$R = 1 - \frac{6\Sigma d^2}{n(n^2 - 1)}$$

where d = difference between the pairs of ranked values
n = number of pairs of rankings.

This will be illustrated by the following example.

Example 2

A group of 8 accountancy students are tested in Quantitative Techniques and Management Accountancy. Their rankings in the two tests were:

Table 4

Student	Q.T. Ranking	M.A. Ranking	d	d^2
A	2	3	−1	1
B	7	6	+1	1
C	6	4	+2	4
D	1	2	−1	1
E	4	5	−1	1
F	3	1	+2	4
G	5	8	−3	9
H	8	7	+1	1
			Σd^2	= 22

The 'd column' is obtained by QT ranking − MA ranking

$$\therefore R = 1 - \frac{6\sum d^2}{n(n^2-1)} = 1 - \frac{6 \times 22}{8(8^2-1)} = +0.74$$

As the Rank Correlation coefficient is +0.74 we are able to say that there is a reasonable agreement between the student's performances in the two types of tests.

Notes

a) **R** can vary, like **r**, between +1 and −1 with similar meanings.

b) As with **r**, care should be taken in any interpretation of the value of **R** whether it is a particularly high or low value.

c) The values in Example 2 are the rankings of the students not the actual marks obtained in the test.

Tied rankings

24. A slight adjustment to the formula is necessary if some students obtain the same marks in a test and thus are given the same ranking.

The adjustment is

$$\frac{t^3 - t}{12}$$

where **t** is the number of tied rankings and the adjusted formula is

$$R = 1 - \frac{6\left(\sum d^2 + \frac{t^3 - t}{12}\right)}{n(n^2 - 1)}$$

For example assume that students E and F achieved equal marks in QT and were given joint third place. The revised data are

Student	Q.T. Ranking	M.A. Ranking	d	d^2
A	2	3	−1	1
B	7	6	+1	1
C	6	4	+2	4
D	1	2	−1	2
E	$3\frac{1}{2}$	5	$-1\frac{1}{2}$	$2\frac{1}{4}$
F	$3\frac{1}{2}$	1	$+2\frac{1}{2}$	$6\frac{1}{4}$
G	5	8	−3	9
H	8	7	+1	1
			Σd^2	$26\frac{1}{2}$

$$\therefore R = 1 - \frac{6\left(\sum d^2 + \frac{t^3 - t}{12}\right)}{n(n^2 - 1)} = 1 - \frac{6\left(26\frac{1}{2} + \frac{2^3 - 2}{12}\right)}{8(8^2 - 1)} \quad \underline{\underline{+0.68}}$$

As will be seen, the **R** value has moved also from +0.74 to 0.68.

Note: It is conventional to show the shared rankings as above, i.e. the shared 3rd place takes up the 3rd and 4th rankings thus it is divided between the two as $3\frac{1}{2}$ each.

Summary

25. a) Correlation measures the interdependence between two sets of numbers.

b) The two main measures of correlation are the Product moment coefficient of correlation (**r**) and the Spearman Rank Coefficient of Correlation (**R**).

c) The formula for **r** is

$$r = \frac{n\Sigma xy - \Sigma x\Sigma y}{\sqrt{n\Sigma x^2 - (\Sigma x)^2} \times \sqrt{n\Sigma y^2 - (\Sigma y)^2}}$$

d) There must be a causal connection between the variables otherwise spurious or nonsense correlation exists.

e) Care should be taken in interpreting both high and low values of the correlation coefficient. A low value indicates little linear relationship but a curvi-linear relationship may exist.

f) The Rank Coefficient of Correlation formula is:

$$R = 1 - \frac{6 \sum d^2}{n(n^2 - 1)}$$

g) A common form of prediction is to calculate a linear regression line using the method of least squares and to use this line for extrapolation and interpolation.

h) The least squares method finds values for the coefficients **a** and **b**, in the equation for a straight line,

$$y = a + bx$$

i) To assess the accuracy of the regression line the coefficient of determination is calculated.

j) Confidence limits based on the standard errors of the estimates and forecasts can be prepared for individual predictions or for the regression line as a whole.

Points to note

26. a) The coefficient, **r**, measures only the strength of a linear relationship in two sets of data. It does not mean that there is a cause–effect relationship.

b) The coefficient of determination, r^2, measures the extent to which movements in **y** are associated with movements in **x**. There is an implied causality from **x** to **y**.

c) **r, a** and **b** are sample statistics used to estimate, ρ, α and β the population parameters.

Self reveiw questions Numbers in brackets refer to paragraph numbers

1.   What is a scatter diagram? (4)
2.   What is meant by correlation? (6)
3.   What is the Product Moment Coefficient of Correlation? (7)
4.   What is a nonsense correlation? (8)
5.   What is the general equation for a straight line? (12)
6.   Distinguish between a positive slope and a negative slope. (12)
7.   Define the Least Squares Method of Linear regression. (14)
8.   What are the Normal Equations for calculating the constants **a** and **b** in the equation **y = a + bx**? (15)
9.   Why is the co-efficient of determination calculated? (17,18)
10. What are the significance tests for the intercept and slope? (20)
11. Define **R**. (23)

Exercises with answers

1. The following data have been collected regarding sales and advertising expenditure.

Sales (£'m)	Advertising expenditure (£'000)
8.5	210
9.2	250
7.9	290
8.6	330
9.4	370
10.1	410

Plot the above data on a scatter diagram and, using judgement, decide whether there is a correlation between sales and advertising expenditure.

2. Calculate **r** for the data in 1 and interpret.

3. Calculate **r²** for the data in 1 and interpret.

4. Find the values of **a** and **b** in the formula **y = a + bx** from the following data relating to costs incurred at various output levels. (Use a calculator and the direct formulae for **a** and **b** given in the chapter).

Output level (units)	Cost incurred (£)
40	812
55	890
68	955
73	948
82	1050
89	1100
94	1160
95	1095
103	1250
110	1380

Answers to exercises

1.

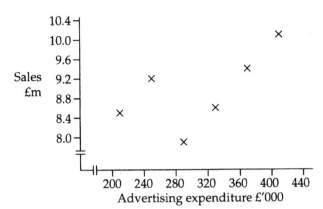

Advertising expenditure £'000

There is a positive correlation between advertising expenditure and sales but it is far from perfect.

2.

$$r = \frac{n\sum xy - \sum x \sum y}{\sqrt{n\sum x^2 - (\sum x)^2}.\sqrt{n\sum y^2 - (\sum y)^2}}$$

x	y	x^2	y^2	xy
8.5	210	72.25	44,100	1,785
9.2	250	84.64	62,500	2,300
7.9	290	62.41	84,100	2,291
8.6	330	73.96	108,900	2,838
9.4	370	88.36	136,900	3,478
10.1	410	102.01	168,100	4,141
53.7	1,860	483.63	604,600	16,833

$$r = \frac{6 \times 16,833 - 53.7 \times 1,860}{\sqrt{6 \times 483.63 - 53.7^2}.\sqrt{6 \times 604,600 - 1,860^2}}$$

$$= \underline{\underline{0.64}}$$

There is a reasonably strong positive correlation between sales advertising expenditure.

3.
$$r^2 = 0.64^2 = 0.41$$

∴ factors other than advertising cause $1 - 0.41 = 59\%$ of variations in sales.

4.

$$b = \frac{10 \times 893,189 - 809 \times 10,640}{10 \times 69,793 - 654,481}$$

$$= 7.46$$

$$a = \frac{10{,}640 - 7.46 \times 809}{10}$$

$$= 460.50$$

Regression line:

$$\mathbf{y} = \pounds460.50 + 7.46\mathbf{x}$$

8 Multiple and non-linear regression

Objectives

1. After studying this chapter you will
    - know what is meant by linear multiple regression;
    - be able to calculate separate regressions, the multiple regressions, and the multiple coefficient of determination;
    - understand the general characteristics of non-linear models;
    - be able to describe exponential and logarithmic functions;
    - understand the principles of learning curves;
    - know how to distinguish between marginal and cumulative average learning curves;
    - be able to calculate the learning coefficient.

Alternatives to a simple linear function

2. There will be occasions when the value of r^2 in the simple linear model, $y = a + bx$, will not be considered satisfactory. This means that the simple linear model will not be a good enough predictor. In such circumstances there are two possible courses of action.

    a) To investigate the possibility that movements in y, the dependent variable, are caused by several independent factors and not just one as in the basic model. For example, changes in demand for a product may depend on
        - the price of the product
        - the price of substitutes
        - the level of incomes
        - consumer tastes and so on

    If linearity can be assumed then a *linear multiple regression model* can be used. These models are dealt with in the first part of the chapter.

    b) Alternatively a *non-linear model* may be considered more appropriate and several of the more important non-linear functions are dealt with later in the chapter.

Multiple regression model development

3. A model which incorporates several independent variables is known as a *multiple regression model*. Because of the lengthy nature of the calculations it would be unlikely that a detailed question on multiple regression would appear in the examinations for which this book is intended. Familiarity with the processes involved and the structure of the model is, however, necessary. The development of the model is shown below:

The basic two variable model (one dependent and one independent variable) is:

$$y = a + bx$$

which can be solved using the Normal Equations thus:

$$\Sigma y = an + b\Sigma x$$

$$\Sigma xy = a\Sigma x + b\Sigma x^2$$

From this can be developed models with more than 2 variables and this is illustrated below using a 3 variable model (one dependent and two independent variables; y, x_1, and x^2).

$$y = a + b_1 x_1 + b_2 x_2$$

which can be solved by the Normal Equations for a three variable model, as follows:

$$\Sigma y = an + b_1\Sigma x_1 + b_2\Sigma x_2$$

$$\Sigma x_1 y = a\Sigma x_1 + b_1\Sigma x_1^2 + b_2\Sigma x_1 x_2$$

$$\Sigma x_2 y = a\Sigma x_2 + b_1\Sigma x_1 x_2 + b_2\Sigma x_2^2$$

The *line* of best fit gives way to a *plane* of best fit. b_1 is the slope of the plane along the x_1 axis, b_2 is the slope along the x_2 axis, and the plane cuts the y axis at 'a'.

The aim of adding to the simple two variable model is to improve the fit of the data. The closeness of fit is measured by the co-efficient of multiple determination R^2 for which the general formula and a useful computational formula are given below:

$$R^2 = \frac{\text{Explained variation}}{\text{Total variation}}$$

$$= \frac{\Sigma(YE - \overline{Y})^2}{\Sigma(Y - \overline{Y})^2}$$

where YE now equals the estimate of Y for each value of x_1 and x_2.

$$R^2 = \frac{a\Sigma y + b_1\Sigma x_1 y + b_2\Sigma x_2 y - \frac{(\Sigma y)^2}{n}}{\Sigma y^2 - \frac{(\Sigma y)^2}{n}}$$

It is not necessarily the case that the value of the co-efficient of determination will improve with the addition of extra variables.

The above models are illustrated by the following examples.

Example of multiple regression

4.

> **Example 1**
>
> The Association of Accountants is investigating the relationship between performance in Quantitative Methods and hours studied per week and the general level of intelligence of candidates. The Association has data on ten students which are:
>
> | Student | Hours | I.Q. | Examination Grade |
> |---------|-------|------|-------------------|
> | | x_1 | x_2 | y |
> | 1 | 9 | 99 | 56 |
> | 2 | 6 | 100 | 45 |
> | 3 | 12 | 119 | 80 |
> | 4 | 14 | 95 | 73 |
> | 5 | 11 | 110 | 71 |
> | 6 | 6 | 117 | 55 |
> | 7 | 19 | 98 | 95 |
> | 8 | 16 | 101 | 86 |
> | 9 | 3 | 100 | 34 |
> | 10 | 9 | 115 | 66 |
>
> Calculate the separate regressions, the multiple regression and the coefficients of determination.

Solution

Part A – Calculation of separate regressions

	y	y^2	x_1	x_1^2	x_2	x_2^2	x_1y	x_2y	x_1x_2
1	56	3,136	9	81	99	9,801	504	5,544	891
2	45	2,025	6	36	100	10,000	270	4,500	600
3	80	6,400	12	144	119	14,161	960	9,520	1,428
4	73	5,329	14	196	95	9,025	1,022	6,935	1,330
5	71	5,041	11	121	110	12,100	781	7,810	1,210
6	55	3,025	6	36	117	13,689	330	6,435	702
7	95	9,025	19	361	98	9,604	1,805	9,310	1,862
8	86	7,396	16	256	101	10,201	1,376	8,686	1,616
9	34	1,156	3	9	100	10,000	102	3,400	300
10	66	4,356	9	81	115	13,225	594	7,590	1,035
	661	46,889	105	1,321	1,054	111,806	7,744	69,730	10,974

Table 1

For Regression y on x_1 (exam. scores: hours studied)

$$b_{x_1} = \frac{n\Sigma x_1 y - \Sigma x_1 \Sigma y}{n\Sigma x_1^2 - (\Sigma x_1)^2}$$

$$= \frac{10 \times 7,744 - 105 \times 661}{10 \times 1,321 - 105^2}$$

$$b_{x_1} = 3.68$$

$$a_{x_1} = \frac{\Sigma y}{n} - \frac{b x_1 \Sigma x_1}{n}$$

$$= \frac{661}{10} - \frac{3.68 \times 105}{10}$$

$$a_{x_1} = 27.46$$

The regression equation for the relationship of hours studied and examination result is

$$y_{x_1} = a_{x_1} + b_{x_1} x_1$$
$$= 27.46 + 3.68$$

The coefficient of correlation for this relationship is

$$r_{x_1} = \frac{n\Sigma x_1 y - \Sigma x_1 \Sigma y}{\sqrt{n\Sigma x_1^2 - (\Sigma x_1)^2} \times \sqrt{n\Sigma y^2 - (\Sigma y)^2}}$$

Note: This formula is the same as that given in Chapter 7 but is easier to work with since all except

$n\Sigma y^2 - (\Sigma y)^2$ is already known.

$$n\Sigma y^2 - (\Sigma y)^2 = 10 \times 46{,}889 - 661^2$$

$$= 468{,}890 - 436{,}921 = 31{,}969$$

$$r_{x_1} = \frac{8{,}035}{\sqrt{2{,}185} \times \sqrt{31{,}969}}$$

$$= 0.9613$$

$\therefore\ r_{x_1}{}^2 = \underline{0.924}$ i.e. coefficient of determination for $y : x_1$.

In a similar manner the regression y on x_2 (exam. scores: IQ scores) is calculated resulting in

$$y_{x_2} = a_{x_2} + b_{x_2} x_2$$

$$= 57.16 + 0.085 x_2$$

$$r_{x_2}{}^2 = 0.001608$$

The scatter diagrams for the two independent variables with the least squares lines of best fit are shown in Figures 8/1 and 8/2.

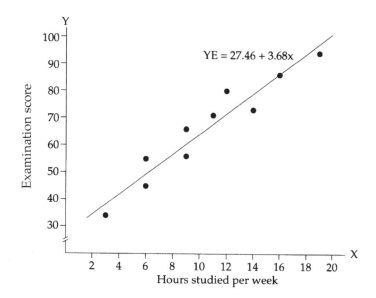

Scatter diagram of examination scores and hours studied ($y : x_1$). Figure 8/1

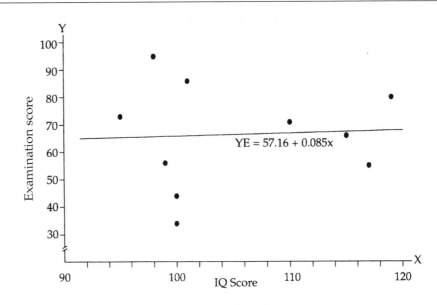

Scatter diagram of examination scores I.Q. scores (y : x_2). Figure 8/2

Solution

Part B – The multiple regression (y : x_1 and x_2)

The multiple regression calculations are carried out using the three variable Normal Equations from Para 3 and the results in Table 1 above, thus:

$$
\begin{aligned}
661 &= 10a + 105b_1 + 1{,}054b_2 \\
7{,}744 &= 105a + 1{,}321b_1 + 10{,}974b_2 \\
69{,}730 &= 1{,}054a + 10{,}974b_1 + 111{,}806b_2
\end{aligned}
$$

Using standard simultaneous equation procedures results in the following values for the coefficients in the equation

$$y = a + b_1 x_1 + b_2 x_2$$
$$y = -38.06 + 3.90x_1 + 0.6x_2$$

This result could be used to predict the examination score for a candidate, given the number of hours worked and IQ. For example, what is the expected score of a candidate who has worked for 13 hours per week and who has an IQ of 102?

$$y = -38.06 + 3.90 \times 13 + 0.6 \times 102$$
$$= 73.84\% \text{ expected examination score}$$

Solution

Part C – Coefficient of multiple determination, R^2

Using the computational formula given in Para 3 and the values calculated above, R^2 can be calculated thus:

$$R^2 = \frac{(-38.06 \times 661) + (3.90 \times 7,744) + (0.6 \times 69,730) - \dfrac{661^2}{10}}{46,889 - \dfrac{661^2}{10}}$$

$$= \underline{0.9995}$$

The various coefficients of determination can now be summarised and interpreted

$$r^2_{x_1} = 0.9243$$

$$r^2_{x_2} = 0.0016$$

$$R^2 = 0.9995$$

$r^2_{x_1}$ – This indicates that about 92% of the variation in examination scores is caused by variation in hours of study, which is obviously a major influence.

$r^2_{x_2}$ – This indicates that only 0.16% of any variation in examination score is caused by variation in IQ score which is a very small influence indeed.

R^2 – This shows the combined effect of the two independent variables and indicates that 99.95% of the movement in examination score is brought about by movements in hours studied and IQ score. This, however, assumes that it is a reasonable hypothesis that examination results are influenced by the intelligence of candidates and how hard they work!

Non-linear models

5. There are many occasions when the relationship between variables cannot be adequately described by linear functions, whether they use a single independent variable or several. In such circumstances some form of non-linear or curvi-linear model is likely to be more suitable and the following paragraphs describe some commonly encountered non-linear models.

Exponential growth

6. A linear trend line as previously described is suited to data which are expected to increase by the same *absolute* amount in each period. Where this relationship does not apply some form of non-linear regression must be used and an important non-linear function relates to what is known as exponential growth, i.e. where the data is expected to grow by the same *proportion* or *percentage* in each period. A typical example of exponential growth was world fertiliser usage between 1938 and 1968 as shown in Figure 8/3.

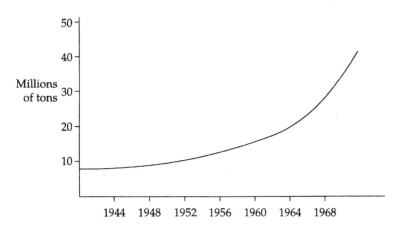

World fertiliser consumption 1938–1968. Figure 8/3

The shape of this graph, a typical exponential growth curve, should be compared with the typical linear trend line shown in Figure 8/1.

The exponential function

7. The exponential function takes the form

$$y = ab^x$$

where **y** is the variable to be predicted (millions of tons in Figure 8/3)

> **a** and **b** are constants

and   x denotes the number of the period (shown on the time axis in Figure 8/3).

It follows that y takes the value of '**a**' at time 0 and is multiplied by a factor of **b**^{x} in each period. If '**b**' exceeds 1 there is exponential growth, if '**b**' is less than 1 there is exponential decay. The variable '**y**' is said to be an exponential function of **x**.

Linear form of the exponential function

8. The exponential function can be reduced to linear form by taking the logarithm of the function thus

$$\log y = \log a + x \log b$$

or

$$\log y = A + Bx$$

where $A = \log a$ and $B = \log b$

The similarity of this expression and the linear regression line previously discussed will be apparent. An interesting feature of the log form of the exponential function is that it is equivalent to fitting a straight line to a graph drawn on semi-logarithmic scale graph paper (i.e. a logarithmic scale on the vertical axis and an ordinary arithmetic scale on the horizontal axis).

Calculation of the exponential function

9. The calculation will be illustrated by using the following data.

Year	Sales £m
0	100
1	150
2	225
3	337.5
4	506.25

Step 1: Find the logs of the sales (y). This produces the following table similar to that previously used for linear regression calculations.

x	log y	x log y	x^2
0	2.0000	0	0
1	2.1761	2.1761	1
2	2.3522	4.7044	4
3	2.5282	7.5846	9
4	2.7045	10.8180	16
$\Sigma x = 10$	$\Sigma y = 11.7610$	$\Sigma xy = 25.2831$	$\Sigma x^2 = 30$

These values can then be substituted into the formulae for **a** and **b** given in Chapter 7.

$$\mathbf{A} = \frac{\Sigma y}{n} - \frac{b\Sigma x}{n}$$

$$\mathbf{B} = \frac{n\Sigma xy - \Sigma x \Sigma y}{n\Sigma x^2 - (\Sigma x)^2}$$

i.e. $\mathbf{\log b} = \dfrac{5 \times 25.2831 - 10 \times 11.7610}{5 \times 30 - 100} = \dfrac{8.8055}{50} = 0.17611$

(The antilog of 0.17611 is 1.5, but as the expression for 'a' utilises **log b**, the log, 0.17611 is used directly.)

$$\mathbf{\log a} = \frac{11.7610}{5} - \frac{0.17611 \times 10}{5} = 2.3522 - 0.3522 = 2.000$$

$$\mathbf{\log a} = 2.000 \text{ gives } \mathbf{a} = 100$$

$$\mathbf{\log b} = 0.17611 \text{ gives } \mathbf{b} = 1.5$$

The exponential function obtained is

$$\mathbf{y = ab^x}$$

$$= \mathbf{100 \times 1.5^x}$$

Thus when $x = 0$ $\qquad$ $y = 100 \times 1.5^0 = 100 \times 1 = \mathbf{100}$

when $x = 1$ $\qquad$ $y = 100 \times 1.5^1 = 100 \times 1.5 = \mathbf{150}$

and so on.

Note: In this artificial example the data were a perfect exponential function, therefore the values of **a** and **b** obtained fit perfectly. This is, of course, unlikely to happen in practice.

Logarithmic functions

10. An alternative non-linear function is what is known as a logarithmic function which has the form of

$$y = ax^b$$

where **y** denotes variable to be predicted, **a** and **b** are constants and **x** generally denotes the time periods.

As with the exponential function, this function can be expressed in a linear form using logarithms thus

$$\log y = \log a + b \log x$$

In this function **y** is said to be a logarithmic function of **x**.

This function is equivalent to fitting a straight line to a graph drawn on log-log paper (i.e. both horizontal and vertical scales being logarithmic).

Calculation of the logarithmic function

11. The calculations will be illustrated using the following data

Period	Sales £ms
1	50
2	200
3	450
4	800
5	1,250

As for the exponential function the linear logarithmic form will be used, i.e.

$$\log y = \log a + b \log x$$

It will be noted that logs of both the periods (**x**) and of the sales (**y**), are required whereas in the exponential calculations only **log y** was necessary.

log x	log y	$(\log x)^2$	log x.log y
0	1.6990	0	0
0.3010	2.3010	0.0906	0.6926
0.4771	2.6532	0.2276	1.2658
0.6021	2.9031	0.3625	1.7479
0.6990	3.0969	0.4886	2.1647
$\Sigma x = \quad 2.0792$	$\Sigma y = \quad 12.6532$	$\Sigma x^2 = \quad 1.1693$	$\Sigma xy = \quad 5.8710$

$$b = \frac{5 \times 5.8710 - 2.0792 \times 12.6532}{5 \times 1.1693 - 2.0792^2} = \frac{3.0465}{1.5234} = \underline{\underline{2.00}}$$

$$\log a = \frac{12.6532}{5} - \frac{2 \times 2.0792}{5} = 1.6989 \text{ the antilog of which is } \underline{\underline{50}}$$

$\therefore$ the values of the logarithmic function are

$$y = ax^b$$
$$= \underline{50x^2}$$

Thus when $x = 1$

$$y = 50 \times 1^2 = 50 \times 1 = 50$$

when $x = 2$

$$y = 50 \times 2^2 = 50 \times 4 = 200$$

when $x = 3$

$$y = 50 \times 3^2 = 50 \times 9 = 450 \text{ and so on}$$

Note: Once again to illustrate the function calculations simply, a perfect set of data has been used which is unlikely to be met in practice.

Learning curves

12. Forecasting is concerned with what we anticipate will happen in the future. Unthinking extrapolation of past conditions is unlikely to produce good forecasts. If we are aware of an expected change in conditions in the future this must be taken into account when preparing the finalised forecast.

A particular example of this relates to what are known as *learning curves* which are a practical application of a non-linear function. The learning curve depicts the way people learn by doing a task and are therefore able to complete the task more quickly the next time they attempt it. Learning is rapid in the early stages and the rate gradually declines until a sufficient number of units or tasks have been completed, when the time taken will become constant. The main practical application is concerned with direct labour times and costs.

Cost predictions, especially those relating to direct labour times and costs, should allow for the effects of the learning process. During the early stages of producing a new part or carrying out a new process, experience and skill is gained, productivity increases and there is a reduction of time taken per unit.

There are two forms of learning curve model, the *cumulative average* model and the *marginal* model. Both models use the same general formula but with some changes in the definitions of the formula elements. The learning curve is a non-linear function for which the general formula is:

$$y = ax^b$$

where **a** = number of labour hours for the first unit

  x = cumulative number of units

  b = the learning coefficient

It is the definition of y which determines the model to be used as follows:

When the *cumulative average model* is used:

  y = cumulative average time per unit for x units.

When the *marginal model* is used:

  y = marginal time for the xth unit.

The two approaches are described in detail below.

Cumulative average learning curves

13. Studies have shown that there is a tendency for the time per unit to reduce at same constant rate as production mounts. For example, a 90% learning curve means that as cumulative production quantities double the cumulative average time per unit falls by 10%. An 80% learning curve would mean that a doubling of production causes a 20% fall in the cumulative average time per unit and so on.

This is illustrated in Table 2 which is based on a product with an 80% learning curve where the first unit of production takes 50 hours.

Cumulative production (units)	Cumulative average time per unit (hours)		Thus, cumulative time taken (hours)
1	50		50
2	40	(50 × 80%)	80
4	32	(50 × 80% × 80%)	128
8	25.6	(50 × 80% × 80% × 80%)	204.8

Illustration of 80% cumulative average learning curve. Table 2

Note that the cumulative average time of 25.6 hours when cumulative production is 8 units is *not* the time taken for the 8th unit; it is the average over all production to date, i.e.

$$\frac{204.8}{8} = 25.6$$

If it is required to find the time the 8th unit takes it is necessary to find the cumulative time for 7 units of cumulative production and deduct it from the cumulative time for 8 units. This is illustrated below after the learning curve formula is explained.

Cumulative average learning curves by formula

14. When the cumulative average model is used the formula is as follows:

$$y = ax^b$$

where y = cumulative average labour hours per unit
$\quad\quad a$ = number of labour hours for the first unit
$\quad\quad x$ = cumulative number of units
$\quad\quad b$ = the learning coefficient

The learning coefficient is calculated as follows:

$$b = \frac{\log (1 - \text{Proportionate decrease})}{\log 2}$$

thus for a 20% decrease (i.e. an 80% learning curve)

$$b = \frac{\log (1 - 0.2)}{\log 2} = \frac{-0.09691}{0.30103} = \underline{\underline{-0.322}}$$

Note: The log values can be found directly from a calculator.

The formula is used to solve questions based on the product in Table 2.

a) What is the cumulative average time per unit when cumulative production is 8 units, 16 units?

b) How long did unit 8 take to manufacture?

Solutions

a) When cumulative production is 8 units

$$y = 50 \times 8^{-0.322}$$

$\therefore$ Cumulative average time = 25.6 hours (confirming the value in Table 2.)

When cumulative production is 16 units

$$y = 50 \times 16^{-0.322}$$

$\therefore$ Cumulative average time = 20.48 hours.

(This can be verified by multiplying 25.6 hours by 80%)

b) Time to manufacture the 8th unit.

As explained above this is found by deducting the cumulative time for 7 units from the cumulative time for 8 units.

The cumulative time for 8 units is already known from Table 2, i.e. 204.8 hours, but it is necessary to calculate the cumulative time for 7 hours, as follows:

Cumulative average time for 7 units $= 50 \times 7^{-0.322}$

$$= 26.72 \text{ hours}$$

Thus cumulative time for 7 units

$$= 26.72 \times 7 = 187.04 \text{ hours}$$

$\therefore$ Time for 8th unit $= 204.8 - 187.04$ hours

$$= \textbf{17.76 hours}$$

Marginal learning curve model

15. This model is based on the assumption that the time for the marginal, or last, unit reduces by a given percentage when cumulative production doubles.

This is illustrated in Table 3 based on the same product used for Table 2, i.e. an 80% learning curve where the first unit takes 50 hours.

Cumulative production (units)	Cumulative time taken (hours)	Time taken for marginal unit (hours)
1	50	50
2	90	40 (50 × 80%)
4	?	32 (50 × 80% × 80%)
8	?	25.6 (50 × 80% × 80% × 80%)

Illustration of the 80% marginal learning curve. Table 3

Although the figures in the right-hand column are the same as in Table 2 their meaning is quite different. In Table 3 the 50, 40, 32, and 25.6 hours are the times the 1st, 2nd, 3rd and 8th unit take to manufacture *not* an overall average. It will also be seen that it is not directly possible to find the cumulative production time at the 4 and 8 unit production levels. This is because we do not yet know the marginal times for the 3rd, 5th, 6th and 7th units. These can be found using the formula and this process is demonstrated below.

Marginal learning curves by formula

16. As already explained the general learning curve formula is applicable but y is defined differently as follows:

$$y = ax^b$$

where y = time for the marginal unit

and all other elements are as previously defined and the learning coefficient is calculated in the same manner.

Once again the formula is used to solve questions based on the product in Tables 2 and 3.

a) How long did unit 8 take to manufacture?

b) What was the cumulative production time when cumulative production is 8 units?

Solutions

a) Time to manufacture 8th unit

$$y = 50 \times 8^{-0.322}$$

$$= 25.6 \text{ hours (confirming the value in Table 3)}$$

b) Cumulative production time when cumulative production is 8 units.

The cumulative production time cannot be found directly. It is necessary to find the times to manufacture each of the units. 1 to 8 separately and total them. The times for units 1, 2, 4 and 8 are already known from Table 3 but it is necessary to calculate the times for units 3, 5, 6 and 7 and then add all the times together.

Time for unit 3 = $50 \times 3^{-0.322}$ = 35.1 hours
Time for unit 5 = $50 \times 5^{-0.322}$ = 29.78 hours
Time for unit 6 = $50 \times 6^{-0.322}$ = 28.08 hours
Time for unit 7 = $50 \times 7^{-0.322}$ = 26.72 hours

Thus the cumulative time for a cumulative production of 8 units is

$$50 + 40 + 35.1 + 32 + 29.78 + 28.08 + 26.72 + 25.6$$

$$= \textbf{267.28 hours}$$

It will be apparent that this process is tedious so that if large quantities were being produced, computer assistance would be required to carry out these calculations if they were thought necessary.

Cumulative average and marginal models compared

17. Although the same general formula is applicable there are substantial differences in results depending on the definition for y in the formula $y = ax^b$. In general, the marginal model predicts that the reduction in time taken for production is *less* than the cumulative average model for any given learning curve. This can be proved by examining the total time for 8 units in the examples above:

With an 80% learning curve:

Total time predicted for 8 units using Cumulative Average model	Total time predicted for 8 units using Marginal model
204.8 hours	267.28 hours

Thus 80% learning curves (or any other values) are not equivalent in the two models.

Great care must therefore be taken to use the correct model. Remember, this is entirely dependent on the meaning given to y in the general formula.

Practical example of learning curve

18. A Production Planning Department is considering the production schedules for Period 9. In particular they wish to calculate the time to be allocated for the manufacture of a batch of 100 of a computer controlled machine tool called ROBO XI. The first ROBO XI took 80 hours to make and it is known from past experience that there is a learning effect. From past records the following information is available.

ROBO XI

Cumulative production (units)	Cumulative time taken (hours)	Time per unit (hours)
600	18,153.6	30.256
1200	32,676	27.23

They calculate that the cumulative production at the beginning of Period 9 will be 3000 units.

Required

a) What type of learning curve model do the records suggest?

b) What value of learning curve do the records show?

c) Calculate the learning coefficient.

d) Calculate the time allowance necessary for the batch of 100 in Period 9.

Solution

a) Type of learning curve model.

From the data supplied it is clear that the time per unit is a cumulative average because the cumulative time is production multiplied by time per unit, e.g.

$$600 \times 30.256 = \underline{18,153.6}$$

$\therefore$ a *cumulative average model* is being used.

b) Value of learning curve.

$$\frac{27.23}{30.256} = 0.9$$

$\therefore$ there is a 90% learning curve.

c) Learning coefficient for a 90% learning curve.

$$b = \frac{\log (1 - \text{Proportional decrease})}{\log 2}$$

$$= \frac{\log (1 - 0.1)}{\log 2}$$

$$= \frac{-0.04576}{0.30103} \qquad = \underline{\underline{-0.152}}$$

d) Time allowance for a batch of 100 in Period 9 when cumulative production at start is 3,000 units.

The time allowance will be the difference between the cumulative time for 3,000 units and the cumulative time for 3,100 units thus

At 3,000 units:

Cumulative average per unit $= 80 \times 3{,}000^{-0.152} = 23.69$ hours

$\therefore$ cumulative time $= 3000 \times 23.69 = 71{,}070$ hours

At 3,100 units:

Cumulative average per unit $= 80 \times 3{,}100^{-0.152} = 23.57$ hours

$\therefore$ cumulative time $= 3{,}100 \times 23.57 = 73{,}067$ hours

Thus the time allowance for the batch of $100 = 73{,}067 - 71{,}070 = \underline{\textbf{1,997 hours}}$

Note that in the above example and in the earlier explanation of learning curves it has been assumed that regular reductions in labour times occur in strict accordance with the theoretical models. In practice, of course, it is very unlikely that this will happen. Although learning does take place and reductions in labour times per unit do occur, the consistency and regularity of the theoretical models are not likely to be encountered. In spite of this, questions using learning curves are popular with examiners.

Summary

19. a) If a linear two variable model is deemed inappropriate then multiple linear regression models or non-linear regression models can be tried.

b) The line of best fit for the two variable model becomes a plane of best fit in a three variable model.

c) Multiple regression uses R^2, the coefficient of multiple determination to assess the improvement in the fit of the model on the introduction of one or more additional causal variables.

d) The exponential growth model deals with situations where the **y** variable is expected to grow by a constant proportion for a given change in the **x** value.

e) There are occasions where a non-linear relationship may be converted to a linear relationship by the use of logarithms. This is known as a logarithmic function and takes the form $\mathbf{y = ax^b}$.

f) The learning curve is an important application of a non-linear function and takes the form $\mathbf{y = ax^b}$

g) There are two forms of learning curve: the marginal and the cumulative average.

h) In the cumulative average learning curve model y in the general formula equals the cumulative average time per unit over all production to date.

i) In the marginal learning curve model y equals the time for the last unit.

Points to note

20. a) The process of plotting the calculated values for a non-linear function onto a scatter diagram is known as curve fitting.

b) Although it is unlikely that full analysis of multiple regression would be required in a time constrained examination understanding of the method is necessary.

Self review questions *Numbers in brackets refer to paragraph numbers*

1. What is multiple regression? (3)
2. Define R^2 and explain its purpose? (3)
3. How does exponential growth differ from linear growth? (6)
4. What is exponential decay? (7)
5. What is the linear form of the exponential function? (8)
6. What is a logarithmic function? (10)
7. What is a learning curve? (14)
8. How is the learning coefficient calculated? (14)
9. What are the key differences between marginal and cumulative average models? (14, 15)

Exercises with answers

1. Analysis of representatives' car expenses shows that the expenses are dependent on the miles travelled (x_1) and the type of journey (x_2). The general form is:

$$y = a + b_1x_1 + b_2x_2$$

Calculations have produced the following values (where y is expenses per month)

$$y = £86 + 0.37x_1 + 0.08x_2$$

$$r_{x_1}^2 = 0.78$$

$$r_{x_2}^2 = 0.16$$

$$R^2 = 0.88$$

Interpret these values.

2. For a given operation a 10% marginal learning curve operates. Assuming that the first unit takes 30 minutes, how long should the 20th unit take?

3. Assume the same problem as for question **2**, but with a 30% cumulative average learning curve.

Answers to exercises

1.
$$r^2_{x_1} = 0.78$$

This is the coefficient of determination of miles travelled to cost and means that 78% of total cost is attributable to mileage.

$$r^2_{x_2} = 0.16 \text{ i.e. 16% of cost is accounted for by the type of journey}$$

$R^2 = 0.88$ is the overall coefficient of determination and indicates that the multiple regression equation accounts for 88% of the total variation in costs.

The coefficients in the equation are:

$$a = £86 = \text{fixed costs}$$
$$b_1 = 0.37 = \text{amount per mile}$$
$$b_2 = 0.08 = \text{influence of the type of journey.}$$

2. With a 10% marginal learning curve the predicted time for the 20th unit can be found directly. Learning curve = ax^b

$$\text{where : } b = \frac{\log (1 - 0.1)}{\log 2} = -0.152$$

$$\therefore \text{ Time for 20th unit} = 30 \times 20^{-0.152}$$

$$= \textbf{19.03 minutes}$$

3. Using a cumulative average model it is necessary to find the time taken in total for 20 units and deduct from this the time for 19 units thus

For a 30% learning curve $\quad b = \dfrac{\log (1 - 0.3)}{\log 2} = -0.5146$

$\therefore$ cumulative average time per unit when

$$20 \text{ units made} = 30 \times 20^{-0.5146}$$
$$= 6.42 \text{ minutes}$$

$\therefore$ overall time for 20 units $= 6.42 \times 20 = \underline{128.4}$

Cumulative average time per unit when 19 units made

$$= 30 \times 19^{-0.5146}$$

$\therefore$ overall time for 19 units $\qquad = 6.59 \times 20 = \underline{131.8}$

$\therefore$ Time for 20th unit $\qquad = 131.8 - 128.4$

$\qquad = \textbf{3.4 minutes}$

9 Forecasting – time series analysis

Objectives

1. After studying this chapter you will
    - understand the purposes of forecasting;
    - know the principles of key qualitative forecasting techniques such as the Delphi method, Market Research and Historical Analogy;
    - be able to calculate a moving average;
    - understand the principles of exponential smoothing;
    - know the distinction between additive and multiplicative time series models;
    - be able to analyse or decompose a time series;
    - understand how to use regression analysis for forecasting.

Forecasting – definition

2. For our purposes forecasting can be defined as attempting to predict the future by using qualitative or quantitative means. In an informal way, forecasting is an integral part of all human activity, but from the business point of view increasing attention is being given to formal forecasting systems which are continually being refined. Some forecasting systems involve very advanced statistical techniques beyong the scope of this book, so are not included.

    All forecasting methodologies can be divided into three broad headings i.e. forecasts based on:

What people have *done*	**What people *say***	**What people *do***
examples:	examples:	examples:
Time Series Analysis	Surveys	Testing Marketing
Regression Analysis	Questionnaires	Reaction tests

The data from past activities are cheapest to collect but may be outdated and past behaviour is not necessarily indicative of future behaviour.

Data derived from surveys are more expensive to obtain and needs critical appraisal – intentions as expressed in surveys and questionnaires are not always translated into action.

Finally, the data derived from recording what people actually do are the most reliable but also the most expensive and occasionally it is not feasible for the data to be obtained.

Forecasting – application

3. Virtually every form of decision making and planning activity in business involves forecasting. Typical applications include:

production planning	demand forecasts
inventory control	advertising planning
investment cash flows	corporate planning
cost projections	budgeting

Qualitative and quantitative techniques

4. A convenient classification of forecasting techniques is between that are broadly qualitative and those that are broadly quantitative. These classifications are by no means mutually exclusive but serve as a means of identification.

Quantitative techniques

5. These are techniques of varying levels of statistical complexity which are based on analysing past data of the item to be forecast, e.g. sales figures, stores issues, costs incurred. However sophisticated the technique used, there is the under-lying assumption that past patterns will provide some guidance to the future. Clearly for many operational items (material usage, sales of existing products, costs) the past does serve as a guide to the future, but there are circumstances for which no data are available, e.g. the launching of a completely new product, where other, more qualitative techniques are required. These techniques are dealt with briefly first and then the detailed quantitative material follows.

Qualitative techniques

6. These are techniques which are used when data are scarce, e.g. the first introduction of a new product. The techniques use human judgement and experience to turn qualitative information into quantitative estimates. Although qualitative techniques are used for both short and long term purposes, their use becomes of increasing importance as the time scale of the forecast lengthens. Even when past data are available, so that standard quantitative techniques can be used, longer term forecasts require judgement, intuition, experience, flair etc, that is, qualitative factors, to make them more useful. As the time scale lengthens, past patterns become less and less meaningful. The qualitative methods briefly dealt with in this book are the *Delphi Method, Market Research* and *Historical Analogy*.

Delphi method

7. This is a technique mainly used for longer term forecasting, designed to obtain expert consensus for a particular forecast, without the problem of submitting to pressure to conform to a majority view. The procedure is that a panel of experts independently answer a sequence of questionnaires in which the responses to one questionnaire are used to produce the next questionnaire. Thus any information available to some experts and not others is passed on to all, so that their subsequent judgements are refined as more information and experience become available.

Market research

8. Market research uses opinion surveys, analyses of market data, questionnaires and other investigations to gauge the reaction of the market to a particular product, design, price, etc. Market research is often very accurate for the relatively short term, but longer term forecasts based purely on surveys are likely to be suspect because peoples' attitudes and intentions change.

Historical analogy

9. Where past data on a particular item are not available, e.g. for a new product, data on similar products are analysed to establish the life cycle and expected sales of the new product. Clearly, considerable care is needed in using analogies which relate to different products in different time periods, but such techniques may be useful in forming a broad impression in the medium to long term.

    Note on qualitative methods:
    a) When past quantitative data are unavailable for the item to be forecast, then inevitably much more judgement is involved in making forecasts.
    b) Some of the qualitative techniques mentioned above use advanced statistical techniques, e.g. some of the sampling methods used in Market Research. Nevertheless, any such method may prove to be a relatively poor forecaster, purely due to the lack of appropriate quantitative data relating to the factor being forecast.

Quantitative forecasting

10. A prerequisite to the use of the technique to be described is data on past usage or demand. The longer a period covered by the data, the more likely that the patterns in the data will be representative of the future. This is the main assumption behind the use of statistical forecasting techniques.

    Nevertheless, however long a period is covered by past data, any extrapolations or forecasts produced from that data by whatever technique should be treated with caution. Conditions can, and do, change quite rapidly. Judgement, experience and a wide knowledge of the market place always play a part in establishing a reliable forecast.

Time series analysis

11. As the name suggests, time series analysis uses some form of mathematical or statistical analysis on past data arranged in a time series, e.g. sales by month for the last ten years. Time series analyses have the advantage of relative simplicity, but certain factors need to be considered.
    a) Are the past data representative? For example, do they contain the results of a recession/boom, a major shift of taste, etc., etc.
    b) On the whole, time series methods are more appropriate where short term forecasts are required. Over the longer term external pressures, internal

policy changes make historical data less appropriate.

c) Time series methods are best suited to relatively stable situations. Where substantial fluctuations are common and/or conditions are expected to change, then time series methods may give relatively poor results.

Time series analysis – moving average

12. If the forecast for next month's sales, say December, was the actual sales for November, then the forecasts obtained would fluctuate up and down with every random fluctuation. If the forecast for the next month's sales was an average of sales for several preceding months then, hopefully, random fluctuations would cancel each other out, i.e. would be smoothed away. This is the simple principle of the moving average method which is one of the smoothing techniques. The method is illustrated by the following example.

Example 1

Past sales of widgets		Forecasts produced by		
Month	Actual Sales (Units)	3 monthly moving average	6 monthly moving average	12 monthly moving average
January	450			
February	440			
March	460			
April	410	450		
May	380	437		
June	400	417		
July	370	397	423	
August	360	383	410	
September	410	377	397	
October	450	380	388	
November	470	407	395	
December	490	443	410	
January	460	470	425	424

Table 1

Any month's forecast is the average of the proceeding **n** months' actual sales. For example, the 3 monthly moving average forecasts were prepared as follows:

$$\text{April's forecast} = \frac{\text{January Sales} + \text{February Sales} + \text{March Sales}}{3}$$

$$= \frac{450 + 440 + 460}{3} = \underline{\underline{450}}$$

$$\text{May's forecast} = \frac{\text{February Sales} + \text{March Sales} + \text{April Sales}}{3}$$

$$= \frac{440 + 460 + 410}{3} = \underline{\underline{437}}$$

Similar logic applies for the 6 and 12 monthly moving averages.

Note 1: A moving average can be used as a forecast as shown above but when graphing moving averages it is important to realise, that being averages, they must be plotted at the mid point of the period to which they relate.

Note 2: Twelve-monthly moving averages or moving annual totals form part of a commonly used diagram, the Z chart. It is called a Z chart because the completed diagram is shaped like a Z. The top part of the Z is formed by the moving annual total, the bottom part by the individual monthly figures and the sloping line by the cumulative monthly figures.

Characteristics of moving averages

13. a)   The different moving averages produce different forecasts.

   b)   The greater the number of periods in the moving average, the greater the smoothing effect.

   c)   If the underlying trend of the past data is thought to be fairly constant with substantial randomness, then a greater number of periods should be chosen.

   d)   Alternatively, if there is thought to be some change in the underlying state of the data, more responsiveness is needed, therefore fewer periods should be included in the moving average.

Limitations of moving averages

14. a)   Equal weighting is given to each of the values used in the moving average calculation, whereas it is reasonable to suppose that the most recent data is more relevant to current conditions.

   b)   An **n** period moving average requires the storage of **n** − **1** values to which is added the latest observation. This may not seem much of a limitation when only a few items are considered, but it becomes a significant factor when, for example, a company carries 25,000 stock items each of which requires a moving average calculation involving say 6 months of usage data to be recorded.

   c)   The moving average calculation takes no account of data outside the period of average, so full use is not made of all the data available.

   d)   The use of the unadjusted moving average as a forecast can cause misleading results when there is an underlying seasonal variation.

Exponential smoothing

15. This is a frequently encountered forecasting technique which largely overcomes the limitations of moving averages. The method involves the automatic weighting of past data with weights that decrease exponentially with time, i.e. the most current values receive the greatest weighting and the older observations receive a decreasing weighting. The exponential smoothing technique is a weighted moving average system and the underlying principle is that the

New Forecast = Old Forecast + a proportion of the forecast error

The simplest formula is

New forecast = Old forecast + α (Latest Observation – Old Forecast)

where α (alpha) is the smoothing constant.

Exponential smoothing illustration

16. The data in Table 1 are reproduced below and forecasts have been prepared using α values of 0.1 and 0.5

Actual		Exponential Forecasts	
Month	Sales (units)	α Values 0.1	α Value 0.5
January	450		
February	440	450	450
March	460	449	445
April	410	450.10	452.50
May	380	445.69	431.25
June	400	439.12	405.63
July	370	435.21	402.82
August	360	428.69	386.41
September	410	421.82	373.21
October	450	420.64	391.61
November	470	423.57	420.81
December	490	428.21	445.41
January	460	434.39	467.71

Table 2

Notes:

a) Because in this example no previous forecast was available, January sales were used as February's forecast.

b) Thereafter the normal formula was used, for example, when α = 0.1

March forecast = February Forecast + 0.1 (Feb Sales – February Forecast)

March Forecast = 450 + 0.1 (440 – 450)

March Forecast = 449

and similarly for all other forecasts.

c) In practice the forecasts would be rounded to the nearest unit. They have been left as calculated so that students may check and compare some of their own calculations.

d) It will be apparent that the higher α value, 0.5, produces a forecast which adjusts more readily to the most recent sales.

The smoothing constant

17.   The value of α can be between 0 and 1. The higher value of α (i.e. the nearer to 1), the more sensitive the forecast becomes to current conditions, whereas the lower the value, the more stable the forecast will be, i.e. it will react less sensitively to current conditions. An approximate equivalent of alpha values to the number of periods' moving average is given below:

α Value	Approximate periods in equivalent moving average
0.1	19
0.25	7
0.3	6
0.5	3

The total of the weights of observations contributing to the new forecast is 1 and the weight reduces exponentially progressively from the alpha value for the latest observation to smaller values for the older observations. For example, if the alpha value was 0.3 and June's sales were being forecast, then June's forecast is produced from averaging past sales weighted as follows.

0.3 (May's Sales) + 0.21 (April's Sales) + 0.147 (March's Sales)

+ 0.1029 (February Sales) + 0.072 (January Sales)

+ 0.050 (December Sales), etc.

From this it will be noted that the weightings calculated approach a total of 1. It should be emphasised that the application of the formula given in Para 15 automatically gives these weightings; further calculations are not required.

The α value is chosen by the analyst usually after experimentation with different values to see what value gives the most realistic forecasts.

Extensions of exponential smoothing

18.   The basic principle of exponential smoothing has been outlined above, but to cope with various problems such as seasonal factors, strongly rising or falling demand, etc., many developments to the basic model have been made. These include double and triple exponential smoothing and correction for trend and delay factors, etc. These are outside the scope of the present book, so are not covered.

Characteristics of exponential smoothing

19.   a)   Greater weight is given to more recent data.

b)   All past data are incorporated there is no cut-off point as with moving averages.

c)   Less data needs to be stored than with the longer period moving averages.

d)   Like moving averages it is an adaptive forecasting system. That is, it adapts continually as new data becomes available and so it is frequently incorporated as an integral part of stock control and production control systems.

e) To cope with various problems (trend, seasonal factors, etc.) the basic model needs to be modified.

f) Whatever form of exponential smoothing is adopted, changes to the model to suit changing conditions can simply be made by altering the α value.

Time series analysis – decomposition

20. Observations taken over time (i.e. a time series) often contain the four following characteristics:

a) **A long-term trend** (denoted by **T**)

i.e. the long-term tendency of the whole series to rise or fall.

b) **Seasonal variations** (denoted by **S**)

i.e. short-term periodic fluctuations in values due to different circumstances, e.g. sales in a shop are usually higher on Fridays and Saturdays, heating fuel sales are higher in the winter, etc. In time series analysis *seasonal* has a wider meaning than just the seasons of the year.

c) **Cyclical variations** (denoted by **C**)

These are medium-term changes caused by factors which apply for a while, then go away, and then come back again in a repetitive cycle. The economic cycle (boom, recession, boom . . .) is an example. Longer-term than seasonal variations.

d) **Random or residual variations** (denoted by **R**)

These are non-recurring random variations, e.g. war, fire, change of government, etc.

To make reasonably accurate forecasts, it is often necessary to quantify the different aspects separately (i.e. **T, S, C** and **R**) from the raw data. This is known as *time series decomposition* or often just *time series analysis*. The separated elements are then combined to produce a forecast.

There are two main Time Series models, known as the *additive* and *multiplicative* models. These are defined as follows:

Additive model

Time Series Value = **T + S + C + R** where **S, C** and **R** are expressed as *absolute* values.

Multiplicative model

Time Series Value = (**T** × **S** × **C** × **R**) where **S, C** and **R** are expressed as *percentages* or *proportions*.

The multiplicative model is commonly used in practice and is more appropriate if the characteristics interact, e.g. where a higher trend value increases the seasonal variation. The additive model is more suitable if the component factors are independent, e.g. where the amount of seasonal variation is not affected by the value of the trend. Figure 9/1 illustrates the appropriateness of the two models where there is an upward trend.

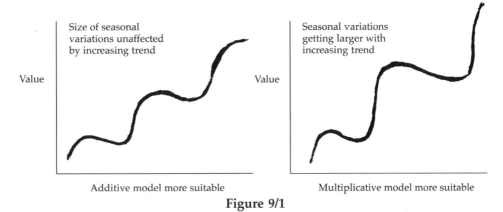

Figure 9/1

Of the four elements the most important are the first two, the trend and seasonal variation, so this book concentrates on these two. The following illustration shows how the trend (T) and seasonal variation (S) are separated out from a time series and how the calculated T and S values are used to prepare a forecast. The process of separating out the trend and seasonal variation is known as *deseasonalising the data*.

There are two approaches to this process: one is based on regression through the actual data points and the other calculates the regression line through moving average trend points. The method using regression through the actual data is demonstrated first followed by the moving average method.

Time series analysis: trend and seasonal variation using regression on the data

21. The following data will be used to illustrate how the trend and seasonal variation are calculated.

Example 1

Sales of widgets in '000s

	Quarter 1	Quarter 2	Quarter 3	Quarter 4
Year 1	20	32	62	29
2	21	42	75	31
3	23	39	77	48
4	27	39	92	53

It will be apparent that there is a strong seasonal element in the above data (low in Quarter 1 and high in Quarter 3) and that there is a generally upward trend.

The steps in analysing the data and preparing a forecast are:

Step 1: Calculate the trend in the data using the least squares method.

Step 2: Estimate the sales for each quarter using the regression formula established in Step 1.

Step 3: Calculate the percentage variation of each quarter's actual sales from the estimates, obtained in Step 2.

Step 4: Average the percentage variations from Step 3. This establishes the average seasonal variations.

Step 5: Prepare forecast based on trend × percentage seasonal variations.

Step 1

Calculate the trend in the data by calculating the least squares linear regression line $y = a + bx$ by the procedure explained in Chapter 7.

	x (quarters)	y (sales)	xy	x^2
Year 1	1	20	20	1
	2	32	64	4
	3	62	186	9
	4	29	116	16
Year 2	5	21	105	25
	6	42	252	36
	7	75	525	49
	8	31	248	64
Year 3	9	23	207	81
	10	39	390	100
	11	77	847	121
	12	48	576	144
Year 4	13	27	351	169
	14	39	546	196
	15	92	1,380	225
	16	53	848	256
	$\Sigma x = 136$	$\Sigma y = 710$	$\Sigma xy = 6{,}661$	$\Sigma x^2 = 1{,}496$

Table 3

Least squares equations

$$\Sigma y = an + b\Sigma x$$

$$\Sigma xy = a\Sigma x + b\Sigma x^2$$

$$710 = 16a + 136b$$

$$6{,}661 = 136a + 1{,}496b$$

$$\therefore 626 = 340b$$

$$b = 1.84 \text{ and substituting we obtain}$$

$$a = 28.74$$

$$\text{Trendline} = y = 28.74 + 1.84x$$

Steps 2 and 3

Use the trend line to calculate the estimated sales for each quarter.

For example, the estimate for the first quarter in year 1 is

$$\text{estimate} = 28.74 + 1.84\,(1) = \underline{30.58}$$

The actual value of sales is then expressed as a percentage of this estimate. For example, actual sales in the first quarter were 20 so the seasonal variation is

$$\frac{\textbf{Actual sales}}{\textbf{Estimate}}\% = \frac{20}{30.58} = \underline{65\%}$$

These calculations are shown in Table 4.

	x (quarters)	y (sales)	Trend	$\dfrac{\text{Actual}}{\text{Trend}}\%$
Year 1	1	20	30.58	65
	2	32	32.42	99
	3	62	34.26	181
	4	29	36.10	80
Year 2	5	21	37.94	55
	6	42	39.78	106
	7	75	41.62	180
	8	31	43.46	71
Year 3	9	23	45.30	51
	10	39	47.14	83
	11	77	48.98	157
	12	48	50.82	94
Year 4	13	27	52.66	51
	14	39	54.50	72
	15	92	56.34	163
	16	53	58.18	91

Trend Estimates and Percentage Variations. Table 4

Step 4

Average the percentage variations to find the average seasonal variations.

	Q1 %	Q2 %	Q3 %	Q4 %
	65	99	181	80
	55	106	180	71
	51	83	157	94
	51	72	163	91
	222	360	681	336
÷ 4 =	56%	90%	170%	84%

Table 5

These then are the average variations expected from the trend for each of the quarters; for example, on average the first quarter *of each year* will be 56% of the value of the trend. Because the variations have been averaged, the amounts *over* 100% (Q3 in this example) should equal the amounts *below* 100%. (Q1, Q2 and Q4 in this example). This can be checked by adding the average variations and verifying that they total 400% thus:

$$56\% + 90\% + 170\% + 84\% = 400\%.$$

On occasions, roundings in the calculations will make slight adjustments necessary to the average variations.

Step 5

Prepare final forecasts based on the trend line estimates from Table 4 (i.e. 30.58, 32.42, etc.) and the averaged seasonal variations from Table 5. (i.e. 56%, 90%, 170% and 84%).

The seasonally adjusted forecast is calculated thus:

Seasonally adjusted forecast = Trend estimate × Seasonal variation%

For example, the forecasts for the first and second quarters in year 1 are:

$$\text{Forecast } \mathbf{Q_1} = 30.58 \times 56\% = 17.12$$

$$\text{Forecast } \mathbf{Q_2} = 32.42 \times 90\% = 29.18$$

The final forecasts are shown in Table 6.

	x (quarters)	y (sales)	Seasonally adjusted forecast
Year 1	1	20	17.12
	2	32	29.18
	3	62	58.24
	4	29	30.32
Year 2	5	21	21.24
	6	42	35.80
	7	75	70.75
	8	31	36.51
Year 3	9	23	25.37
	10	39	42.43
	11	77	83.27
	12	48	42.69
Year 4	13	27	29.49
	14	39	49.05
	15	92	95.78
	16	53	48.87

Seasonally adjusted Forecasts. Table 6

The forecasts based on the analysis as shown in Table 6 are compared with the actual data to get some idea of how good extrapolated forecasts might be. With further analysis they enable us to quantify the residual variations.

Extrapolation using the trend and seasonal factors

Once the formulae above have been calculated, they can be used to forecast (extrapolate) future sales. If it is required to estimate the sales for the next year (i.e. Quarters 17, 18, 19 and 20 in our series) this is done as follows:

Quarter 17   Basic Trend $= 28.74 + 1.84(17)$

$$= 60.02$$

Seasonal adjustment for a first quarter $= 56\%$

$$\text{Adjusted Forecast} = 60.02 \times 56\%$$

$$= \mathbf{33.61}$$

A similar process produces the following figures:

Adjusted forecasts   Quarter 18 = **55.67**

19 = **108.29**

20 = **55.05**

Notes:

a) Time series decomposition is not an adaptive forecasting system like moving averages and exponential smoothing.

b) Forecasts produced by such an analysis should always be treated with caution. Changing conditions and changing seasonal factors make long term forecasting a difficult task.

c) The above illustration has been an example of a *multiplicative model*. This is because the seasonal variations were expressed in percentage or proportionate terms. Similar steps would have been necessary if the additive model had been used except that the variations from the trend in Table 4 would have been the absolute values. For example, the first two variations would have been

$$Q_1 : 20 - 30.58 = \text{absolute variation} = -10.58.$$

$$Q_2 : 32 - 32.42 = \text{absolute variation} = -0.42.$$

and so on.

The absolute variations would have been averaged in the normal way to find the average absolute variation, whether + or −, and these values would have been used to make the final seasonally adjusted forecasts.

Figure 9/2 shows the actual sales, the trend line and the seasonally adjusted forecast.

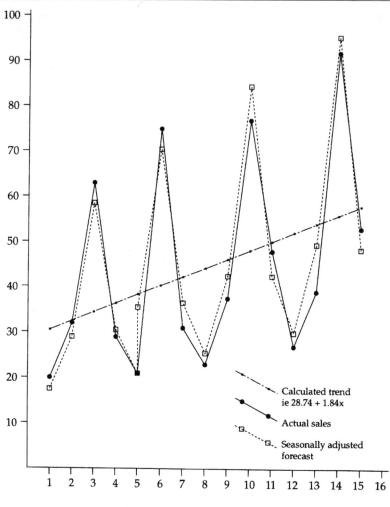

Time Series Analysis. Figure 9/2

Times series analysis: trend and seasonal variation using moving averages

22.   When the correlation coefficient is low the method of calculating the regression line through the actual data points as shown in para 21 should not be used. This is because the regression line is too sensitive to changes in the data values.

In such circumstances, calculating a regression line through the moving average trend points is more robust and stable.

Example 1 is reworked below using this method and, because there are many similarities to the earlier method, only the key stages are shown.

Table 7 summarises the process.

x	y	3 point moving average (1)	Trendline (2)	$\dfrac{\text{Actual}}{\text{Trend}}\%$
1	20		34.38	58
2	32	38	35.70	90
3	62	41	37.02	167
4	29	37.3	38.34	76
5	21	30.7	39.66	53
6	42	46	40.98	102
7	75	49.3	42.30	177
8	31	43	43.62	71
9	23	31	44.94	51
10	39	46.3	46.26	84
11	77	54.7	47.58	162
12	48	50.7	48.90	98
13	27	38	50.22	54
14	39	52.7	51.54	76
15	92	61.3	52.86	174
16	53		54.18	98

Trend estimates and percentage variations utilising moving averages. Table 7

Notes:

1. The first 3 point moving average is calculated thus:

$$\frac{20 + 32 + 62}{3} = 38 \text{ which is entered opposite Period 2}$$

The next is calculated:

$$\frac{32 + 62 + 29}{3} = 41, \text{and so on}$$

2. The regression line $y = a + bx$ of the Moving Average values is calculated in the normal manner and results in the following:

$$y = 33.06 + 1.32x$$

This is used to calculate the trend line:

e.g.

$$\text{For Period 1 } y = 33.06 + 1.32(1) = 34.38$$

$$\text{For Period 2 } y = 33.06 + 1.32(2) = 35.70$$

etc.

The percentage variations are averaged as previously shown, resulting in the following values:

	Q_1	Q_2	Q_3	Q_4
Average seasonal variations %	54	89	170	86

The trend line and the average seasonal variations are then used in a similar manner to that previously described.

For example, to extrapolate future sales for the next year (i.e. quarters 17, 18, 19 and 20) is as follows:

Quarter 17

$$\text{Forecast sales} = (33.06 + 1.32(17)) \times 0.54 = \mathbf{29.97}$$

A similar process produces the following figures:

$$\text{Quarter } 18 = \mathbf{50.57}$$
$$19 = \mathbf{98.84}$$
$$20 = \mathbf{51.13}$$

Measuring forecast errors

23. Differences between actual results and predictions may arise for many reasons. They may arise from random influences, normal sampling errors, choice of the wrong forecasting system or alpha value or simply that future conditions turn out to be radically different from the past. Whatever the cause(s) management wish to know the extent of the forecast errors and various methods exist to calculate these errors.

    A commonly used technique, appropriate to time series, is to calculate the *mean squared error of the deviations* between forecast and actual and choose the forecasting system and/or parameters which gives the lowest value of mean squared errors, i.e. akin to the 'least squares' method of establishing a regression line.

Illustration of mean squared error calculation

24. The forecasts produced for July to January by the 3 monthly and 6 monthly moving average calculations in Table 1 will be used as an example.

	Actual sales (units)	3 monthly moving average	Forecast error	Squared error	6 monthly moving average	Forecast error	Squared error
July	370	397	+27	729	423	+53	2,809
August	360	383	+23	529	410	+50	2,500
September	410	377	−33	1,089	397	−13	169
October	450	380	−70	4,900	388	−62	3,844
November	470	407	−63	3,969	395	−75	5,625
December	490	443	−47	2,209	410	−80	6,400
January	460	470	+10	100	425	−35	1,225
				13,525			22,572

Table 8 (Extract from Table 1)

Mean Squared Errors are

$$\text{3 monthly M.A.} = \frac{13,525}{7-1} = \underline{\underline{2,254}}$$

$$\text{and 6 monthly M.A.} = \frac{22,572}{7-1} = \underline{\underline{3,762}}$$

On the basis of the few readings available the 3 monthly moving average would be preferred to the 6 monthly.

Note: The denominator of the above calculations will be recognised as the degrees of freedom, i.e. $n-1$.

Longer-term forecasting

25. Moving averages, exponential smoothing and decomposition methods tend to be used for short to medium term forecasting. Longer term forecasting is usually less detailed and is normally concerned with forecasting the main trends on a year by year basis. Any of the techniques of regression analysis described in the preceding chapters could be used depending on the assumptions about linearity or non-linearity, the number of independent variables and so on. The least squares regression approach is often used for trend forecasting and is illustrated below.

Forecasting using least squares

26.

Example 2

Data have been kept of sales over the last seven years.

Year	1	2	3	4	5	6	7
Sales (in '000 units)	14	17	15	23	18	22	27

It is required to forecast the sales for year 8 and to calculate the coefficient of determination.

Solution

The data are drawn on a time series graph where **x**, the independent variable representing time, is represented on the horizontal axis of Figure 9/3. Note that, unlike a scatter diagram, the points are joined. The least squares line of best fit will become the linear trend when plotted on the graph.

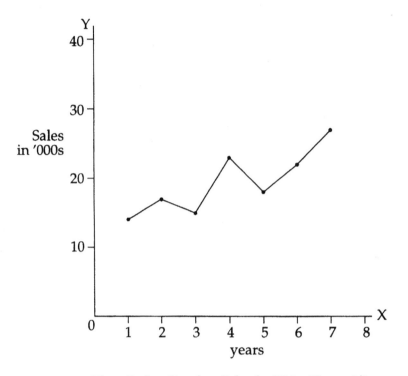

Time Series Graph – Sales in '000s. Figure 9/3

The calculations for the regression line are given opposite and follow the principles already explained in Chapter 7.

Example 3

As there are 7 pairs of readings **n** = 7 the data are set out as follows:

Years (x)	Sales (y)	xy	x^2
1	14	14	1
2	17	34	4
3	15	45	9
4	23	92	16
5	18	90	25
6	22	132	36
7	27	189	49
$\Sigma x = 28$	$\Sigma y = 136$	$\Sigma xy = 596$	$\Sigma x^2 = 140$

(All calculations to two decimal places) $136 = 7a + 28b$

$$596 = 28a + 140b$$

which reduces to $52 = 28b$

$$\therefore b = \underline{\mathbf{1.86}}$$

and substituting in one of the equations we obtain

$$a = \underline{\mathbf{12}}$$

$\therefore$ Regression line = **y = 12 + 1.86x**

or, in terms of the problem above:

Sales (in '000s of units) = 12.00 + 1.86 (no of years)

To use this expression for forecasting, we merely need to insert the number of the year required.

For example, 8th year sales $= 12 + 1.86\,(8)$

$$= \underline{\mathbf{26.88,\ i.e.\ 26,888\ units.}}$$

Coefficient of determination for Example 2

27.   This is calculated as in Chapter 7.

$$r^2 = \frac{\Sigma(YE - \bar{Y})^2}{\Sigma(y - \bar{Y})^2}$$

$$y = 12 + 1.86x$$

$$\therefore \bar{Y} = \frac{136}{7} = 19.43$$

x(years)	y(sales)	YE	(YE − Ȳ)	(YE −Ȳ)²	y − Ȳ	(y −Ȳ)²
1	14	13.86	−5.57	31.02	−5.43	29.48
2	17	15.72	−3.71	13.76	−2.43	5.90
3	15	17.58	−1.85	3.42	−4.43	19.62
4	23	19.44	0.01	0	3.57	12.74
5	18	21.30	1.87	3.49	−1.43	2.04
6	22	23.16	3.73	13.91	2.57	6.60
7	27	25.02	5.59	31.24	7.57	57.30
	Σy = 136			Σ(YE − Ȳ²= 96.84		Σ(y − Ȳ²) = 133.68

Table for Example 2

$$r^2 = \frac{\sum(YE - \bar{Y})^2}{\sum(y - \bar{Y})^2} = \frac{96.84}{133.68} = \underline{72.44\%} = \underline{72\%}$$

This can be interpreted that in the example given 72% of the variations of the actual values of **y** (sales) may be predicted by changes in the actual values of **x** (years). In other words, factors other than changes in the value of **x** influence **y** to the extent of $(100 - 72)\%$, i.e. 28%.

Summary

28. a) Time series analysis is based on data which are arranged in regular time periods, e.g. sales per month.

   b) The moving average system may be based on any number of periods. Typically 3, 6 or 12 month moving averages are used. One of the most important uses of moving averages is a 4 point moving average based on quarters.

   c) Exponential smoothing is less cumbersome than moving averages and requires much less data storage. It is an adaptive forecasting system, variants of which are often included in inventory control systems.

   d) The key factor in an exponential smoothing system is the choice of the smoothing constant, α. The higher the value, the more responsive is the system to current conditions. Typical values are from 0.3 to 0.5.

   e) Time series decomposition, or time series analysis, separates out from the data, the trend, seasonal factors and the cyclical movement.

   f) To enable the best forecasting system to be chosen, it is necessary to calculate the forecast errors. A common method is to calculate the mean squared error of the deviations.

   g) The techniques of least squares regression are frequently used for time series analysis.

Points to note

28. a) The basic forecasting methods and principles have been described in the last three chapters, but many hybrid systems exist in practice.

b) Questions involving forecasting could occur in virtually any syllabus, perhaps on their own or as part of some other topic.

c) For short term forecasting single and double exponential smoothing systems are the most commonly used and most computer based stock control and forecasting systems incorporate these methods. For longer term forecasting the use of trend curves developed by mathematical regression, are the most useful in terms of reliability.

Self review questions *Numbers in brackets refer to paragraph numbers*

1. What is meant by forecasting? (2)
2. Distinguish between quantitative and qualitative forecasting techniques (5,6,7)
3. What is the Delphi Technique? (7)
4. What is Time Series Analysis? (11)
5. What is the Moving Averages system of forecasting? (12)
6. What are the limitations of the basic moving averages system and how does exponential smoothing overcome these? (14 and 15)
7. What are the main characteristics of exponential smoothing? (15 and 16)
8. Into what factors can a Time Series be separated? (20)
9. What are the steps in Time Series Decomposition? (21)
10. How can forecast errors be calculated? (23)
11. How many correlation and regression techniques be applied to time series forecasting? (25)

Exercises with answers

1. Calculate the 3 and 6 monthly averages of the following data.

	Sales
January	1,200
February	1,280
March	1,310
April	1,270
May	1,190
June	1,290
July	1,410
August	1,360
September	1,430
October	1,280
November	1,410
December	1,390

2. Using the January sales as the old forecast and a smoothing constant of 0.3 (α value) calculate the forecast for February onwards using Exponential smoothing.

3. Calculate the forecast errors for the 3 and 6 monthly moving averages calculated in question 1.

Answers to exercises

1.

	Sales	3 Month M.A.	6 Month M.A.
January	1,200		
February	1,280		
March	1,310		
April	1,270	1,263	
May	1,190	1,287	
June	1,290	1,257	
July	1,410	1,250	1,257
August	1,360	1,297	1,292
September	1,430	1,353	1,305
October	1,280	1,400	1,325
November	1,410	1,357	1,327
December	1,390	1,373	1,363

2. Exponential smoothing forecasts for February onwards $\alpha = 0.3$

February	1,200
March	1,224
April	1,250
May	1,256
June	1,233
July	1,250
August	1,283
September	1,327
October	1,358
November	1,335
December	1,357

3.

	Sales	3 Month MA	Forecast Error	Squared	6 Months MA	Forecast Error	Squared
April	1,270	1,263	7	49			
May	1,190	1,287	97	9,409			
June	1,290	1,257	33	1,089			
July	1,410	1,250	160	25,600	1,257	153	23,409
August	1,360	1,297	63	3,969	1,292	68	4,624
September	1,430	1,353	77	5,929	1,305	125	15,625
October	1,280	1,400	120	14,400	1,325	45	2,025
November	1,410	1,357	53	2,809	1,327	83	6,889
December	1,390	1,373	17	289	1,363	27	729
Sum July–December				52,996			53,301

Thus over the same comparative period the 3 month M.A., is slightly the better forecast but only just.

Assessment and revision
Chapters 4 to 9

Examination questions with answers

Answers commence on page 523

A1. A business software company provides several types of support for its customers, one of which is advice by telephone. It is now investigating the costs of these support systems. From past experience, the company has found that its average cost of advice per telephone call is about £20.

The number of telephone support calls given each day during the last 64-day quarter, which is representative, is shown below:

Telephone support calls

Number of calls:	0–9	10–19	20–29	30–39	40–49	50–59	60–69	70+	Total
Frequency:	Nil	5	10	20	15	10	4	Nil	64

You are required to find

a) the mean and standard deviation number of telephone support calls during the last quarter.

b) 95% confidence limits for the annual total cost of the telephone support system, and to comment on your answer.

$$\left[S^2 = \frac{\sum FX^2}{\sum F} - \bar{x}^2 \right]$$

CIMA Quantitative methods

A2. The managers of an import agency are investigating the length of time that customers take to pay their invoices, the normal terms for which are 30 days net.

They have checked the payment record of 100 customers chosen at random and have compiled the following table:

Payment in:	Number of customers
5 to 9 days	4
10 to 14 days	10
15 to 19 days	17
20 to 24 days	20
25 to 29 days	22
30 to 34 days	16
35 to 39 days	8
40 to 44 days	3

Required:

a) Calculate the arithmetic mean.

b) Calculate the standard deviation.

c) Construct a histogram and insert the modal value.

d) Estimate the probability that an unpaid invoice chosen at random will be between 30 and 39 days old.

CIMA Cost Accounting and Quantitative Methods

A3. a) For a sample of items drawn from any population describe in words:

1) the general relationship between the standard error and the standard deviation of the population;

2) how and why, the standard error changes with the sample size.

b) A sample of 50 units is taken from each batch of product manufactured by a company.

The volume of the sample units taken from a particular batch averages 0.25 litres per unit and the sample standard deviation is 0.004 litres.

You are required to:

Find the limits of the mean unit volume of the batch with 95% confidence.

ACCA Management Information

A4. A researcher in a large retail organisation wishes to study sickness absence amongst its employees. The organisation has a large number of branches, each of which keeps full records of sickness leave. A random sample of ten such branches produced the following data showing the number of days of sickness per branch in the past year:

$$18 \quad 23 \quad 26 \quad 30 \quad 32 \quad 35 \quad 39 \quad 45 \quad 48 \quad 54$$

Required:

a) Using the above data

1) calculate a 95% confidence interval for the mean amount of sickness days per branch,

2) estimate the number of branches that should be included in a simple random sample so that a 95% confidence interval for the mean number of days sickness should not have a width greater than four days.

b) After the sample had been collected, it became apparent that the branches fell into three natural groups in terms of sales – small, medium and large. From her data on all of the branches, the researcher found that of 210 randomly selected staff, 90 worked in small branches, 36 in medium-sized branches and the rest worked in large branches. In total, 96 of the selected staff had no days off for sickness, of which 52 worked in small branches, and 29 worked in large-sized branches.

1) Form a table showing the information clearly.

2) Carry out an appropriate statistical test to investigate whether size of branch influences the occurrence of sickness absence. Interpret your results clearly.

ACCA Decision Making Techniques

A5. The quality controller, Mrs Brooks, at Queensville Engineers has become aware of the need for an acceptance sampling programme to check the quality of bought-in components. This is of particular importance for a problem the company is currently having with batches of pump shafts bought in from a local supplier. Mrs Brooks proposes the following criteria to assess whether or not to accept a large batch of pump shafts from this supplier.

From each batch received take a random sample of 50 shafts, and accept that batch only if no more than two defectives are found in the sample.

Mrs Brooks needs to calculate the probability of accepting a batch P_a, when the proportion of defectives in the batch, p, is small (under 10%, say).

Required:

a) Explain why the Poisson distribution is appropriate to investigate this situation.

b) Using the Poisson distribution, determine the probability of accepting a batch P_a, containing p = 2% defectives if the proposed method is used.

c) Determine P_a for p = 0%, 5%, 10%, 15% and hence plot a graph showing the relationship between P_a (on the vertical axis) and p (on the horizontal axis).

Use this graph to estimate

1) The proportion of defectives which must not be exceeded if the supplier needs to be 95% certain that the batch will be accepted.

2) The proportion of defectives in the batch for which there is only a 5% probability of acceptance by Queensville.

Briefly outline the importance of these values to the supplier.

ACCA Decision Making Techniques

A6. A research assistant working for a large company wanted to investigate attitudes towards a proposed management buy-out. Thus she asked a random sample of the company's workforce whether or not they were in favour of the buy out. The research assistant was particularly interested in finding out whether the workers' views on the buy out were related to the classification of their job (manual or non-manual). Altogether 120 workers were interviewed by her, of these 70 were manual workers and 50 were nonmanual workers. Of the manual workers 30 were in favour of the buy out, whereas 25 of the non-manual workers were in favour.

Required:

a) Draw up a 2 × 2 table showing this information and carry out a χ^2 test of association to see if job classification affects attitude to the proposed management buy-out.

b) The research assistant was also interested in the question of whether the workers' views on the buy out were related to their gender. A computer analysis was carried out for each gender separately and the results are shown below.

Expected frequencies are written below observed frequencies

	Males				Females		
	Manual	Non-manual	Total		Manual	Non-manual	Total
In favour	26	22	48	In favour	4	3	7
	31.2	16.8			1.75	5.25	
Not in favour	39	13	52	Not in favour	1	12	13
	33.8	18.2			3.25	9.75	
Total	65	35	100	Total	5	15	20

Chi-square test statistic = 3.89   Chi-square test statistic = 3.59
Degrees of freedom = 1   Degrees of freedom = 1

Criticise the use of the chi-square test for one gender and interpret the results for the other gender.

c) Write a short report about the relationship between gender, job classification and attitude towards the buy-out for this company's workers.

ACCA Decision Making Techniques

A7. The managers of a company are preparing revenue plans for the last quarter of 1993/94 and for the first three quarters of 1994/5. The data below refer to one of the main products:

Revenue	April–June Quarter 1	July–Sept Quarter 2	Oct–Dec Quarter 3	Jan–March Quarter 4
£'000	£'000	£'000	£'000	
1990/91	49	37	58	67
1991/92	50	38	59	68
1992/93	51	40	60	70
1993/94	50	42	61	–

Required:

a) Calculate the four-quarterly moving average trend for this set of data.

b) Calculate the seasonal factors using either the additive model or the multiplicative model, but not both.

c) Explain, but do not calculate, how you would use the results in parts (a) and (b) of this question to forecast the revenue for the last quarter of 1993/4 and for the first three quarters of 1994/95.

CIMA Cost Accounting and Quantitative Methods.

A8. A company has a fleet of vehicles and is trying to predict the annual maintenance costs per vehicle. The following data have been supplied for a sample of vehicles:

Vehicle number	Age in years	Maintenance cost per annum £ × 10
	(x)	(y)
1	2	60
2	8	132
3	6	100
4	8	120
5	10	150
6	4	84
7	4	90
8	2	68
9	6	104
10	10	140

Required:

a) Using the least squares technique, calculate the values of a and b in the equation y = a + bx, to allow managers to predict the likely maintenance cost, knowing the age of the vehicle.

b) Prepare a table of maintenance costs covering vehicles from 1 to 10 years of age, based on your calculations in (a).

c) Estimate the maintenance costs of a 12-year-old vehicle and comment on the validity of making such an estimate.

CIMA Cost Accounting and Quantitative Methods

A9. A housing consultant believes that the number of houses sold in a region for a given year is related to the mortgage rate in that period. He collects the following relevant data.

Year	Mortgage interest rate, X	Housing sales index, Y
1982	12	80
1983	10	90
1984	8	105
1985	6	115
1986	7	125
1987	8	120
1988	10	115
1989	12	100
1990	14	85
1991	13	70
1992	11	80

Required:

a) The correlation coefficient between X and Y is −0.81.

Use the test statistic

$$T = \frac{r\sqrt{n-2}}{\sqrt{1-r^2}} \sim t_{n-2}$$

to test its significance. Interpret your answer clearly.

b) Having collected the data, the housing consultant observes that the number of houses sold in a year may be better related to the mortgage interest rate in the previous year.

To investigate this possibility

i) plot a scatter diagram,

ii) calculate a relevant correlation coefficient,

and hence comment on the appropriateness of this model.

c) Determine a regression equation to predict the housing sales index based on the mortgage rate in the previous year. Hence predict the housing sales index in 1993.

d) Comment on the degree of accuracy provided by your prediction.

ACCA Decision Making Techniques

A10. 'The learning curve is a simple mathematical model but its application to management accounting problems requires careful thought.'

Required:

Having regard to the above statement,

a) Explain the 'cumulative average time' model commonly used to represent learning curve effects.

b) sketch two diagrams to illustrate, in regard to a new product, the relative impacts of 70%, 80% and 90% learning curves on

- cumulative average hours per unit,

- cumulative hours taken.

c) explain the use of learning curve theory in budgeting, budgetary control and project evaluation; explain the difficulties that the management accountant may encounter in such use.

d) compare and contrast the learning curve with the experience curve; explain the circumstances when each may be most relevant.

CIMA Management Account Applications

A11. Gourmet Cater-Fish Limited specialises in high-quality seafood products for restaurants and hotels.

One of its most popular lines is scampi fillets covered in batter. There are 21 separate fillets in a packet with a nominal weight of 1 kg. Typically the company sells 100,000 packets in a period.

Each individual scampi fillet is covered in batter, cooked and frozen. Because of the nature of the material and the cooking process, there is an unavoidable variation in the finished weight of each fillet. On average the weight of each individual prepared fillet is 50 g with a standard deviation of 5 g.

Assume a Normal distribution.

Required:

a) Calculate the average weight and standard deviation of a packet of completed fillets.

b) What is the probability that a packet will be below the nominal weight of 1 kg?

c) The company has a policy of giving £50 to each customer who finds that they have been supplied with an underweight packet. From experience, Gourmet Cater-Fish Limited knows that only 50% of customers who receive underweight packets actually complain and receive the £50 payment.

d) Advise management whether it would be worthwhile to put 22 fillets in each packet to reduce the number of underweight payments. Each individual fillet costs 50p to buy in.

Note: Ignore any legal consequences which might arise from selling underweight packages.

CIMA Cost Accounting and Quantitative Methods

A12. A company manufactures Product XX on two different machines, Machine 1 and Machine 2, in order to satisfy demand. 70% of production is on Machine 2.

Samples of Product XX (100 units from each machine) have been tested in order to measure the proportion of production falling within the targeted quality standard. The results of the testing are as follows:

	Machine 1	Machine 2
Within targeted standard	93 units	97 units
Outside targeted standard	7 units	3 units

Required:

Based upon the above sample testing results:

(a) Calculate the probability that a manufactured item will:

   (i) have been manufactured on Machine 2 and be within the targeted quality standard;

   (ii) be within the targeted quality standard.

b) Calculate the limits of the proportion of Product XX from Machine 1 that will be outside the targeted quality standard, with 95% confidence.

c) Explain how the required sample size should be determined in order to establish, within given limits and to a required degree of confidence, the proportion of items falling within the targeted quality standard.

ACCA Management Information

A13. A company is building a model in order to forecast total costs based on the level of output. The following data are available for last year:

Month	Output '000 units (x)	Costs £000 (y)
January	16	170
February	20	240
March	23	260
April	25	300
May	25	280
June	19	230
July	16	200
August	12	160
September	19	240
October	25	290
November	28	350
December	12	200

Required:

a) State two possible reasons for the large variation in output per month.

b) Plot a graph of output and costs, and comment on the relationship observed.

c) Using the least square technique, calculate the values of **a** and **b** in the equation $y = a + bx$ in order to predict costs given the output, and explain the meaning of the calculated values.

d) Calculate the correlation coefficient between output and costs, and comment on the value.

e) Prepare a forecast of the costs for the next two months, when output will be 20,000 and 40,000 units respectively. Discuss the validity of your forecasts.

CIMA Business Mathematics

A14. A retailer sells computer games. Tabulated below are the unit sales of one of its games for the last 13 quarters, together with the computer-generated trend equation and average seasonal variations. For planning purposes, forecasts of sales are required for the remaining three quarters of this year. (The brand was introduced in 1997.)

Sales	Q_1	Q_2	Q_3	Q_4
1998	105	95	150	250
1999	80	70	110	180
2000	50	50	70	110
2001	30			

Trend equation: $S = 180 - 10T$

where S = sales units and

T = time period in quarters

(For example, $T = 1$ in Q_1 of 1998, $T = 2$ in Q_2 of 1998, $T = 3$ in Q_3 of 1998, $T = 5$ in Q_I of 1999, and so on.)

Average seasonal variations for the years 1998–2001

	Q₁	Q₂	Q₃	Q₄
	-40%	-40%	0	$+80\%$

Required:

a) Plot a time series graph of the unit sales, and include the trend on the same graph.

b) Forecast the unit sales for the second, third and fourth quarters of 2001, and comment on the reliability of these forecasts.

c) Interpret the key factors in this game's unit sales, 1998–2001.

CIMA Business Mathematics

Examination questions without answers

B1. In April 1995 a survey of customer transaction values was carried out by a retail business. The following data was collected:

Transaction value	No of customers
below £1.00	137
£1.00 to £1.99	259
£2.00 to £2.99	297
£3.00 to £3.99	378
£4.00 to £4.99	193
£5.00 to £5.99	84
£6.00 and over	52

Required:

a) For the above data calculate:

i) The mean transaction value (stating clearly assumptions made.)

ii) the standard deviation.

b) A similar survey had been carried out in October 1994. The resulting data gave a mean transaction value of £2.79 and a standard deviation of £1.43.

Comment on how the spending pattern of customers has changed between the two months.

c) Outline the advantages and disadvantages of simple random sampling.

ACCA Management Information

B2. The managers of a Sales Department have recorded the number of successful sales made by their 50 tele-sales persons for one week, and the raw scores are reproduced below;

20	10	17	22	35	43	29	34	12	24
24	32	34	13	40	22	34	21	39	12
10	49	32	33	29	26	33	34	34	22
24	17	18	34	37	32	17	36	32	43
12	27	43	32	35	26	38	32	20	21

Sales persons who achieve fewer than 20 sales are required to undertake further training.

Required:

a) Classify the raw data into class intervals of five units wide, i.e. 10–14, 15–19, etc.

b) Draw a cumulative frequency diagram to show the classified information from (a). Insert the median and the upper and lower quartiles.

c) Calculate the arithmetic mean for the classified information. Briefly comment on why the mean differs from the median in (b).

d) What proportion of the sales force will need extra training?

CIMA Cost Accounting and Quantitative Methods

B3. In May 1993 the sales manager of Brighton Electronics won a contract for 5,000 shortrange transmitters. The contract stipulated 100% inspection prior to delivery to ensure that all transmitters in the consignment were operating within given specifications. One of the critical components in the transmitter was an impedance sub-assembly which contained eight tuning coils.

The sales manager had arranged with the company's purchasing manager to obtain trail batches of tuning coils from two suppliers, York Ltd and Zeus Ltd, from which Brighton Electronics had been purchasing electronic components for several years. A careful analysis of their coils indicated that for York, 6% of coils could be expected to lie outside the design specifications and for Zeus, the proportion of rejects was 2%. Their unit prices are 45 pence and 50 pence respectively.

The design engineer looked very closely at the design requirements and concluded that if in any sub-assembly not more than one of the eight coils is out of specification then the circuit will still operate in an acceptable manner. If two of the eight coils are outside the specification then the problem is overcome by fitting a supplementary adapter, costing £2.25 each. If, however, three or more of the eight coils are out of specification the subassembly has to be scrapped, costing £8.50 per sub-assembly.

Required:

a) Explain why we can use the binomial distribution in this problem.

b) If the coils are bought from York Ltd, how many of 5,000 sub-assemblies would you expect to contain no defective tuning coils, how many one, and how many two defective coils? What will these figures be if, instead, Zeus is chosen as the supplier?

c) How many sub-assemblies would you expect to be scrapped if York Ltd was selected as supplier? How many for Zeus Ltd?

d) What is the expected total number of tuning coils required to enable 5,000 acceptable sub-assemblies to be made for each of the two alternatives?

e) What are the expected total costs if the coils were bought from York, or, alternatively, from Zeus? From which supplier should the tuning coils be purchased?

ACCA Decision Making Techniques

B4. A department store has recently carried out a series of studies of its Saturday customers. Over a period of several weeks a very large representative sample was asked about buying intentions as they entered, and about actual purchases as they left. (The incentive to co-operate was a free canvas shopping bag.) Some of the results for the Sports Department, based on 1,000 people, were as follows:

Number of items	0	1	2	3	4	5+	Whole sample
Stated intention to buy this number of items (%)	19	51	22	5	2	1	100
Actual purchases (%)	48	25	11	10	4	2	100

From till records, the average (mean) value of Saturday purchases from the Sports Dept was found to be £20, with a standard deviation of £10. The average (mean) number of Saturday customers was 2,000, with a negligible standard deviation.

You are required

a) using averages and percentages, to comment on any differences between intention to buy and actual purchases made in the Sports Department.

b) to find 95% confidence limits for the percentage of "customers" expected to enter the Sports Department on Saturdays and to buy nothing;

c) to find 95% confidence limits for Saturday's total takings, stating any assumptions.

CIMA Quantitative Method.

B5. Annual sales of Brand Y over the last eleven years have been as follows.

Unit sales of Brand Y, 1983 – 1993 (thousands)

1983	1984	1985	1986	1987	1988	1989	1990	1991	1992	1993
50	59	46	54	65	51	60	70	56	66	76

You are required

a) to calculate a three-year moving average trend;

b) to plot the series and the trend on the same graph;

c) to produce a sales forecast for 1994, stating any assumptions.

CIMA Quantitative Methods

B6. A book publishing company wishes to develop a model that it can use to help predict textbook sales for books it is considering for future publication. The marketing department has collected data on a number of variables from a sample of ten books. These data are listed below.

Copies sold (thousands), Y	Pages X_1	Advertising budget, (£000) X_2	Number of competing books, X_3	Cost (£), X_4
16	126	27	5	15
10	204	30	7	19
33	411	33	7	14
15	176	25	3	15
77	600	41	9	17
59	333	19	1	13
75	483	40	8	11
57	400	26	3	12
88	504	51	12	9
26	302	58	15	14

A statistical package was used to carry out a multiple regression routine of Y based on the four independent variables X_1, X_2, X_3, X_4.

Correlation matrix

	Y	X_1	X_2	X_3
X_1	.89			
X_2	.31	.44		
X_3	*	.35	.98	
X_4	−.64	−.41	−.27	−.13

The regression equation is

$Y = 29.9 + 0.148X_1 + 0.79X_2 - 3.11X_3 - 2.87X_4$

$R^2 = 90.5\%$

Required:

a) Calculate the product-moment correlation coefficient between Y and X_3, and interpret its meaning.

b) Use the test statistic

$$T = \frac{r\sqrt{(n-2)}}{\sqrt{1-r^2}} \quad \text{(which has } t_{n-2} \text{ distribution under } H_0\text{)}$$

to investigate whether there is a significant correlation between copies sold, Y, and each of the four dependent variables individually.

c) Interpret the value R^2 in the context of the data.

d) Use the regression equation to predict the number of copies sold for a book with 350 pages, having an advertising budget of £35,000, with 6 competing books and costing £16.

e) Criticise the use of this model for predicting Y. Explain how the model could be improved.

ACCA Decision Making Techniques

B7. a) Distinguish between

i) variability due to common causes, and

ii) variability due to special causes when monitoring production processes.

Winchester Electronics manufactures two types of cathode ray tube, type A tubes and type B tubes, on a mass production basis.

b) The following data were collected over a 12-day period and are based on samples of 6 tubes per day. The measurements record the strength of tube type A.

Day	1	2	3	4	5	6	7	8	9	10	11	12
X̄	162	163	155	160	165	160	160	157	155	158	164	161
R	12	14	19	16	11	15	12	17	18	14	19	13

For type A tubes, calculate the mean of sample means and the mean range, and hence establish the '3σ' upper and lower control limits for

i) sample means,

ii) sample ranges.

Hence discuss whether or not the production process for type A cathode ray tubes is operating satisfactorily.

c) The following table contains data from 15 days and shows the number of rejected cathode ray tubes (type B) out a total of 120 inspected each day.

Number rejected:	9	10	12	7	11	6	9	8	11	10	13	9	7	0	4

i) Establish upper and lower '3σ' control limits for the proportion of defective type B tubes, and assess whether the production process is operating normally.

ii) Explain why it might be sensible to stop the production process when a result falls below the lower control limit of a defective chart.

ACCA Decision Making Techniques

B8. The sales of Yelesol ice cream for the last twelve quarters have been as follows:

		Sales (thousand units)	Moving annual total (thousand units)
1991	Q4	14	–
1992	Q1	16	–
	Q2	30	100
	Q3	40	104
	Q4	18	104
1993	Q1	16	108

	Q$_2$	34	112
	Q$_3$	44	108
	Q$_4$	14	108
1994	Q$_1$	16	112
	Q$_2$	38	120
	Q$_3$	52	–
	Q$_4$		–

You are required

a) to calculate a centred moving average;

b) to plot a graph of the data and the centred moving average;

c) to find the average quarterly seasonal variations;

d) to calculate sales forecasts for the next two quarters.

CIMA Quantitative Methods

10 Calculus

Objectives

1.   After studying this chapter you will
   - know why calculus is used;
   - understand the principles and uses of differentiation;
   - be able to calculate the derivatives of common expressions;
   - be able to use differentiation to solve typical problems;
   - know how to use differentiation to find the maximum and minimum points of various functions;
   - understand the general principles of partial differentiation;
   - know the basics of integration;
   - be able to distinguish between definite and indefinite integrals.

Why is calculus used?

2.   Frequently it is possible to represent relationships by simple linear functions. For example, a simple linear function for total cost might have the form:

$$y = a + bx$$

where y = total cost, the dependent variable.

   x = output or activity, the independent variable and a and b are constants representing fixed cost and variable (or marginal) cost respectively.

Such a function is shown in Figure 10/1.

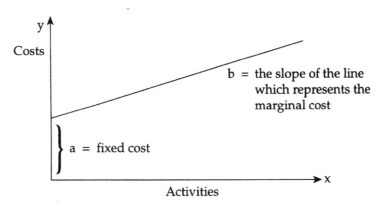

Graph of a simple linear cost function. Figure 10/1

In such a function the rate of change of cost (represented by "b", the gradient of the line) is constant at all levels of activity and will not increase or decrease at any

205

level of activity. This is, of course, is what is meant by a linear function. The value of **b** can be easily found by simple arithmetic without recourse to more sophisticated techniques.

However, there are many occasions when a linear function is not an accurate representation of reality and some form of curvi-linear function is required.

Some examples are given in Figure 10/2.

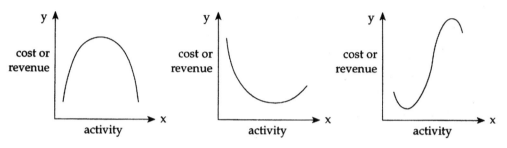

Examples of curvi-linear functions. Figure 10/2

Examination of curvi-linear functions shows that the slope or gradient changes at various activity levels and that at some maximum or minimum value – or both – there is a turning point. For many business applications it is essential to know the *rate of change* of a function (representing say, marginal cost or marginal revenue) and also the point of zero gradient, the *turning point* (representing, say, maximum revenue or minimum cost). The process of *differentiation* provides a ready means of finding the rates of change of curvi-linear functions and of their turning points and it can thus be used as a simple means of optimising.

In addition, it is sometimes necessary to be able to calculate the total amount of revenue or cost between two activity levels on a curvi-linear function. This is done by the process of *integration* which provides the means of finding the area under the curve of a function.

Differentiation

3.   The process of differentiation establishes the slope of a function at a particular point. Alternatively this can be described as establishing the rate of change of the dependent variable (say, cost) with respect to an infinitesimally small increment in the value of the independent variable (say, activity).

For illustration consider the following two functions:

$$\text{Function I} \qquad y = x$$

$$\text{Function II} \qquad y = x^2$$

Function I is a linear function whereas, because x is raised to the power of 2 (i.e. x^2) in Function II, that function is curvi-linear.

Assume now that the independent variable, x, is altered by a very small amount, m. What is the rate of change in y caused by the change in the value of x?

Function I

If x is at some value A, then y is also at the value A. If x is altered by m then its value becomes x = A + m and, as a direct consequence, the value of the dependent variable becomes

$$y = A + m$$

It follows therefore that:

$$\text{The rate of change of } \mathbf{y} \text{ with } \mathbf{x} = \frac{\text{Change in value of } \mathbf{y}}{\text{change in value of } \mathbf{x}} = \frac{m}{m} = 1$$

This has the obvious meaning that the rate of change (i.e. slope) is constant and equal to one so that y changes by exactly the same amount as x, regardless of the level of activity or the amount of the change.

Function II

What is the consequence of x changing from x = A to x = A + m, along the function $y = x^2$?

$$\text{When } x = A \quad y = A^2$$

$$\text{When } x = A + m \quad y = (A + m)^2 = A^2 + 2mA + m^2$$

Thus the change in value of y caused by the increase of m in the value of x is

$$(A^2 + 2mA + m^2) - A^2$$

which reduces to $2mA + m^2$

In a similar fashion to that outlined above for Function I it follows that:

$$\text{The rate of change of } \mathbf{y} \text{ with } \mathbf{x} = \frac{\text{Change in value of } \mathbf{y}}{\text{change in value of } \mathbf{x}} = \frac{2mA}{m} + \frac{m^2}{m} = \underline{2A + m}$$

If the value of the small change, m, tends to zero the rate of change becomes 2A.

This means that at any value on the function $y = x^2$, the rate of change in the value of y with respect to x is 2x. This is known as the *derivative* or *differential coefficient*.

Thus the derivative for the function $y = x^2$ is 2x. The following tables gives some numerical examples for this function.

Value of x (independent variable)	Value of y (dependent variable) $y = x^2$	Rate of Change of y i.e. the value of the derivative 2x
1	1	1
2	4	4
3	9	6
4	16	8
5	25	10
6	36	12
7	49	14

Notes:

a) The small change in value denoted above as **m**, is conventionally known as Δx (**delta x**). As this value tends towards zero, i.e. $\Delta x \to 0$, the comparison of the changes in value becomes:

$$\text{Limit} \frac{\Delta y}{\Delta x}$$

$$\Delta x \to 0$$

generally written as $\frac{dy}{dx}$, which means the derivative of a function when Δx tends towards zero.

Thus, Original function: $\qquad y = x^2$

Derivative: $\qquad \frac{dy}{dx} = 2x$

b) The derivative of a function gives the *exact* rate of change at a *point* and only gives an approximate result when used over a finite range. An example of the effect of this is where differentiation is used to obtain the marginal cost from a curvi-linear cost function. Accountants frequently define marginal cost as the increase in total cost due to an increase in output of one unit. A whole unit is a finite range as far as differentiation is concerned and consequently the marginal cost obtained from the derivative is only an approximation. This is a technical point which is considered not to be of great practical significance.

Rules for finding derivatives

4. In a similar fashion to that outlined above the derivative of any type of function could be found from first principles. However, this would be a tedious, lengthy process and, once the general ideas of differentiation is understood, it is much simpler to follow a few simple rules which are shown in the following paragraphs.

Derivatives – the basic rule

5.   Where the function is $y = x_n$

$$\text{The derivative}\frac{dy}{dx} = nx^{n-1}$$

Examples

Function	Derivative
$y = x^2$	$\frac{dy}{dx} = 2x$
$y = x^{10}$	$\frac{dy}{dx} = 10x^9$
$y = -x^2$	$\frac{dy}{dx} = -2x$
$y = \sqrt{x}$ (i.e. $x^{-1/2}$)	$\frac{dy}{dx} = -\frac{1}{2}x^{-1/2}$ or $\frac{1}{2\sqrt{x}}$

Note: All the other rules are merely extensions of the basic method so it must be understood at this stage. All that is necessary is to multiply x by its original index and to realise that the new index of x is one less than the original value.

Derivatives – where the function has a coefficient

6.   Where $y = kx^n$ (**k** is the coefficient)

$$\frac{dy}{dx} = nkx^{n-1}$$

Examples

Function	Derivative
$y = 3x^2$	$\frac{dy}{dx} = 6x$
$y = 8x^4$	$\frac{dy}{dx} = 32x^3$
$y = -7x^5$	$\frac{dy}{dx} = -35x^4$
$y = \frac{1}{2}x^9$	$\frac{dy}{dx} = 4.5x^8$
$y = 3x$	$\frac{dy}{dx} = 3$

Derivatives – where the function contains a constant

7. Where $y = x^n + c$ (c is the constant)

$$\frac{dy}{dx} = nx^{n-1}$$

Examples

Function	Derivative
$y = x^3 + 8$	$\frac{dy}{dx} = 3x^2$
$y = 6x^4 + 27$	$\frac{dy}{dx} = 24x^3$
$y = \frac{1}{3}x^3 - 5$	$\frac{dy}{dx} = x^2$

Note: On differentiation the constant disappears. This is to be expected because the derivative measures the rate of change and a constant (e.g. fixed cost) by defintion does not change.

Derivatives – where the function is a sum

8. Where $y = x^n + x^m$

$$\frac{dy}{dx} = nx^{n-1} + mx^{m-1}$$

Examples

Function	Derivative
$y = x^2 + 6x^4$	$\frac{dy}{dx} = 2x + 24x^3$
$y = \frac{1}{6}x^2 - 12x^5$	$\frac{dy}{dx} = \frac{1}{3}x - 60x^4$
$y = 3x^4 + 10x^2 + 9x^3$	$\frac{dy}{dx} = 12x^3 + 20x + 27x^2$

Derivatives – where the function is a product

9. Let **m** and **n** represent function of x and $y = mn$ then

$$\frac{d}{dx}(mn) = m\frac{dn}{dx} + n\frac{dm}{dx}$$

To illustrate this consider the function

$$y = (8x + 4)(2x^3 + 6)$$

$$(8x + 4) \text{ represents } \mathbf{m} \text{ and}$$

$$(2x^3 + 6) \text{ represents } \mathbf{n}$$

$$\therefore \frac{dm}{dx} = 8$$

$$\frac{dn}{dx} = 6x^2$$

$$\therefore \frac{dy}{dx} = (8x + 4)\, 6x^2 + (2x^3 + 6)8$$

$$= 48x^3 + 24x^2 + 16x^3 + 48$$

$$= 64x^3 + 24x^2 + 48$$

Examples

Function	Derivative
$y = (10x^2 + 5)(3x^3 + 2)$	$\dfrac{dy}{dx} = (10x^2 + 5).9x^2 + (3x^3 + 2).20x$
$y = (x^2 + 4)(6x^{1/2} + 3)$	$\dfrac{dy}{dx} = (x^2 + 4).3x^{-1/2} + (6x^{1/2} + 3).2x$

Derivatives – where the function is a quotient

10. Let $\mathbf{m}$ represent the function of x which is the numerator and $\mathbf{n}$ represents the function which is the denominator, then:

$$y = \frac{m}{n} \text{ and } \frac{dy}{dx} = \frac{n\dfrac{dm}{dx} - m\dfrac{dn}{dx}}{n^2}, \ n \neq 0$$

As an illustration consider the function

$$y = \frac{4x^3 + 2}{x^6}, \text{ so } \frac{dm}{dx} = 12x^2$$

$$\text{and } \frac{dn}{dx} = 6x^5$$

$$\therefore \frac{dy}{dx} = \frac{x^6.12x^2 - (4x^3 + 2)\, 6x^5}{(x^6)^2}$$

$$= \frac{12x^8 - 24x^8 - 12x^5}{x^{12}}$$

$$= -\left(\frac{12x^5 + 12x^8}{x^{12}}\right)$$

$$= -\left(\frac{12 + 12x^3}{x^7}\right)$$

Examples

Function	Derivative
$y = \dfrac{3 - 2x}{3 + 2x}$	$\dfrac{dy}{dx} = \dfrac{-12}{(3 + 2x)^2}$
$y = \dfrac{x}{1 - 4x^2}$	$\dfrac{dy}{dx} = \dfrac{1 + 4x^2}{(1 - 4x^2)^2}$

Derivatives – functions of a function

11. Where $y = (2x + 6)^3$ and the expression in the brackets is a differentiable function say, m, i.e. $m = 2x + 6$, the whole expression can be written as $y = m^3$.

In such cases the rule for differentiation is $\dfrac{dy}{dx} = \dfrac{dy}{dm} \times \dfrac{dm}{dx}$ which is known as the *chain rule*.

Thus to differentiate $y = (2x + 6)^3$, let $m = 2x + 6$, **then** $y = m^3$

$$\text{and } \frac{dy}{dm} = 3m^2 \text{ and } \frac{dm}{dx} = 2$$

$$\therefore \frac{dy}{dx} = 3(2x + 6)^2 \times 2$$

$$= 6(2x + 6)^2$$

Examples

Function	Derivative
$y = (1 - 6x)^6$	$\dfrac{dy}{dx} = -36(1 - 6x)^5$
$y = (8 + 4x - x^2)^{1/2}$	$\dfrac{dy}{dx} = (2 - x)(8 + 4x - x^2)^{-1/2}$
	or $\dfrac{2 - x}{y}$

Note: The derivative $\dfrac{dy}{dx}$ is the inverse of the derivative $\dfrac{dx}{dy}$, i.e.

$$\frac{dy}{dx} = \frac{1}{\left(\dfrac{dx}{dy}\right)}$$

A practical example of differentiation

12. Now that the idea of differentiation has been explained and the rules given for differentiating common functions, a practical example can be considered.

Example

A firm has analysed their operating conditions, prices and costs and have developed the following functions.

Revenue £(R) = 400Q − 4Q² and **Cost £(C) = Q² + 10Q + 30**

where **Q** is the number of units sold

The firm wishes to maximise profit and wishes to know

a) What quantity should be sold?

b) At what price?

c) What will be the amount of profit?

Note: Previous examples have used **y** and **x** which results in the derivative $\dfrac{dy}{dx}$.

This example uses **R**, **C** and **Q** and results in the derivatives $\dfrac{dR}{dQ}$ and $\dfrac{dC}{dQ}$.

Solution

From basic economic theory it will be recalled that profit is maximised when Marginal Cost = Marginal Revenue and, as explained in this chapter, differentiating a function gives the rate of change of that function which is equivalent to the marginal cost or revenue.

$$£R = 400Q - 4Q^2 \text{ and } \frac{dR}{dQ} = 400 - 8Q = \textbf{Marginal Revenue}$$

$$\text{and } £C = Q^2 + 10Q + 30 \text{ and } \frac{dC}{dQ} = 2Q + 10 = \textbf{Marginal Cost}$$

Point of profit maximisation is when

$$\textbf{MR} = \textbf{MC} \text{ or } \frac{dR}{dQ} = \frac{dC}{dQ}$$

$$\text{i.e. } 400 - 8Q = 2Q + 10$$

$$\therefore Q = \underline{\underline{39}} \qquad \qquad \text{answer (a)}$$

$$\text{Total Revenue} = 400(39) - 4(39^2),$$

$$= £9,516$$

and, as 39 will be sold, the price will be $\dfrac{9,516}{39} = \underline{\underline{£244}}$ each   answer (b)

$$\text{Total Profit} = \text{Revenue} - \text{Cost}$$

$$\text{Revenue} = £9,516 \text{ from above and}$$

$$\text{Cost} = (39)^2 + 10(39) + 30 = 1,941$$

$$\therefore \text{ Profit} = \underline{\underline{£7,575}}$$ answer (c)

Turning points

13.   Some functions have turning points, i.e. points of local minima or maxima, and these points are of particular interest in many business applications because they represent points of minimum cost or maximum profit or revenue.

Figure 10/3 shows two such turning points.

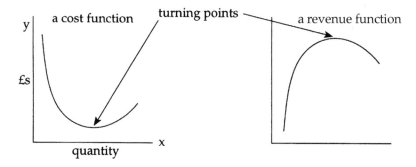

Figure 10/3

These points are points of zero slope or gradient and can be likened to the exact top of a hill. One climbs up to the summit and once it has been passed one starts to go down. On the exact top-most point (or bottom-most point) one is not going up or down so there is zero gradient. At that point, if a tangent is drawn it will be parallel to the x axis, i.e. horizontal.

Using differentiation to find turning points

14.   The derivative of a function shows the slope or rate of change of the function. If the derivative is calculated and equated to zero this will show the turning point of the function.

For many functions this turning point will be the required maximum or minimum point. However, there are certain functions which have two (or indeed more than two) turning points so that it is necessary to test whether the calculated turning point is at a maximum or minimum value.

This is done by calculating the *second derivative*, designated as $\dfrac{d^2y}{dx^2}$ (d two y, dx squared).

The second derivative is found by differentiating the first derivative,

$$\text{i.e.} \ \dfrac{dy}{dx}$$

using the normal rules of differentiation. If the second derivative is *negative* the turning point is a *maximum*, if the second derivative is *positive* the turning point is a *minimum*.

This procedure is summarised in the following table.

To find the maximum or minimum value of a function

Step 1 Find first derivative of the function

$$\text{i.e.} \ \dfrac{dy}{dx}$$

Step 2 Set $\dfrac{dy}{dx}$ to zero and calculate the turning point where, say x = **a**

Step 3 Find the second derivative, i.e. $\dfrac{d^2y}{dx^2}$, by differentiating the first derivative.

Step 4 If $\dfrac{d^2y}{dx^2}$ is *negative* at the point x = **a** the turning point is a *maximum*

If $\dfrac{d^2y}{dx^2}$ is *positive* at the point x = **a** the turning point is a *minimum*

Example

Find the point of maximum value of the Revenue function in Paragraph 12.

i.e. $R = 400Q - 4Q^2$

Step 1 $\qquad\qquad\qquad\qquad \dfrac{dr}{dQ} = 400 - 8Q$

Step 2 At the turning point $\dfrac{dr}{dQ} = 0$

$$\therefore \ 400 - 8Q = 0$$

$$\therefore Q = \underline{50}$$

Step 3
$$\frac{dr}{dQ} = 400 - 8Q$$

$$\therefore \frac{d^2r}{dQ^2} = -8 \text{ i.e. negative for all values of } Q.$$

Step 4 As $\frac{d^2r}{dQ^2}$ is negative, the turning point when $Q = 50$ is a maximum and the revenue at that point is

$$\pounds\, R = 400Q - 4Q^2$$
$$= 400(50) - 4(50^2)$$
$$= \underline{\pounds 10,000}$$

Notes:

a) When the turning point is at a maximum the second derivative is negative and this can be thought of as the gradient going downhill, which is to be expected at the 'top of the hill'. Alternatively at the bottom of the valley the gradient will be going up hill from that point which is shown by the second derivative being positive.

b) On occasions the second derivative will be found to be zero. This denotes what is known as an inflexional point, which is a bend in the curve, and is not a true turning point.

c) The inverse relationship between $\frac{dy}{dx}$ and $\frac{dx}{dy}$ does *not* apply to the second derivatives, i.e.

$$\frac{d^2y}{dx^2} \neq \frac{1}{\left(\frac{d^2x}{dy^2}\right)}$$

d) Notes (a) and (b) above are generalisations which are broadly accurate but there are specialised circumstances in which they do not apply.

Derivatives with more than one independent variable

15. So far in this chapter the functions have contained a single independent variable, for example:

$$y = x^2 + 6x^4$$

where y is dependent upon the value of the independent variable, x. The rules of differentiation covered earlier provide a method of finding the rates of change (slopes of the curve) for any value of x, the single independent variable.

However there are occasions where the functions contains two or more independent variables as, for example, when the cost function of a firm depends on both labour hours and machine hours. If it is required to find the rates of change in these circumstances a process known as *partial differentiation* is used.

Partial differentiation

16.   Assume that a cost function is as follows:

$$y = 10x^2 + 5z^2 - 4xz + 12$$

where y = total cost (the dependent variable)
$\quad\quad x$ = labour hours (an independent variable)
$\quad\quad z$ = machine hours (an independent variable)

With such a function there is a multi-dimensional cost surface with several slopes thus:

a)   the slope when x changes but z is held constant
b)   the slope when z changes but x is held constant
c)   the slope when both x and z are changing

Each of these slopes has a special derivative. (a) and (b) are known as *partial derivatives* and (c) is known as a *total derivative*.

Here we are concerned with the partial derivatives and these are written as follows:

$\quad\quad$ *Partial derivative (a) above*, i.e. when x changes and z is constant $= \dfrac{\partial y}{\partial x}$

(this is called 'the partial derivative of y with respect to x')

$\quad\quad$ *Partial derivative (b) above*, i.e. when z changes and x is constant $= \dfrac{\partial y}{\partial z}$

(this is called 'the partial derivative of y with respect to z')

Note: Learn the difference between the symbol used for partial differentiation, ∂, and the symbol used previously, d.

The derivative $\dfrac{dy}{dx}$ calculated in the early part of this chapter is an example of a total derivative.

Rules for partial derivatives

17.   Fortunately, the rules already given for ordinary derivatives also apply to partial derivatives. These are illustrated using the cost function given earlier thus:

$$y = 10x^2 + 5z^2 - 4xz + 12$$

The problem is to find the partial derivatives, $\dfrac{\partial y}{\partial x}$ and $\dfrac{\partial y}{\partial z}$.

This is done by differentiating in the normal way for one of the independent variables, say **x**, whilst at the same time treating the other variable (**z** in this case) as a constant, as follows:

$$\frac{\partial y}{\partial x} = 20x - 4z$$

Note: As explained in the early part of the chapter, any part of the expression which does not contain the variable being differentiated (**x** in this case) disappears. Thus $5z^2$ and 12 disappear. In the case of the mixed element $(-4xz)$ this is treated as $(-4z)x$, which becomes $-4z$ when differentiated with respect to **x**.

In a similar manner the other partial derivative is derived thus:

$$\frac{\partial y}{\partial z} = 5z - 4x$$

Practical example of partial differentiation

18. DIY Ltd supply tool kits for the home handyman. Each tool kit comprises a standard plastic box which contains a variable number of tools depending on the type of tools, the market, and the wholesalers requirements. The firm has derived a profit function which shows that their profits are dependent both on the number of tool kits supplied and the number of tools in each kit. The profit function is as follows:

$$P = 8K - 0.0001K^2 + 0.05KT - 77.5T^2 - 10,000$$

where **P** = Profit in £s
K = No of kits
T = No. of tools in each kit

How many tool kits containing how many tools should be sold?

Solution

$$\frac{\partial P}{\partial K} = 8 - 0.0002K + 0.05T = 0 \text{ at maximum}$$

$$\frac{\partial P}{\partial T} = 0.05K - 155T \qquad = 0 \text{ at maximum}$$

(*Note*: The second derivatives of each are negative, -0.0002 and -155, respectively indicating maxima)

∴ solving for **T** and substituting gives

$$0.05K = 155T$$

$$\therefore T = \frac{0.05K}{155} = 0.0003226K$$

and substituting as follows:

218

$$8 - 0.0002\mathbf{K} + 0.05(0.0003226\mathbf{K}) = 0$$

$$\therefore 8 - 0.00018387\mathbf{K} = 0$$

$$\therefore \mathbf{K} = \frac{8}{0.00018387}$$

$$= \underline{\mathbf{43{,}509}} \text{ tool kits.}$$

This value can be substituted into the partial derivative $\dfrac{\partial \mathbf{P}}{\partial \mathbf{T}}$ thus :

$$0.05\mathbf{K} - 155\mathbf{T} = 0$$

$$0.05(43509) - 155\mathbf{T} = 0$$

$$\therefore \mathbf{T} = \frac{0.05(43509)}{155}$$

$$\approx \underline{\mathbf{14\ tools}}$$

Thus profit will be maximised by the sale of 43,509 kits each containing 14 tools. This gives a profit of **£63,735.**

Integration

19. For our purposes integration can be regarded as the reverse of differentiation. Differentiation establishes the slope of the function at a point whereas integration can be defined as the procedure for finding the area under the curve of a function.

As integration is the reverse of differentiation it follows that:

Original function	Derivative	Original function	Integral
x^4	$4x^3$	$4x^3$	x^4

The integral can be written as

$$\int 4x^3 dx = x^4$$

which can be described as

'the function which gives $4x^3$ when differentiated with respect to x is x^4

However the integral above is not complete because it will be recalled from paragraph 7 that $x^4 + 10$, or $x^4 + 50$, or $x^4 + \mathbf{c}$, where $\mathbf{c}$ is any constant, also have $4x^3$ as their derivative so that is is essential to recognise this possibility by writing

$$\int 4x^3 dx = x^4 + \mathbf{c}$$

The integral including the undetermined constant is known as the indefinite integral.

The value of the constant can, in some instances, be inferred to be zero or it may have a value when additional information known as an initial condition is supplied. Examples of both are given later in the chapter.

Basic rule for integration

20. As integration is the reverse of differentiation it follows that whereas differentiation reduced the index of x by 1 and used the old index as the coefficient of x, integration requires us to increase the index of x by 1 and then to divide through by the value of the new index.

i.e. where $y = kx^n$

$$\int y dx = \frac{kx^{n+1}}{n+1} + c \quad (n \neq -1)$$

Examples

Functions	Integral

$y = 10x^4$

$$\int 10x^4 dx = \frac{10x^5}{5} + c = 2x^5 + c$$

$y = 6x^3 + 2x + 3$

$$\int (6x^3 + 2x + 3)dx = \frac{6}{4}x^4 + \frac{2}{2}x^2 + 3x + c$$

$$= 1.5x^4 + x^2 + 3x + c$$

Notes:

a) It will be seen that a constant is added to each integral.

b) Since differentiation is a more straight forward process is is useful to check the integral by differentiating it to see that it comes back to the original expression. For instance differentiating $2x^5 + c$ gives $10x^4$ which is the first example above.

The value of the integration constant

21. Before numeric results can be obtained from the integral, which is the typical requirement in business applications, it is often necessary to establish the value of c, the constant. The value of c depends entirely on the particular situation in which the function is being used. Two examples are given below.

Example 1

Assume that a marginal revenue function is

$$y = 3x + 10 \text{ where x is sales in units.}$$

$$\text{The integral is}: \int (3x + 10)dx = \frac{3}{2}x^2 + 10x + c$$

which is the Total Revenue function. Now by inference when sales are zero (i.e. $x = 0$) revenue is zero.

thus, when $x = 0$

$$R = \frac{3}{2}x^2 + 10x + c = 0$$

$\therefore$ c must equal zero as this is the only value which satisfies this equation.

Example 2

Assume that a marginal profit function is

$y = 100 - 2x$ when y is the £s and x is sales in units.

You have also found that the company breaks even on sales of 5 units. (This is the additional information known as an initial condition). What are the fixed costs of the company?

Solution

The first step is to find the Total Profit function which is found by integrating the marginal function thus

$$\int (100 - 2x)dx = 100x - x^2 + c$$

In addition we know that the company breaks even (i.e. total profits are zero) when $x = 5$

$$\therefore P = 100x - x^2 + c = 0 \text{ when } x = 5$$

$$\therefore 100(5) - 5^2 + c = 0$$

$$\therefore c = -\underline{475}$$

c represents the fixed costs of the company so the answer is £475.

(Alternatively this can be expressed as follows: What would be the value of the profit function at zero activity, i.e. $x = 0$? This is a loss of £475 which represents the fixed costs of the organisation.)

Definite integrals

22. When we require a numeric result, say, the total revenue between two activity levels, the expression is termed a definite integral and is written thus:

$$\int_a^b ydx, \text{ where a is a definite value and b is some larger definite value.}$$

For example, what is the increase in profit in moving from an activity level of 10 units to 15 units when the marginal profit function is as given in Example 2 above, i.e. $y = 100 - 2x$?

The definite integral of this function is written thus

$$\int_{10}^{15} (100 - 2x)dx = [100 - x^2]_{10}^{15}$$

$$= [100(15) - 15^2] - [100(10) - 10^2]$$

$$= \underline{\underline{£375}}$$

Note: It will be seen that the constant, c, has been omitted. This is because the value of c (in this case -475) is the same at both the higher and lower limits and therefore it cancels out and is not a factor when we are seeking the increase in profit by moving from one activity level to another.

Summary

23. a) Differential calculus is used to find rates of change and turning points of functions.

b) The differential coefficient or derivative is the rate of change of y (the dependent variable) with respect to an infinitesimally small change in x (the independent variable).

c) Where the function is $y = x^n$ the derivative is

$$\frac{dy}{dx} = nx^{n-1}$$

d) Differentiation rules exist where the function has a constant, is a sum, is a product, is a quotient, is a function of a function and for many other types of more complex functions.

e) Major practical application of differentiation include finding marginal revenue, marginal cost, marginal profit and thereby establishing output levels which maximise profit.

f) Some functions have turning points and these can represent points of maximum revenue or minimum cost.

g) By calculating the derivative of a function and setting this to zero the turning point can normally be found.

h) The turning point should always be tested to see if it is a maximum or minimum. This is done by finding the second derivative, i.e.

$$\frac{d^2y}{dx^2}$$

If this is negative the turning point is a *maximum*, if *positive* it is a *minimum*.

i) Partial differentiation is used when there is more than one independent variable.

j) The normal rules for differentiation apply with one of the independent variables being treated as a constant whilst the other is differentiated.

k) integration is the reverse of differentiation is denoted thus

$$\int y \, dx$$

l) The basic integration rule is

$$\int kx^n \, dx = \frac{kx^{n+1}}{n+1} + c \quad (n \neq -1)$$

m) The value of the constant depends on circumstances and may be found by inference or by some additional information known as an initial condition.

n) When a numeric answer is required the expression becomes a definite integral.

Points to note

24. a) Differentiation and integration presuppose continuous functions. Business functions may be stepped or discontinuous so it may not be possible to apply the techniques in every case.

b) Frequently the total cost or revenue functions are known. If so, total values or values between two activity levels can be found easily and directly without using integration. Integration becomes necessary when only the marginal curves are known.

c) A word of caution. Calculus is a vast, complex subject which this chapter barely introduces. Many of the rules given are operational rules only and exceptions can occur in particular circumstances. Nevertheless the chapter provides sufficient material to cope with the examinations for which it is intended.

d) As explained in this chapter, calculus is frequently used to find information about economic and business models which, by their nature, are normally approximations so that any results obtained need to be used with caution.

Self review questions *Numbers in brackets refer to paragraph numbers*

1. In what circumstances is calculus used? (2)
2. What is the objective of differentiation? (3)
3. What is the basic rule for finding derivatives? (4)
5. How can differentiation be used to find a turning point? (14)
6. What is the second derivative? (14)
7. What is a partial derivative? (16)

8.　How is a partial derivative found? (17)

9.　What is the relationship between integration and differentiation? (19)

10.　What is the basic integration rule? (20)

11.　Distinguish between a definite and indefinite integral (22)

Exercises with answers

1.　Find the derivative of

a) $y = 6x - x$

c) $y = \sqrt{1 + 2x}$

b) $y = \dfrac{1}{x^2}$

d) $y = \dfrac{1}{\sqrt{x}}$

2.　A cost function is

$$£(c) = Q^2 - 30Q + 200$$

where Q = quantity of units produced

Find the point of minimum cost.

3.　A firm selling a Trade Directory has developed a profit function as follows:

$$P = 9D - 0.0005D^2 + 0.06DA - 80A^2 - 5,000$$

where D = number of directories sold and
A = number of advertising pages.

How many directories containing how many advertising pages should be sold to maximise profits?

Answers to exercises

1.　a) 6　　b) $\dfrac{-2}{x^3}$　　c) $\dfrac{1}{\sqrt{1+2x}}$　　d) $-\dfrac{1}{2x\sqrt{x}}$

2.　$\dfrac{dc}{dQ} = 2Q - 30$

　　$\therefore Q = 15$ at minimum

　　　　Note: $\dfrac{d^2C}{dQ^2} = 2$, which is positive indicating a minimum value.

3.

$$\frac{\partial P}{\partial D} = 9 - 0.001D + 0.06A$$

$$\frac{\partial P}{\partial A} = 0.06D - 160A$$

$$\therefore D = \frac{160A}{0.06} = 2,667A$$

$$\therefore\ 9 - 0.001(2,667\mathbf{A}) + 0.06\mathbf{A}$$

$$9 = 2.727\mathbf{A}\ \therefore\ \mathbf{A} = \underline{\textbf{3.3 pages}}$$

and substituting

$$9 - 0.001\mathbf{D} + 0.06(3.3)$$

$$9.198 = 0.001\mathbf{D}$$

$$\therefore\ \mathbf{D} = \underline{\textbf{9,198 copies}}$$

ventory control – introduction
rminology

Objectives

1. After studying this chapter you will
   - have been introduced to the key principles of inventory control which are developed in the following three chapters;
   - know the reasons why stocks are held;
   - understand the four costs associated with inventory;
   - be able to define common inventory, control terms such as Lead time, Physical and Free Stock, Economic Ordering Quantity;
   - know what is meant by Pareto or ABC analysis.

Types of inventory

2. A convenient classification of the types of inventory is as follows:
   a) Raw materials – the materials, components, fuels, etc., used in the manufacture of products.
   b) Work-in-Progress – (W-I-P) – partly finished goods and materials, sub-assemblies, etc., held between manufacturing stages.
   c) Finished goods – completed products ready for sale or distribution.

   The particular items included in each classification depend on the particular firm. What would be classified as a finished product for one company might be classified as raw materials for another. For example, steel bars would be classified as a finished product for a steel mill and as raw material for a nut and bolt manufacturer.

Reasons for holding stocks

3. The main reasons for holding stocks can be summarised as follows:
   a) to ensure that sufficient goods are available to meet anticipated demand;
   b) to absorb variations in demand and production;
   c) to provide a buffer between production processes. This is applicable to work-in-progress stocks which effectively de-couple operations;
   d) to take advantage of bulk purchasing discounts;
   e) to meet possible shortages in the future;
   f) to absorb seasonal fluctuations in usage or demand;
   g) to enable production processes to flow smoothly and efficiently;
   h) as a necessary part of the production process, e.g. the maturing of whiskey;

i)   as deliberate investment policy particularly in times of inflation or possible shortage.

Alternative reasons for the existence of stocks

4.   The reasons given in Para. 3. above are the logical ones based on deliberate decisions. However, stocks accumulate for other, less praiseworthy reasons, typical of which are the following:

a)   Obsolete items are retained in stock.

b)   Poor or non existent inventory control resulting in over-large orders, replenishment orders being out of phase with production, etc.

c)   Inadequate or non-existent stock records.

d)   Poor liaison between the Production Control, Purchasing and Marketing departments.

e)   Sub-optimal decision making, e.g. the Production Department might increase W-I-P stocks unduly so as to ensure long production runs.

Stock costs

5.   Whether as a result of deliberate policy or not, stock represents an investment by the organisation. As with any other investment, the costs of holding stock must be related to the benefits to be gained. To do this effectively, the costs must be identified. There are four categories: costs of holding stock; costs of obtaining stock; stockout costs; and the costs of the stock itself.

Costs of holding stock

6.   These costs, also known as carrying costs, include the following:

a)   Interest on capital invested in the stock.

b)   Storage charges (rent, lighting, heating, refrigeration, air conditioning, etc.).

c)   Stores staffing, equipment maintenance and running costs.

d)   Handling costs.

e)   Audit, stocktaking or perpetual inventory costs.

f)   Insurance, security.

g)   Deterioration and obsolescence.

h)   Pilferage, vermin damage.

Costs of obtaining stock

7.   These costs, sometimes known as ordering or procurement costs, include the following:

a)   The clerical and administrative costs associated with the Purchasing, Accounting, and Goods Received departments.

b)   Transport costs.

c) Where goods are manufactured internally the set up and tooling costs associated with each production run.

Note: Some students consider ordering costs to include only those costs associated with ordering external to the firm. However, internal ordering (i.e. own manufacture) may involve high costs for production planning, set-up.

Stockout costs

8. These are the costs associated with running out of stock. The avoidance of these costs is the basic reason why stocks are held in the first instance. These costs include the following:

   a) Lost contribution through the lost sale caused by the stockout.

   b) Loss of future sales because customers go elsewhere.

   c) Loss of customer goodwill.

   d) Cost of production stoppages caused by stockouts of W-I-P or raw materials.

   e) Labour frustration over stoppages.

   f) Extra costs associated with urgent, often small quantity, replenishment purchases.

   Clearly many of these costs are difficult to quantify, but they are often significant.

Costs of stock

9. These costs are the buying in prices or the direct costs of production. These costs need to be considered when:

   a) discounts are available for bulk purchases.

   b) Savings in production costs are possible with longer batch runs.

Objective of inventory control

10. The overall objective of inventory control is to maintain stock levels so that the combined costs, detailed in Paras. 6, 7 and 8 above, are at a minimum. This is done by establishing two factors, *when to order* and *how many to order*. These decisions are the subject of the following chapters, but before they can be explained in detail some terms used in Inventory Control are defined.

Inventory control terminology

11. Brief definitions of common inventory control terms are given below and are illustrated in Para 12.

   a) Lead or Procurement time. The period of time, expressed in days, weeks, months, etc., between ordering (either externally or internally) and replenishment, i.e. when the goods are available for use.

   b) Demand. The amount required by sales, production, etc. Usually expressed as a rate of demand per week, month, etc. Estimates of the rate of demand during the lead time are critical factors in inventory control systems. This is dealt with in more detail in Chapter 14.

c) Economic Ordering Quantity (EOQ) or Economic Batch Quantity (EBQ). This is a calculated ordering quantity which minimises the balance of costs between inventory holding costs and re-order costs. The rationale of EOQ and derivation of the EOQ formulae are dealt with in Chapter 13.

d) Physical stock. The number of items physically in stock at a given time.

e) Free stock. Physical stock plus outstanding replenishment orders minus unfulfilled requirements.

f) Buffer Stock or Minimum Stock or Safety Stock. A stock allowance to cover errors in forecasting the lead time or the demand during the lead time. Buffer stock is further explained in Chapter 14.

g) Maximum Stock. A stock level selected as the maximum desirable which is used as an indicator to show when stocks have risen too high.

h) Re-order level. The level of stock at which a further replenishment order should be placed. The re-order level is dependent upon the lead time and the demand during the lead time.

i) Re-order Quantity. The quantity of the replenishment order. In some types of inventory control systems this is the EOQ, but in some other systems a different value is used. This aspect is dealt with in detail in Chapter 14.

A simple stock situation illustrated

12. The following diagram shows a stock situation simplified by the following assumptions: regular rates of demand, a fixed lead time, and replenishment in one batch.

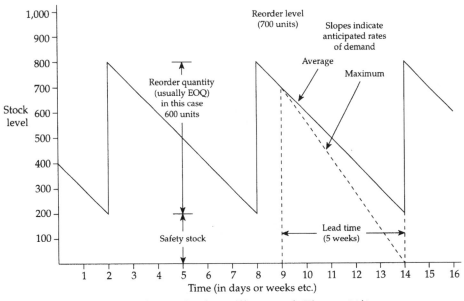

Stock terminology illustrated. Figure 11/1

Notes:

a) It will be seen from Figure 11/1 that the safety stock in this illustration is needed to cope with periods of maximum demand during the lead time.

b) The lead time as shown is 5 weeks, the safety stock 200 units, and the re-order quantity 600 units.

c) With constant rates of demand, as shown, the average stock is the safety stock plus $\frac{1}{2}$ Re-order quantity; for example, in Figure 11/1 the average stock is

$$200 + \tfrac{1}{2}(600) = \underline{\underline{\textbf{500 units}}}$$

Pareto analysis

13. Detailed stock control uses time and resources and can cost a considerable amount of money. Because of this it is important that the effort is directed where it can be most cost effective – there is little point in elaborate and costly recording and control procedures for an item of insignificant value.

Because of this it is worthwhile carrying out a so called Pareto or ABC analysis. It is often found that a few items account for a large proportion of the value and accordingly should have the closest monitoring. A typical analysis of stock items could be as follows

Class A items – 80% of value in 20% of items – Close day-to-day control

Class B items – 15% of value in 30% of items – Regular review

Class C items – 5% of value in 50% of items – Infrequent review.

Such a review can help to ensure that resources are used to maximum advantage.

Detailed, selective control will be more effective than a generalised approach which treats all items identically.

Deterministic and stochastic models

14. Like most other quantitative techniques, two types of models can be used for inventory control – deterministic and stochastic (or probabilistic) models.

- a deterministic model is one which assumes complete certainty. The values of all factors (e.g. demand, usage, lead time, carrying cost, etc.) are known exactly and there is no element of risk and uncertainty.

- a stochastic model exists where some or all of the factors are not known with certainty and can only be expressed in probabilistic or statistical terms. For example, if the usage rate or lead time was specified as a probability distribution, this would be a stochastic or probabilistic model.

Both types of model are covered in the chapters which follow.

Summary

15. a) Stocks may conveniently be classified into Raw Materials, Work-in-Progress, and Finished Goods.

b) The classification of an item depends upon the nature of the firm.

c) Stocks are held to satisfy demands quickly, to allow unimpeded production, to take advantage of bulk purchasing, as a necessary part of the production process, and to absorb seasonal and other fluctuations.

d) Stocks accumulate unnecessarily through poor control methods, obsolescence, poor liaison and sub-optimal decision making.

e) The costs associated with stock are: holding costs, costs of obtaining stock, and stockout costs.

f) The overall objective of inventory control is to maintain stock at a level which minimises total stock costs.

g) Inventory control has its own terminology, the basic contents of which are given in this chapter. The various definitions should be thoroughly learned.

Points to note

16. a) Inventory control is an important area of financial control which is often neglected.

b) A few per cent saving on inventory costs would represent millions of pounds on a national scale.

c) The stocks recorded by the official system are not necessarily the only stocks held. Unofficial stocks – often as unofficial WIP – often occur.

d) Pareto analysis is sometimes known as the 80 : 20 rule.

e) All stocks represent an investment so they should be kept to an absolute minimum. Some of the newer production systems, e.g. the Just-in-Time System, seek to maintain even production flows with minimum or zero inventory.

Self review questions *Numbers in brackets refer to paragraph numbers.*

1. How may inventories be classified? (2)
2. Why are stocks held? (3)
3. What are the major categories of cost associated with stocks? (5 to 8)
4. What is the overall objective of stock control systems? (9)
5. Define at least six common terms used in Inventory Control. (10)
6. What is ABC analysis? (12)

12 Inventory control – types of control system

Objectives

1. After studying this chapter you will
   - have been introduced to the two main inventory control systems;
   - know the principles of the re-order level or two-bin system;
   - be able to calculate the key control levels: re-order level, minimum level and maximum level;
   - understand the characteristics of the periodic review system.

Re-order level system

2. This system is also known as the two-bin system. Its characteristics are as follows:
   a) A predetermined re-order level is set for each item.
   b) When the stock level falls to the re-order level, a replenishment order is issued.
   c) The replenishment order quantity is invariably the EOQ.
   d) The name 'two-bin system' comes from the simplest method of operating the system whereby the stock is segregated into two bins. Stock is initially drawn from the first bin and a replenishment order issued when it becomes empty.
   e) Most organisations operating the re-order level system maintain stock records with calculated re-order levels which trigger off the required replenishment order.

   Note: The illustration in Chapter 11 was of a simple re-order level system. The re-order level system is widely used in practice and is the subject of frequent examination questions.

Illustration of a simple manual re-order level system

3. The following data relate to a particular stock item.

Normal usage	110 per day
Minimum usage	50 per day
Maximum usage	140 per day
Lead Time	25–30 days
EOQ (Previously calculated)	5,000

Using this data the various control levels can be calculated.

Re-order Level = Maximum Usage × Maximum Lead Time

= 140 × 30

$$= \underline{\textbf{4,200 units}}$$

Minimum Level = Re-order Level − Average Usage for Average Lead Time

$$= 4,200 - (110 \times 27.5)$$

$$= \underline{\textbf{1,175 units}}$$

Maximum Level = Re-order Level + EOQ − Minimum Anticipated Usage in Lead Time

$$= 4,200 + 5,000 - (50 \times 25)$$

$$= \underline{\textbf{7,950 units}}$$

Notes:

a) In a Manual system the three levels would be entered on a Stock record card and comparisons made between the actual stock level and the control levels each time an entry was made on the card.

b) The re-order level is a definite action level, the minimum and maximum points are levels at which management would be warned that a potential danger may occur.

c) The re-order level is calculated so that if the worst anticipated position occurs, stock would be replenished in time.

d) The minimum level is calculated so that management will be warned when demand is above average and accordingly buffer stock is being used. There may be no danger, but the situation needs watching.

e) Maximum level is calculated so that management will be warned when demand is the minimum anticipated and consequently the stock level is likely to rise above the maximum intended.

f) In a manual system if warnings about maximum or minimum level violations were received, then it is likely that the re-order level and/or EOQ would be recalculated and adjusted. In a computer based system such adjustments would take place automatically to reflect current and forecast future conditions.

g) A critical factor in establishing re-order levels and for calculating EOQs is the forecast of expected demand. Forecasting has been covered earlier in the book.

Periodic review system

4. This system is sometimes called the constant cycle system. The system has the following characteristics:

    a) Stock levels for all parts are reviewed at fixed intervals, e.g. every fortnight.

    b) Where necessary a replenishment order is issued.

    c) The quantity of the replenishment order is not a previously calculated EOQ, but is based upon; the likely demand until the next review, the present stock level and the lead time.

d) The replenishment order quantity seeks to bring stocks up to a predetermined level.

e) The effect of the system is to order *variable quantities* at *fixed intervals* as compared with the re-order level system, described in Para. 2, where *fixed quantities* are ordered at *variable intervals*.

Illustration of a simple manual periodic review system

5. A production control department maintains control of the 500 piece parts used in the assembly of the finished products by the periodic review system. The stock levels of all 500 parts are reviewed every 4 weeks and a replenishment order issued to bring the stock of each part up to a previously calculated level. This level is calculated at six-monthly intervals and is based on the anticipated demand for each part.

Based on the above system, the following graph shows the situation for one of the piece parts, part No.1101x, over a period of 16 weeks.

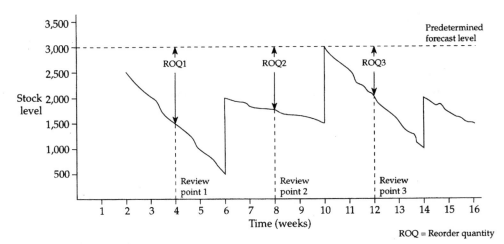

Stock Levels of Part No. 1101. Figure 12/1

Notes:

a) The re-order quantities, based on the agreed system are 1,500 units, 1,200 units, and 1,000 units.

b) It will be seen that the rates of usage are assumed to be variable and the lead time has been assumed to be 2 weeks.

c) The above illustration is merely one way of operating a periodic review system and many variants exist, particularly relating to the method of calculating the re-order quantities.

Periodic review system

6.   **Advantages:**

a)   All stock items are reviewed periodically so that there is more chance of obsolete items being eliminated.

b)   Economies in placing orders may be gained by spreading the purchasing office load more evenly.

c)   Larger quantity discounts may be obtained when a range of stock items are ordered at the same time from a supplier.

d)   Because orders will always be in the same sequence, there may be production economies due to more efficient production planning being possible and lower set up costs. This is often a major advantage and results in the frequent use of some form of periodic review system in production control systems in firms where there is a preferred sequence of manufacture, so that full advantage can be gained from the predetermined sequence implied by the periodic review system.

Disadvantages:

a)   In general larger stocks are required, as re-order quantities must take account of the period between reviews as well as lead times.

b)   Re-order quantities are not at the optimum level of a correctly calculated EOQ.

c)   Less responsive to changes in consumption. If the rate of usage change shortly after a review, a stockout may well occur before the next review.

d)   Unless demands are reasonably consistent, it is difficult to set appropriate periods for review.

Re-order level system

7.   **Advantages:**

a)   Lower stocks on average.

b)   Items ordered in economic quantities via the EOQ calculation.

c)   Somewhat more responsive to fluctuations in demand.

d)   Automatic generation of a replenishment order at the appropriate time by comparison of stock level against re-order level.

e)   Appropriate for widely differing types of inventory within the same firm.

Disadvantages:

a)   Many items may reach re-order level at the same time, thus overloading the re-ordering system.

b)   Items come up for re-ordering in a random fashion so that there is no set sequence.

c)   In certain circumstances (e.g. variable demand, ordering costs, etc.), the EOQ calculation may not be accurate. This is dealt with in more detail in Chapter 15.

Hybrid systems

8. The two basic inventory control systems have been explained above but many variations exist in practice. A firm may develop a system to suit their organisation which contains elements of both systems. In stable conditions of constant demand, lead times, and costs, both basic approaches are likely to be equally effective.

Summary

9. a) There are two basic inventory control systems, the re-order level or two-bin system and the periodic review system.

    b) The re-order level system usually has three control levels; re-order level, maximum level and minimum level.

    c) In the re-order level system the usual replenishment order quantity is the EOQ.

    d) The re-order level system results in fixed quantities being ordered at variable intervals dependent upon demand.

    e) The periodic review system means that all stocks are reviewed at fixed intervals and replenishment orders issued to bring stock back to predetermined level.

    f) The replenishment order quantity is based upon estimates of the likely demand until the next review period.

    g) The periodic review system results in variable quantities being ordered at fixed intervals.

Points to note

10. a) Many examination questions seem implicitly to assume that all inventory control systems are Re-order Level systems. This is not so.

    b) The importance of forecasting in inventory control cannot be over emphasised. Adaptive forecasting systems, already described are frequently an integral part of inventory control systems, particularly those which are computer based.

    c) There is a tendency for the stock level to increase in line with **n**, where **n** is the number of items in stock. Accordingly, standardisation of products, parts and materials can help to reduce stock levels. However, gains from reduced stocks may be offset by loss of marketability so care needs to be taken with standardisation policies.

Self review questions *Numbers in brackets refer to paragraph numbers*

1. What are the characteristics of the Re-order Level System of Inventory Control? (2)

2. What are the three levels commonly calculated for use in re-order level systems? (3)

3. Define the Periodic Review system and explain how it differs from the Re-order Level system? (4)

4. What are the major advantages and disadvantages of the Periodic Review system? (6)

5. What are the major advantages and disadvantages of the Re-order Level system? (7)

Exercises with answers

1. The following data relate to a given stock item.

Normal usage	1,300 per day
Minimum usage	900 per day
Maximum usage	2,000 per day
Lead time	15–20 days
EOQ	30,000

    Calculate the various controls levels.

2. What are the advantages and disadvantages of the Periodic review System of Inventory Control?

Answers to exercises

1.

$$\text{Reorder Level} = 2{,}000 \times 20$$

$$= \underline{\mathbf{40{,}000}}$$

$$\text{Minimum Level} = 40{,}000 - (1{,}300 \times 17\tfrac{1}{2})$$

$$= \underline{\mathbf{17{,}250}}$$

$$\text{Maximum Level} = 40{,}000 + 30{,}000 - (900 \times 15)$$

$$= \underline{\mathbf{56{,}500}}$$

2. a) The main advantage of a periodic review system are:

    i)   All stock holdings are reviewed periodically. Therefore obsolete stocks can be eliminated.

    ii) Specialist labour can be gainfully employed.

    iii) Where there is a cost advantage in placing a batch of orders, normally a number of items from one supplier, it can be exploited.

    iv) Cost advantages may be gained from certain production sequences. For example, where changing from one production run to another is easier than resetting.

    The main disadvantages are:

    v)   It requires additional labour to review items of stock at times other than when receipts and issues are being posted.

vi)   Larger stocks are often required, as re-order levels must take account of the period between reviews, as well as lead time.

vii) Slow to respond to changes in consumption. Should the rate of usage change shortly after a review a stockout may well occur before the next review.

viii) Difficult to set periods for review. Fluctuating, spasmodic usage may render it impossible.

13 Inventory control – economic order quantity

Objectives

1. After studying this chapter you will
   - understand the assumptions necessary to calculate the basic Economic Order Quantity (EOQ);
   - be able to find the EOQ graphically and by formula;
   - know how to calculate the EOQ with gradual replenishment and where stockouts are permitted;
   - understand how to find the best order quantity when price discounts are possible;
   - know the importance of marginal costs in EOQ calculations.

EOQ assumptions

2. The EOQ has been previously defined as the ordering quantity which minimises the balance of cost between inventory holding costs and re-order costs. To be able to calculate a basic EOQ certain assumptions are necessary:

   a) that there is a known, constant stockholding cost;

   b) that there is a known, constant ordering cost;

   c) that rates of demand are known;

   d) that there is a known, constant price per unit;

   e) that replenishment is made instantaneously, i.e. the whole batch is delivered at once;

   f) no stockouts allowed.

   Notes:

   a) It will be apparent that the above assumptions are somewhat sweeping and they are a good reason for treating any EOQ calculation with caution.

   b) Some of the above assumptions are relaxed later in this chapter.

   c) The rationale of EOQ ignores buffer stocks which are maintained to cater for variations in lead time and demand.

A graphical EOQ example

3. The following data will be used to develop a graphical solution to the EOQ problem.

Example 1

A company uses 50,000 widgets per annum which are £10 each to purchase. The ordering and handling costs are £150 per order and carrying costs are 15% of purchase price per annum, i.e. it costs £1.50 p.a. to carry a widget in stock (£10 × 15%).

To graph the various costs involved the following calculations are necessary:

$$\text{Total Costs p.a.} = \text{Ordering Cost p.a.} + \text{Carrying Cost p.a}$$

where $\qquad$ Ordering Cost p.a. = No. of orders × £150 and

$$\text{the No. of orders} = \frac{\text{Annual Demand}}{\text{Order Quantity}}$$

(For example, if the order quantity was 5,000 widgets,

$$\text{the no. of orders} = \frac{50,000}{5,000} = 10 \text{ and the ordering cost p.a.}$$

$$= 10 \times £150 = £1,500)$$

and carrying cost p.a. = average stock level × £1.5

$$\text{and the average stock} = \frac{\text{order quantity}}{2}$$

(For example if the order quantity is $5,000$, carrying costs p.a. are $\dfrac{5,000}{2} \times £1.5 = £3,750$)

Based on the above principles, the following table gives the costs for various order quantities.

Column I	II	III	IV	V	VI
Order quantity	Average no. of orders p.a.	Annual ordering cost	Average stock	Stock holding cost p.a.	Total cost
	$\dfrac{50,000}{\text{Col.I}}$	Col.II × £150	$\dfrac{\text{Col.I}}{2}$	Col.IV × £1.5	Col.III + Col.V
		£	£	£	
1,000	50	7,500	500	750	8,250
2,000	25	3,750	1,000	1,500	5,250
3,000	$16\frac{2}{3}$	2,500	1,500	2,250	4,750
4,000	$12\frac{1}{2}$	1,875	2,000	3,000	4,875
5,000	10	1,500	2,500	3,750	5,250
6,000	$8\frac{1}{3}$	1,250	3,000	4,500	5,750

Ordering and stock holding costs for various order quantities. Table 1

The costs in Table 1 can be plotted in a graph and the approximate EOQ ascertained.

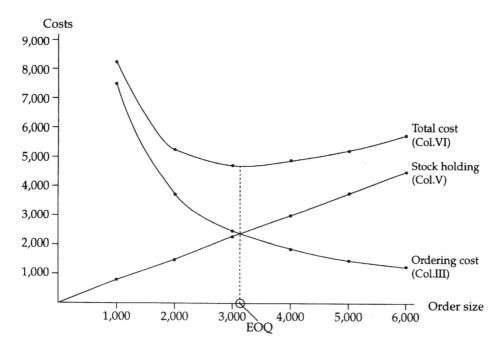

Graph of Data in Table 1. Figure 13/1

From the graph it will be seen that the EOQ is approximately 3,200 widgets, which means that an average of slightly under 16 orders will have to be placed a year.

Notes:

a) From a graph closer accuracy is not possible and is unnecessary anyway.

b) It will be seen from the graph that the bottom of the total cost curve is relatively flat, indicating that the exact value of the EOQ is not too critical. This is typical of most EOQ problems.

The EOQ formula

4. It is possible, and more usual, to calculate the EOQ using a formula. The formula method gives an exact answer, but do not be misled into placing undue reliance upon the precise figure. The calculation is based on estimates of costs, demands, etc., which are, of course, subject to error. The EOQ formula is given below and should be learned. The mathematical derivation is given in Appendix 1 of this chapter.

Basic EOQ formula

$$EOQ = \sqrt{\frac{2.Co.D}{Cc}}$$

where **Co** = ordering cost per order

D = Demand per annum

Cc = Carrying cost per item per annum

Using the data from Example 1, the **EOQ** can be calculated:

$$Co = £150$$

$$D = 50,000 \text{ widgets}$$

$$Cc = £10 \times 15\% \text{ or } £1.5 \text{ per widget.}$$

$$EOQ = \sqrt{\frac{2 \times 150 \times 50,000}{1.5}}$$

$$= \sqrt{10,000,000}$$

$$= \underline{\mathbf{3,162 \text{ widgets}}}$$

Notes:

a) The closest value obtainable from the graph was approximately 3,200 which is very close to the exact figure.

b) Always take care that demand and carrying costs are expressed for the same time period. A year is the usual period used.

c) In some problems the carrying cost is expressed as a percentage of the value whereas in others it is expressed directly as a cost per item. Both ways have been used in this example to provide a comparison.

EOQ with gradual replenishment

5. In example 1 above the assumption was that the widgets were ordered externally and that the order quantity was received as one batch, i.e. instantaneous replenishment as shown in the following diagram.

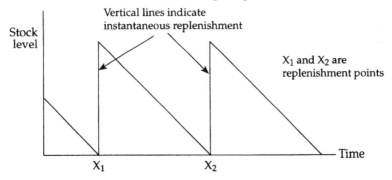

Stock Levels showing Instantaneous Replenishment. Figure 13/2

If however, the widgets were manufactured internally, they would probably be placed into stock over a period of time resulting in the following pattern:

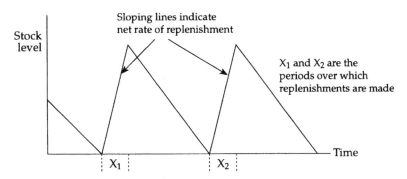

Stock level

Sloping lines indicate net rate of replenishment

X_1 and X_2 are the periods over which replenishments are made

Time

X_1 X_2

Stock Levels showing Non-instantaneous Replenishment. Figure 13/3

The net rate of replenishment is determined by the rate of replenishment and the rate of usage during the replenishment period. To cope with such situations, the basic EOQ formula needs modification thus:

$$\text{EOQ with gradual replenishment} = \sqrt{\frac{2.Co.D}{Cc\left(1 - \dfrac{D}{R}\right)}}$$

where **R** = Production rate per annum, i.e. the quantity that would be produced if production of the item was carried on the whole year.

All other elements in the formula have meanings as previously defined in Para.4.

Note: The derivation of the above formula is given in Appendix 2 of this chapter.

Example of EOQ with gradual replenishment

6.

> **Example 2**
>
> Assume that the firm described in Example 1 has decided to make the widgets in its own factory. The necessary machinery has been purchased which has a capacity of 250,000 widgets per annum. All other data are assumed to be the same.
>
> $$\text{EOQ with gradual replenishment} = \sqrt{\frac{2 \times 150 \times 50,000}{1.5\left(1 - \dfrac{50,000}{250,000}\right)}}$$
>
> $$= \sqrt{\frac{15,000,000}{1.5(.8)}}$$
>
> $$= \underline{\underline{\textbf{3,535 widgets}}}$$

Notes:

a) The value obtained above is larger than the basic EOQ because the usage during the replenishment period has the effect of lowering the average stock holding cost.

b) As pointed out in Chapter 11, Para. 7., the ordering costs for internal ordering usually include set up and tooling costs as well as paper work and administration costs.

EOQ where stockouts are permitted

7. It will be recalled that the overall objective of stock control is to minimise the balance of the three main areas of cost, i.e. holding costs, ordering costs and stockout costs.

Stockout costs are difficult to quantify but nevertheless may be significant and the avoidance of these costs is the main reason why stocks are held in the first place. Where stockout costs are known then they can be incorporated into the EOQ formula which thus becomes:

EOQ (where stockouts are permitted and stock out costs are known)

$$= \sqrt{\frac{2.\text{Co}.\text{D}}{\text{Cc}}} \times \sqrt{\frac{\text{Cc} + \text{Cs}}{\text{Cs}}}$$

where **Cs** = stockout costs per item and the other symbols have the meanings previously given.

Note: It will be seen that the formula is the basic EOQ formula multiplied by a new expression containing the stockout cost. The derivation of the EOQ formula where stockouts are permitted is given in Appendix 3 of this chapter.

Example of EOQ where stockouts are permitted

8. Assume the same data as in Para. 4 except that stockouts are now permitted. When a stockout occurs and an order is received for widgets the firm has agreed to retain the order and, when replenishments are received, to use express courier service for the delivery at a cost of £0.75 per widget. Other administrative costs associated with stockouts are estimated at £0.25 per unit. What is the EOQ?

$$\text{Co} = £150$$

$$\text{D} = 50,000$$

$$\text{Cc} = £1.5$$

$$\text{Cs} = £0.75 + £0.25 = £1$$

$$\text{Thus EOQ (with stockouts)} = \sqrt{\frac{2 \times 150 \times 50,000}{1.5}} \sqrt{\frac{1.5 + 1}{1}}$$

$$= \underline{\textbf{5,000}}$$

EOQ with discounts

9. A particularly unrealistic assumption with the basic EOQ calculation is that the price per item remains constant. Usually some form of discount can be obtained by ordering increased quantities. Such price discounts can be incorporated into the EOQ formula, but it becomes much more complicated. A simpler approach is to consider the costs associated with the normal EOQ and compare these costs with the costs at each succeeding discount point and so find the best quantity to order.

Financial consequences of discounts

10. Price discounts for quantity purchases have three financial effects, two of which are beneficial and one adverse.

Beneficial Effects:

Savings come from

a) Lower price per item

b) The larger order quantity means that fewer orders need to be placed so that total ordering costs are reduced.

Adverse Effect:

Increased costs arise from the extra stockholding costs caused by the average stock level being higher due to the larger order quantity.

Example 3 – Example of EOQ with Discounts

A company uses a special bracket in the manufacture of its products which it orders from outside suppliers. The appropriate data are

Demand = 2,000 per annum

Order cost = £20 per order

Carrying cost = 20% of item price

Basic item price = £10 per bracket

The company is offered the following discounts on the basic price:

For order quantities		
400–799	less 2%
800–1,599	less 4%
1,600 and over	less 5%

It is required to establish the most economical quantity to order.

This problem can be answered using the following procedure:

a) Calculate the EOQ using the basic price.

b) Compare the savings from the lower price and ordering costs and the extra stock-holding costs at each discount point (i.e. 400, 800 and 1,600) with the costs associated with the basic EOQ, thus

$$\text{Basic EOQ} = \sqrt{\frac{2 \times 2,000 \times 20}{10 \times .2}}$$

$$= 200 \text{ brackets}$$

Based on this EOQ the various costs and savings comparisons are given in the following table:

Order Quantity	200 (EOQ)	400	800	1,600	Line No.
Discount	–	2%	4%	5%	1
Average No. of Orders p.a.	10	5	$2\frac{1}{2}$	$1\frac{1}{4}$	2
Average No. of Orders saved p.a.	–	5	$7\frac{1}{2}$	$8\frac{3}{4}$	3
Ordering cost Savings p.a.	–	$(5 \times 20) =$ £100	$(7\frac{1}{2} \times 20)$ £150	$(8\frac{3}{4} \times 20) =$ £175	4 ‹
Price saving per item per annum	– –	20p $(2,000 \times 20p)$ $= 400$	40p $(2,000 \times 40p)$ $= 800$	50p $(2,000 \times 50p)$ $= 1,000$	5
Total gains		£500	£950	£1,175	6
Stockholding Cost p.a.	$(100 \times 10 \times .2)$ $= £200$	$(200 \times 9.8 \times .2)$ $= £392$	$(400 \times 9.6 \times .2)$ $= £768$	$(800 \times 9.5 \times .2)$ $= £1,520$	7
Additional costs incurred by increased order	–	$(£392 - £200)$ $= £192$	$(£768 - £200)$ $= £568$	$(£1,520 - £200)$ $= £1,320$	8
Net gain/loss	–	£308	£382	£145	9

Cost/savings comparisons EOQ to discount points. Table 2

From the above table it will be seen that the most economical order quantity is 800 brackets, thereby gaining the 4% discount.

Notes:

a) Line 2 is $\dfrac{\text{Demand of } 2,000}{\text{Order quantity}}$

b) Line 7 is the cost of carrying the average stock, i.e.

$\dfrac{\text{Order quantity}}{2} \times \text{Cost per item} \times \text{Carrying cost percentage}$

c) Line 9 is Line 6 minus Line 8.

Marginal costs and EOQ calculations

11.   It cannot be emphasised too strongly that the costs to be used in EOQ calculations must be true marginal costs, i.e. the costs that alter as a result of a further order or carrying another item in stock. It follows therefore that fixed costs should not be used in the calculations. In the examples used in this chapter the costs have been clearly and simply stated. In examination questions this is not always the case and considerable care is necessary to ensure that the appropriate costs are used.

Summary

12.   a)   The EOQ is the order quantity which minimises the total costs involved which include holding costs and order costs.

   b)   The basic EOQ calculation is based on constant ordering and holding costs, constant demand and instantaneous replenishment.

   c)   The basic EOQ formula is

$$\sqrt{\frac{2.\text{Co}.\text{D}}{\text{Cc}}}$$

   d)   Where replenishment is not instantaneous, e.g. in own manufacture, the formula becomes

$$\sqrt{\frac{2.\text{Co}.\text{D}}{\text{Cc}\left(1-\frac{\text{D}}{\text{R}}\right)}}$$

   e)   Where replenishment is not instantaneous, the EOQ calculated is larger than the basic EOQ.

   f)   Where stockouts are permitted and stock-out costs are known, the formula becomes

$$\sqrt{\frac{2.\text{Co}.\text{D}}{\text{Cc}}} \times \sqrt{\frac{\text{Cc}+\text{Cs}}{\text{Cs}}}$$

   g)   Where larger quantities are ordered to take advantage of price discounts stock-holding costs increase, but savings are made in the price reductions and reduced ordering costs.

   h)   The costs to be used in EOQ calculations must be marginal costs. Fixed costs should not be included.

Self review questions *Numbers in brackets refer to paragraph numbers*

1.   What are the assumptions usually made in EOQ calculations? (2)

2.   What is the basic EOQ formula? (4)

3.   How is the formula modified to take account of gradual replenishment? (5)

4.   What is the formula when stockouts are permitted? (7)

5.   What are the cost effects of buying larger quantities to obtain a quantity discount? (9)

6.   What type of costs should be used in EOQ calculations? (11)

Appendix I – Derivation of basic EOQ formula

Let D = annual demand

O = order quantity

Co = Cost of ordering for one order

Cc = Carrying cost for one item p.a.

$$\text{Average Stock} = \frac{Q}{2}$$

$$\text{Total annual stock holding cost} = \frac{QCc}{2}$$

$$\text{Number of orders per annum} = \frac{D}{Q}$$

$$\text{Annual ordering costs} = \frac{DCo}{Q}$$

$$\text{Total cost} = \frac{QCc}{2} + \frac{D}{Q}Co$$

The order quantity which makes the total cost (Tc) at a minimum is obtained by differentiating with respect to Q and equating the derivative to zero.

$$\frac{dTc}{dQ} = \frac{Cc}{2} - \frac{DCo}{Q^2}$$

and when $\dfrac{dTc}{dQ} = 0$ costs are at a minimum,

$$\text{i.e. } 0 = \frac{Cc}{2} - \frac{DCo}{Q^2}$$

and to find Q

$$\frac{DCo}{Q^2} = \frac{Cc}{2}$$

$$2DCo = Q^2Cc$$

$$\frac{2DCo}{Cc} = Q^2$$

$$Q \text{ (i.e. the EOQ)} = \sqrt{\frac{2.Co.D}{Cc}}$$

Appendix 2 – Derivation of EOQ with gradual replenishment

It will be recalled that gradual replenishment results in a stock level profile of the following shape where replenishment takes over time, **t**, at rate **R**.

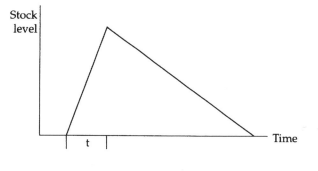

as No. of batches $\times$ **R** $\times$ **t** = **D**

i.e. $\dfrac{D}{Q} \times R \times t = D$

$$\therefore\ t = \frac{Q}{R}$$

The average stock is half the height of the triangle and the height is determined by the rate of replenishment less the demand over the replenishment time.

$$\text{Average stock} = \frac{t(R-D)}{2}$$

Substituting for **t**

$$\text{Average stock} = \frac{\dfrac{Q}{R}(R-D)}{2}$$

or

$$\frac{Q(1-\dfrac{D}{R})}{2}$$

$$\text{Total annual stockholding cost} = \frac{Q\left(1-\dfrac{D}{R}\right)Cc}{2}$$

which expression can be substituted for the corresponding expression in Appendix 1, and the identical steps following resulting in a modified EOQ formula thus

$$\textbf{EOQ (with gradual replacement} = \sqrt{\frac{2.Co.D}{Cc\left(1-\dfrac{D}{R}\right)}}$$

Appendix 3 – Derivation of EOQ where stockouts are permitted

Let all symbols have the meanings previously described.

$$\text{Total cost} = \text{Ordering Cost} + \text{Carrying Cost} + \text{Stockout Cost}$$

$$\text{where Annual ordering cost} = \frac{DCo}{Q}$$

To find the carrying and stockout costs it is necessary to find how much of the time the firm carries stock or is out of stock thus.

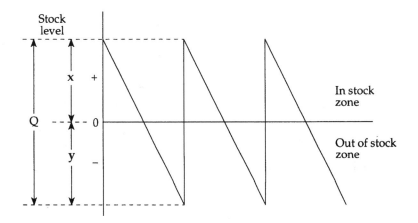

$$\therefore \text{Proportion of time in stock} = \frac{x}{Q}$$

$$\text{proportion of time out of stock} = \frac{y}{Q}$$

and as average stock is ½ the maximum level

$$\text{Carrying Cost (i.e. relating to 'instock' zone)} = \frac{1}{2} \times Cc \frac{x}{Q}$$

$$\text{Out of stock cost (i.e. relating to 'stockout' zone)} = \frac{1}{2} y\, Cs \frac{y}{Q}$$

$$\text{Thus Total Cost} = \frac{DCo}{Q} + \frac{1}{2} \times Cc \frac{x}{Q} + \frac{1}{2} yCs \frac{y}{Q}$$

and substituting for **y**

$$\text{Total cost} = \frac{DCo}{Q} + \frac{\frac{1}{2} x^2 Cc}{Q} + \frac{\frac{1}{2}(Q-x)^2}{Q} Cs$$

This is the total cost function and, as usual, to find the minimum it is necessary to differentiate and set the result to zero.

However as there are two independent variables x and Q it is necessary to partially differentiate with respect to x and Q and set the results to zero.

$$\frac{\partial Tc}{\partial x} = \frac{xCc}{Q} - \frac{(Q-x)Cs}{Q}$$

$$\frac{\partial Tc}{\partial Q} = \frac{-DCo}{Q^2} - \frac{\frac{1}{2}x^2Cc}{Q^2} + \left(\frac{Q(Q-x)Cs - \frac{1}{2}(Q-x)^2}{Q^2}\right)Cs$$

Setting to zero and solving

$$\frac{1}{Q}(xCc - (Q-x)Cs) = 0$$

$$\therefore x = Q\frac{Cs}{Cc+Cs}$$

$$-\frac{1}{Q^2}(DCo + \frac{1}{2}x^2Cc - \frac{1}{2}(Q^2 - x^2)Cs) = 0$$

$$Q^2 = \frac{2DCo}{Cs} + \frac{x^2Cc}{Cs} + x^2$$

$$= \frac{2DCo}{Cs} + x^2\frac{Cc+Cs}{Cs}$$

and as $x = Q\dfrac{Cs}{Cc+Cs}$ this can be substituted as follows

$$Q^2 = \frac{2DC}{Cs} + \frac{Q^2Cs}{Cc+Cs}$$

$$\therefore Q^2\left(1 - \frac{Cs}{Cc+Cs}\right) = \frac{2DCo}{Cs}$$

$$Q^2\left(\frac{Cc}{Cc+Cs}\right) = \frac{2DCo}{Cs}$$

$$\therefore Q^2 = \frac{2DCo(Cc+Cs)}{CcCs}$$

$$\therefore Q = \sqrt{\frac{2DCo}{Cc}} \times \sqrt{\frac{Cc+Cs}{Cs}}$$

13 Inventory control – economic order quantity

Exercises with answers

1. Calculate the various control levels given the following information:

Normal usage	560 per day
Minimum usage	240 per day
Maximum usage	710 per day
Lead Time	15–20 days
EOQ	10,000

2. a) A company uses 100,000 units per year which cost £3 each. Carrying costs are 1% per month and ordering costs are £250 per order.

   What is the EOQ?

   b) What would be the EOQ if the company made the items themselves on a machine with a potential capacity of 600,000 units per year?

3. Demand is 5,000 units per year. Ordering costs are £100 per order and the basic unit price is £5. Carrying costs are 20% p.a.

   Discounts are available thus:

1,200–1,399	less 10%
1,400–1,499	less 15%
1,500 and over	less 20%

   What is the most economical quantity to order?

Answers to exercises

1. Reorder level $= 710 \times 20 = 14{,}200$

   Minimum level $14200 - (560 \times 17.5) = 4{,}400$

   Maximum level $14200 + 10{,}000 - (240 \times 15) = 20{,}600$

2.
$$EOQ = \sqrt{\frac{2 \times 250 \times 100{,}000}{3 \times 0.12}}$$

$$= \underline{\underline{11{,}785}}$$

$$EOQ \text{ with gradual replenishment} = \sqrt{\frac{2 \times 250 \times 100{,}000}{3 \times 0.12\left(1 - \frac{100{,}000}{600{,}000}\right)}}$$

$$= \underline{\underline{12{,}909}}$$

3.
$$EOQ = \sqrt{\frac{2 \times 5{,}000 \times 100}{5 \times 0.2}} = 1{,}000$$

Order qty	1,000	1,200	1,400	1,500
Average orders p.a.	5	4.16	3.57	3.33
Order cost savings	–	84	143	167
Price savings p.a.	–	2,500	3,750	5,000
Extra stock holding cost	–	40	95	100
Total gain		2,544	3,798	5,067

$\therefore$ Order in 1,500 lots.

14 Inventory control – safety stocks and re-order levels

Objectives

1. After studying this chapter you will
    *   understand that safety or buffer stocks are necessary to cope with variations in demand and/or lead time;
    *   know how to calculate the safety stock by tabulating costs;
    *   be able to calculate the safety stock by statistical methods for various risk levels;
    *   understand how sensitivity analysis can be applied to inventory control.

Setting re-order levels in conditions of certainty

2. Where the rate of demand and the lead time is known with certainty, the re-order level is the rate of demand times the lead time. This means that, regardless of the length of the lead time or of the rate of demand, no buffer stock is necessary. This results in a situation as follows

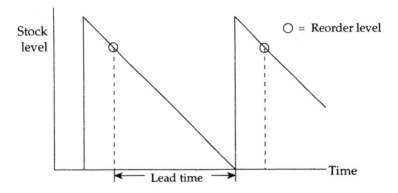

Re-order level in conditions of certainty. Figure 14/1

Re-order level and safety stock relationship

3. It will be seen from Figure 14/1 that, in conditions of certainty, the re-order level can be set so that stock just reaches zero and is then replenished. When demand and/or lead time vary, the re-order level must be set so that, on average, some safety stock is available to absorb variations in demand and/or lead time. In such circumstances the re-order level calculation can be conveniently considered in two parts.

a) the normal or average rate of usage times the normal or average lead time (i.e. as the re-order level calculation in conditions of certainty) *plus*

b) the safety stock

Safety stock calculation by cost tabulation

4.   The amount of safety stock is the level where the total costs associated with safety stock are at a minimum. That is, where the safety stock holding cost plus the stock out cost is lowest. (it will be noted that this is a similar cost position to that described in the EOQ derivation described in Chapter 13). The appropriate calculations are given below based on the following illustration.

Example 1

An electrical company uses a particular type of thermostat which costs £5. The demand averages 800 p.a. and the EOQ has been calculated at 200. Holding costs are 20% p.a. and stock out costs have been estimated at £ 2 per item that is unavailable. Demand and lead times vary, but fortunately the company has kept records of usage over 50 lead times as follows:

(a) Usage in lead time	(b) Number of times recorded	(c) Probability $\frac{b}{50}$
25–29 units	1	0.02
30–34 units	8	0.16
35–39 units	10	0.20
40–44 units	12	0.24
45–49 units	9	0.18
50–54 units	5	0.10
55–59 units	5	0.10
Total	50	1.00

Table 1

From the above the re-order level and safety stock should be calculated.

Solution

Step 1 – Using the mid point of each group calculate the average usage in the lead time.

x	t	tx
27	1	27
32	8	256
37	10	370
42	12	504
47	9	423
52	5	260
57	$\underline{5}$	$\underline{285}$
	$\underline{50}$	2,125

Table 2

$$\text{Average usage} = \frac{2,125}{50} = \underline{\underline{42.5}}$$

Step 2 – Find the holding and stock out costs for various re-order levels.

A	B	C	D	E	F	G	H
			Possible		No of orders		
Re-order level	Safety stock	Holding cost	shortages (mid-points)	Probability (from	p.a. $\left(\frac{800}{200}\right)$	Shortage cost	Total cost
	(A − 42.5)	(B × £1)	(Table 2 − A)	Table 1)		(D × E × F × £2)	(C + G)
		£				£	£
45	2.5	2.5	2	0.18	4	2.88	
			7	0.10	4	5.6	
			12	0.10	4	9.6	20.58
50	7.5	7.5	2	0.10	4	1.6	
			7	0.10	4	5.6	14.7
55	12.5	12.5	2	0.10	4	1.6	14.1
60	17.5	17.5					17.5

Table 3

From Table 3 it will be seen that the most economical re-order level is **55** units. This re-order level, with the average demand in the lead time of 42.5, gives a safety stock of 12.5, say **13** units.

Safety stock calculation by statistical methods

5. The previous method of calculating Safety Stock was based on relative holding and stock out costs, but on occasions these costs, particularly the effects of stock outs, are not known. In such circumstances management may decided upon a particular risk level they are prepared to accept and the safety stock and re-order level are based upon this risk level. For example, management may have decided that they are prepared to accept a 5% possibility of a stock out.

Illustration of a safety stock calculation by statistical methods

6.

Example 2

Using the data from Example 1, it was found that the average demand during the lead time was 42½ units. The company has carried out further analysis and has found that this average lead time demand is made up of an average demand (**D**) of 3.162 units per day over an average lead time (**L**) of 13.44 days. Both demand and lead time may vary and the company has estimated that the standard deviation of demand (σ_D) is 0.4 units and the standard deviation of lead time (σ_L) is 0.75 days. The company are prepared to accept a 5% risk of a stock out and wish to know the safety stock required in the following three circumstances:

i)   Where demand varies and lead time is constant

ii) Where the lead time varies and demand is constant

iii) Where both demand and lead time vary.

In each of these cases the *average usage* is **D** × **L**, i.e. 3.162 × 13.44 = **42½** units.

Solution

a)   Safety stock given variable demand and constant lead time.

From normal area tables it will be found that 5% of the area lies above the mean +1.64σ.

$\therefore$ Safety stock = 1.64 × standard deviation of demand for 13.44 days

$$= 1.64 \times (0.4 \times \sqrt{13.44})$$
$$= \underline{\underline{2.40}}$$

Note: The standard deviation of daily demand, 0.4 is multiplied by $\sqrt{13.44}$ because standard deviations are not additive.

b)   Safety Stock given variable lead time and constant demand.

$\therefore$ Safety stock = 1.64 × standard deviation of lead time for a demand of 3.162

$$= 1.64 \times (0.75)$$
$$= \underline{\underline{1.23}}$$

c)   This is a combination of the two previous sections and is the sum of the separate safety stocks already calculated.

$$= 2.40 + 1.23$$
$$= \underline{\underline{3.63}}$$

Notes:

a)   Safety stock calculations based on risk levels are commonly used.

b) Where lead time and demand vary it can be expensive to maintain sufficient safety stocks for low stock out risks.

The above example uses the properties of the Normal Distribution and values obtained from Normal Area tables. It is but one further application of the statistical principles covered earlier in the book using continuous probability distributions. Discrete probabilities and expected values can also be used to incorporate variability.

Inventory control and sensitivity analysis

7. When an inventory control value has been calculated; for example, the EOQ, the re-order level, the total stockholding cost etc. management may wish to know how sensitive the value is to changes in the factors used to calculate it.

For example, is the EOQ greatly affected when ordering cost changes? The process by which this is done is known as *sensitivity analysis*. The following example shows a typical analysis.

Example 3

Henderson Ltd made the following estimates for a component they use:

Annual usage 1,125

Ordering costs £50 per order

Carrying costs per year £5 per component.

Based on these estimates, an EOQ of 150 and expected total stock costs of £750 p.a. were calculated, as follows:

$$EOQ = \sqrt{\frac{2 \times 50 \times 1,125}{5}} = \underline{\underline{150}}$$

Expected total stock costs = Ordering Cost p.a. + Holding Cost p.a.

$$= \left(\frac{1,125}{150}\right) \times £50 + \frac{£5 \times 150}{2} = \underline{\underline{£750}}$$

During a year the EOQ of 150 was used for re-ordering, but the actual usage of components turned out to be 20% higher at 1,350.

a) Calculate the actual total stock costs.

b) Calculate what the total stock costs would have been if the correct EOQ had been used.

c) Find out how sensitive are total costs to errors in the usage estimates.

Solution

a) Actual total stock costs $= \left(\frac{1,350}{150}\right) \times £50 + \frac{5 \times 150}{2} = \underline{\underline{£825}}$

b) Correct EOQ

$$= \sqrt{\frac{2 \times 50 \times 1,350}{5}} = \underline{\underline{164}}$$

Total costs if an EOQ of 164 was used $= \left(\frac{1,350}{164}\right) \times £50 + \frac{£5 \times 164}{2} = \underline{\underline{£822}}$

c) Thus £3 (i.e. £825 − 822) extra costs were incurred by using the 150 EOQ based on the incorrect usage.

This means that a 20% increase in demand has caused only approximately $\{\frac{1}{2}\%\}$ change in total costs. We conclude that total costs are insensitive to errors in demand estimates.

From management's viewpoint, this is good news. It means that even if errors are made in the estimates, which is all too likely, it doesn't make much difference to the final result.

Summary

8. a) Safety stocks are necessary because of demand and/or lead time variations.

b) Re-order level is the average demand over the average lead time plus safety stock.

c) The safety stock level can be established by comparing the safety stock holding cost and the stock out cost at various re-order levels.

d) Where it is necessary to calculate the safety stock level to cater for a given risk of stock out, statistical methods, based on areas under the normal curve, can be used.

e) The means and standard deviations of demand and lead time must be calculated or estimated and the number of standard deviations appropriate to the risk level added to the average demand and/or lead time before multiplying together to establish the re-order level and thus the safety stock.

Points to note

9. a) This chapter concerns the effects of uncertainty (i.e. variability) on inventory control. It will be recalled from Chapter 1 that including risk and uncertainty is a necessary part of all OR studies. As previously mentioned, models which include uncertainty are stochastic models.

b) Although some OR techniques have little practical application, this is definitely not true of inventory control.

Self review questions *Numbers in brackets refer to paragraph numbers*

1. How is the re-order level set when demand and lead times are known with certainty? (2)

2. Why is the establishment of safety stocks essentially a cost balancing exercise? (4)

3. What are the appropriate costs to be considered in establishing an appropriate Safety Stock Level? (4)

4.   How can safety stocks be estimated if all the relative costs are not known? (5 and 6)

Exercises with answers

1.   A company uses 2,000 components per annum and the cost is £6 per component. Holding costs are £2 per component p.a. and stock out costs are £3 per component per item unavailable. The EOQ is 500 and demand is variable as follows:

Usage in lead time	Probability
80	0.2
90	0.5
100	0.3

What is the most economical re-order level: 90, 95 or 100?

2.   The demand for an item is normally distributed with a mean of 50 per day and a standard deviation of 5 units. Given a lead time of 20 days what is the re-order level and safety stock to meet 90% of demands?

Answers to exercises

1.   Expected usage $= (80 \times 0.2) + (90 \times 0.5) + (100 \times 0.3) = 91$

Cost tabulation

Order level	Safety stock	Holding cost £	Shortages	P	Orders p.a.	Cost of shortages + holding cost		£
90			1	0.5	4	$1 \times 0.5 \times 3 \times 4$	=	6
			10	0.3	4	$10 \times 0.3 \times 3 \times 4$	=	36
							=	42
95	4	8	5 $\times$	0.3	4	$5 \times 0.3 \times 3 \times 4 + 8$	=	26
100	9	18	–				=	18

∴ 100 is most economical order quantity

2.   Average demand in lead time $= 50 \times 20 = 1,000$

∴ Safety stock given 10% of area is above the mean $+ 1.286\sigma$

$= 1.28 \times 5 \times \sqrt{20}$

$= \underline{\mathbf{28.62}}$

Suggested reorder level is 1,028.62

Note: An alternative way of dealing with the variability of demand is to combine the variances of demand.

s.d. of demand $= 5$ ∴ variance $= 5^2 = 25$

∴ variance of demand over lead time $= 20 \times 25 = \underline{\underline{500}}$

$\therefore$ s.d. of demand over lead time $= \sqrt{500} = 22.36$

$\therefore$ Safety stock giving 90% confidence $= 1.28 \times 22.36 = \underline{\underline{28}}$

5 Simulation

Objectives

1.   After studying this chapter you will

   - be able to define simulation and know why it is used;
   - understand what is meant by a model and how models are developed;
   - know what is meant by Monte Carlo Simulation;
   - be able to describe the typical variables used in simulation;
   - be able to construct a simulation model;
   - know the advantages and disadvantages of simulation.

Simulation definition

2.   Simulation is the process of experimenting or using a model and noting the results which occur. In a business context the process of experimenting with a model usually consists of inserting different input values and observing the resulting output values. For example, in a simulation of a queueing situation the input values might be the number of arrivals and/or service points and the output values might be the numbers and/or times in the queue.

Why is simulation used?

3.   Simulation is used where analytical techniques are not available or would be overly complex. Typical business examples are: most queueing systems other than the very simplest, inventory control problems, production planning problems, corporate planning, etc., etc. Simulation often provides an insight into a problem which would be unobtainable by other means. By means of simulation the behaviour of a system can be observed over time and, because only a model of the system is used the actual time span can be compressed.

Thus for example, the simulation of an inventory system's performance over 30 months could be carried out using a computer in as many minutes. In addition, the analyst can manipulate or experiment with the system at will. He could for example, alter the frequencies of receipts and issues, the decision rules governing re-order levels and re-order quantities and so on to observe the effects of these changes on the system being simulated or imitated. Fundamental to simulation is the concept of *a model*.

Business models

4.   A model is any representation (physical or abstract) of a real thing, event or circumstances. Physical models, such as aircraft models for wind tunnel testing, models of space capsule controls, architectural layout models, are well known, but for business planning and decision making purposes are rarely appropriate.

Instead, *symbolic models* are generally used. These are models which represent reality in numeric, algebraic or graphical form. The most useful models for business simulation are discrete models. These are models which enable the system being simulated to be observed or sampled at selected time intervals. Thus an inventory simulation might show stock levels on a daily basis. Alternatively a queuing simulation for supermarket checkouts might use a five minute interval to assess performance. The interval chosen should be that appropriate to the operational reality of the system being simulated.

Model construction

5. The success of a simulation exercise is related to the predictive quality of the underlying model, so that considerable care should be taken with model construction. Important factors in model development are:

   a) Objective oriented. The model should be constructed with some definite purpose and the model results must be directly related to this purpose.

   b) Critical variables and relationships. Modelling buildings is an iterative, creative process with the aim of identifying those variables and relationships that must be included in the model. It is not essential or indeed desirable to include all variables in the model.

   c) Simplicity. The best model is the simplest that has adequate predictive qualities.

   d) Management involvement. To construct good models there must be a thorough understanding of actual operations. Only management have this knowledge, so they must be involved. If model building is left entirely to OR or computer specialists, over-elaborate and sophisticated models may result which do not accurately represent reality. However, it must be recognised that getting accurate, comprehensive information from management is a difficult task in itself.

   e) Model development. If a model is to be used more than once, e.g. a corporate planning model used each quarter, care must be taken to modify and refine the model characteristics so that it continues to represent reality.

   Figure 15/1 provides an overview of model development and use.

Assessing a model's suitability

6. Models are intelligent simplifications of reality and there is no way of proving that a model gives adequate prediction in unknown circumstances. However, many models are developed to help forecast what might happen when an existing operation is enlarged or has extra demands made upon it. In such circumstances, when the model is constructed, values are input that correspond to current operational levels and the output values calculated. If these values correspond to the present position, then there are reasonable grounds for believing that the model is a fair representation of reality.

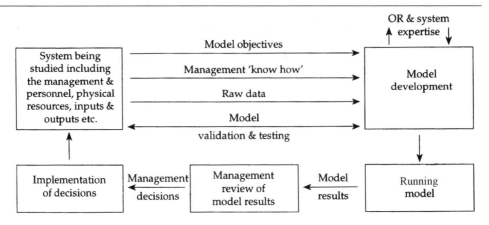

Model development and use. Figure 15/1

Monte Carlo simulation

7.   Any realistic business problem contains probabilistic or random features, e.g. arrivals at a store may average 80 per day, but the actual arrival patterns is likely to be highly variable; in a corporate planning exercise forecasts of the likely sales will obviously vary according to different circumstances, etc. Because models must behave like the system under observation, the model must contain these probabilistic elements. Simulations involving such random elements are sometimes termed *Monte Carlo simulations*.

Random selection

8.   To carry out a realistic simulation involving probabilistic elements, it is necessary to avoid bias in the selection of the values which vary. This is done by selecting randomly (using the term in its statistical sense) using one of the following methods:

a)   Random number generation by computer.

b)   Random number tables. These consist of a table of randomly selected numbers in which bias does not exist.

c)   Lottery selection, e.g. put all the numbers in a bag, shake well, draw out a number.

d)   Roulette wheel. The wheel is spun and a ball dropped in to select a number (hence the name 'Monte Carlo simulation').

e)   Dice or cards. These can also be used, although with cards the card drawn should be replaced and the pack reshuffled before another card is selected.

The repeated random selection of input values and the logging of the resultant outputs is the very essence of simulation. In this way an understanding is gained of the likely pattern of results so that a more informed decision can be taken.

Note: As all operational simulation uses computers, method (a) above is by far the most common.

Variables in a simulation model

9. A business model usually consists of linked series of equations and formulae arranged so that they 'behave' in a similar manner to the real system being investigated. The formulae and equations use a number of factors or variables which can be classified into 4 groups.

   a) *Input or exogenous variables*

   b) *Parameters*

   c) *Status variables*

   d) *Output or endogeneous variables*

   These are described below.

Input variables

10. These variables are of two types – *controlled* and *non-controlled*.

    Controlled variables: These are the variables that can be controlled by management. Changing the input values of the controlled values, and noting the change in the output results, is the prime activity of simulation. For example, typical controlled variables in an inventory simulation might be the re-order level and re-order quantity. These could be altered and the effect on the system outputs noted.

    Non-controlled variables: These are input variables which are not under management control. Typically these are probabilistic or stochastic variables, i.e. they vary but in some uncontrollable, probabilistic fashion.

    For example, in a production simulation the number of breakdowns would be deemed to vary in accordance with a probability distribution derived from records of past break-down frequencies. In an inventory simulation, demand and lead time would also be generally classed as non-controlled, probabilistic variables.

Parameters

11. These are also input variables which, for a given simulation, have a constant value. Parameters are factors which help to specify the relationships between other types of variables. For example in a production simulation a parameter (or constant) might be the time taken for routine maintenance; in an inventory simulation a parameter might be the cost of a stock-out.

Status variables

12. In some types of simulation the behaviour of the system (rates, usages, speeds, demand and so on) varies not only according to individual characteristics but also according to the general state of the system at various times or seasons. As an example, in a simulation of supermarket demand and checkout queueing, demand will be probabilistic and variable on any given day but the general level

of demand will be greatly influenced by the day of the week and the season of the year. Status variables would be required to specify the day(s) and season(s) to be used in a simulation.

Note: On occasions, status variables and parameters would both be termed just parameters, although strictly there is a difference between the two concepts.

Output variables

13. These are the results of the simulation. They arise from the calculations and tests performed in the model, the input values of the controlled values, the values derived for the probabilistic elements and the specified parameters and status values. The output variables must be carefully chosen to reflect the factors which are critical to the real system being simulated and they must relate to the objectives of the real system. For example, output variables for an inventory simulation would typically include:

   - cost of stock holding
   - number of stockouts
   - number of unsatisfied orders
   - number of replenishment orders
   - cost of the re-ordering, and so on.

Constructing a simulation model

14. Some broad guidelines for constructing a simulation model are given below. These will be found useful for dealing with examination questions but in this area especially, practice is vital.

   Step 1: Identify the objective(s) of the simulation.

   A detailed listing of the results expected from the simulation will help to clarify step 5 – the output variables.

   Step 2: Identify the input variables. Distinguish between controlled and non-controlled variables.

   Step 3: Where necessary determine the probability distribution for the non-controlled variables.

   Step 4: Identify any parameters and status variables.

   Step 5: Identify the output variables.

   Step 6: Determine the logic of the model.

   This is the heart of the simulation construction. The key questions are: how are the input variables changed into output results? what formulae/decision rules are required? how will the probabilistic elements be dealt with? how should the results be presented?

   To illustrate these steps a simple problem follows together with a solution using the six step approach given above.

   Note: To show the stages clearly within the constraints of a printed format the example that follows has been carried out manually. In practice, a computer would be used for simulations with either a specialised program or, for small simulations, a spreadsheet package such as Excel or Lotus 1.2.3. It will be noted how the table layouts that follow naturally transform into the cells of a spreadsheet. The model logic of step 6 above is the stage where the programming of the spreadsheet cells takes place and the necessary formulae are entered in the cells.

A simple inventory simulation

15.

Example 1

A wholesaler stocks an item for which demand is uncertain. He wishes to assess two re-ordering policies i.e. order 10 units at a reorder level of 10, or order 15 units at a reorder level of 15 units, to see which is most economical over a 10 day period.

The following information is available:

Demand per day (units)	Probability
4	0.10
5	0.15
6	0.25
7	0.30
8	0.20

Carrying costs £15 per unit per day. Ordering costs £50 per order. Loss of goodwill for each Unit out of stock £30. Lead time 3 days. Opening Stock 17 units.

The probability distribution is to be based on the following random numbers

| 41 | 92 | 05 | 44 | 66 | 07 | 00 | 00 | 14 | 62 |
| 20 | 07 | 95 | 05 | 79 | 95 | 64 | 26 | 06 | 48 |

Note: The reorder level is physical stock plus any replenishment orders outstanding.

Solution

Step 1: Objectives of simulation.

To simulate the behaviour of two ordering policies – order 15 at reorder level of 15 and order 10 at reorder level of 10 – to establish the cheaper policy.

Step 2: Identify the input variables.

Controlled variables:

Order quantity
Reorder level

Non-Controlled variable:

Probabilistic demand

Step 3: Determine probability distribution.

The random numbers are allocated to the demand as follows

Demand	Probability	Cumulative probability	Random numbers
4	10%	10%	00–09
5	15%	25%	10–24
6	25%	50%	25–49
7	30%	80%	50–79
8	20%	100%	80–99

The two random number sequences supplied can then be used for the two runs of the simulation with each pair of digits used to select a demand. For example, for the 15 order quantity simulation, the first two digits, 41, give a Day 1 demand of 6 units. The next two digits, 92, give a Day 2 demand of 8 units and so on.

Step 4: Identify Parameters and Status Variables

Parameters	
Opening Stock	17 units
Carrying costs	£15 per unit per day
Ordering costs	£50 per order
Loss of goodwill	£30
Lead time	3 days

There are no status variables in this simple example.

Step 5: Identify the output variables

The main output variable is Total Cost. However ancillary output variables which arise from the simulation include:

- Number of orders placed
- Number of stockouts
- Cost of stockouts
- Total carrying cost
- Total order cost,

Step 6: Determine model logic.

In this simple example the logic and rules required are nothing more than the basics of inventory control thus:

Reorder level	= Physical stock + Replenishment Orders outstanding
Closing stock	= Opening stock − Demand
Carrying cost per day	= Stock × £15 per unit
Goodwill costs	= Stock shortfall × £30 per unit
Total cost	= Goodwill costs + carrying costs + ordering costs.

The information, values and rules are then used to simulate the two ordering policies. The results of these simulations are shown in the schedules in Figure 15/2 and 15/3 from which it will be seen that the simulation shows that the 'order 15' policy is more economical over the 10 days simulated.

It must be emphasised that in practice the simulations would be carried out over many more cycles than 10 days in order to obtain a truly representative picture.

Day	Opening stock	Demand	Closing stock	Order costs	Carrying costs	Stock out costs	Total costs
1	17	6	11	£50	£165		£215
2	11	8	3		£ 45		45
3	3	4	–		–	£30	30
4	+ 15	6	9	£50	£135		185
5	9	7	2		£ 30		30
6	2	4	–		–	£60	60
7	+ 15	4	11	£50	£165		215
8	11	4	7		£105		105
9	7	5	2		£ 30		30
10	+ 15 + 2	7	10	£50	£150		200
				£200	£825	£90	£1,115

Results of simulation using order quantity and re-order level of 15 units.
Figure 15/2

Notes on Figure 15.2

- The daily demand is found by using the random numbers in Para 15 (i.e. 41 92 05, etc.) and determining the demand from the probability allocation in Step 3 above. Thus 41 is in the 25–49 group and corresponds to a demand of 6, 92 is in the 80–99 group corresponding to a demand of 8 and so on
- An order for 15 is placed whenever physical stock and replenishment orders outstanding ≤ 15. This occurs on Days 1, 4, 7 and 10.

Day	Opening stock	Demand	Closing stock	Order costs	Carrying costs	Stock out costs	Total costs
1	17	5	12		£180		£180
2	12	4	8	£50	120		170
3	8	8	–		–	–	–
4	–	5	–		–	£150	150
5	+10	7	3	50	45		95
6	3	8	–		–	150	150
7	–	7	–		–	210	210
8	+10	6	4	50	60		110
9	4	4	–		–		–
10	–	6	–		–	180	180
				£150	£405	£690	£1,245

Results of simulation using order quantity and re-order level of 10 units.
Figure 15/3

Further example of simulation

16. A common application of simulation is to examine the behaviour of queues in circumstances where the use of queueing formulae is not possible. The following example is typical.

Example 2

A filling station is being planned and it is required to know how many attendants will be needed to maximise earnings. From traffic studies it has been forecast that customers will arrive in accordance with the following table:

   Probability of 0 customers arriving in any minute 0.72

   Probability of 1 customer arriving in any minute 0.24

   Probability of 2 customer arriving in any minute 0.03

   Probability of 3 customer arriving in any minute 0.01

From past experience it has been estimated that service times vary according to the following table

Service time in minutes	1	2	3	4	5	6	7	8	9	10	11	12
Probability	0.16	0.13	0.12	0.10	0.09	0.08	0.07	0.06	0.05	0.05	0.05	0.04

If there are more than two customers waiting, in addition to those being serviced, new arrivals drive on and the sale is lost.

A petrol pump attendant is paid £40 per 8 hour day, and the average contribution per customer is estimated to be £4.

How many attendants are needed?

Solution

Step 1: Objectives of simulation

To find the number of attendants to maximise earnings

Step 2: Input variables

Controlled   Number of attendants

Non-Controlled   Customer arrival rate

          Service time

Step 3: Probability distribution

As previously a random number table is used and an extract is given in Table 1.

5053225496	9565241457	7354776952	2149630416	5579018342
7245174840	2275698645	8416549348	4676463101	2229367983
6749420382	4832530032	5670984959	5432114610	2966095680
7164238934	7666237259	5263097712	9999089966	7544056852
4192054466	0700014629	5169439659	8408705169	1074373131
9697426117	6488888550	4031652526	8123543276	0927534537
2007950579	9564268448	3457415998	1531027886	7016633739
4584768758	2389278610	3859431781	3643768456	4141314518
3840145867	9120831830	7228567652	1267173884	4020651657

Table 1

The arrival pattern estimated is reproduced below with random numbers assigned.

		Random nos assigned
Probability of 0 customers arriving in any minute	= 0.72	01–72
Probability of 1 customer arriving in any minute	= 0.24	73–96
Probability of 2 customer arriving in any minute	= 0.03	97–99
Probability of 3 customer arriving in any minute	= 0.01	00

Similarly for the service pattern we assign the random numbers thus:

Service time minutes	1	2	3	4	5	6	7	8	9	10	11	12
Likelihood	0.16	0.13	0.12	0.10	0.09	0.08	0.08	0.06	0.05	0.05	0.05	0.04
Random numbers assigned	01–16	17–29	30–41	42–51	52–60	61–68	69–75	76–81	82–86	87–91	92–96	97–00

The random number table is read in any direction in groups of two digits and, according to the digits, the appropriate arrival pattern or service time is selected.

For example, assume that for the arrival pattern the table is read from left to right starting from the first row.

Minute no.	Random digits from table	No of arrivals
1	50	0
2	53	0
3	22	0
4	54	0
5	96	1
6	95	1
7	65	0
8	24	0
9	14	0
10	57	0
11	73	1
12	54	0

13	77	1
14	69	0
15	52	0
16	21	0
17	49	0
etc.		0

Table 2

Selecting on this basis over say, a week's operations, will result in a random selection reflecting the estimated probabilities.

Step 4: Parameters

Attendant cost £40 per day

Average contribution per customer £4

Step 5: Output variables

Average contributions per day

Attendant costs per day

No. of unsatisfied customers

Step 6: Logic of simulation

In this case the logic is shown using a flowchart Figure 15/4 and the results of the simulation would be entered on a simple worksheet, Figure 15/5.

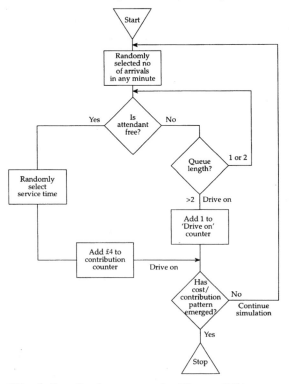

Simulation logic – example. Figure 15/4

The results of such a simulation could be entered on a simple worksheet as follows:

Minute	No. of arrivals	Queue length	Attend-ant	Attend-ant	Contri-bution earned	No. of unsatisfied customers
1						
2						
3						
4						
5						
6						
7						
8						
etc						

Simulation worksheet – for Example 2. Figure 15/5

Minute	No. of arrivals	Queue length	Attendant A	Contri-bution earned	No. of unsatisfied customers
1	0	0			
2	0	0			
3	0	0			
4	0	0			
5	1	0	engaged		
6	1	1	engaged		
7	0	1	engaged		
8	0	1	engaged		
9	0	1	engaged		
10	0	1	engaged		
11	1	2	engaged		
12	0	2	engaged		
13	1	2	engaged		1
14	0	2	engaged	£4	
15	0	1	engaged (and another random time selection made)		

Simulation worksheet with one attendant. Figure 15/6

Carrying out the simulation

17. Now that the objectives, variables, logic and so on have been established the simulation is carried out for a number of iterations each representing 1 minutes operations; first with 1 attendant, then 2 attendants, then 3 and so on until a cost/contribution pattern emerges.

As the flowchart, Figure 15/4, is worked through every minute, the number of arrivals would be randomly selected as in Table 2. If an attendant was free, a random selection of service time would be made and the appropriate number of minutes logged against the attendant. For example, in Table 2 an arrival occurred in minute 5. If the random selection of service time was, say 10 minutes, and the simulation was dealing with one attendant then that attendant would be engaged from minute 5 to minute 14.

In such circumstances the worksheet would be as Figure 15/6.

It will be apparent that such a simulation is simple to set up and use but becomes very tedious indeed to repeat for hundreds of iterations.

Results of simulation

18. The above simulation has been worked through for several days' operation with 1,2,3 and 4 attendants and the results obtained are tabulated below.

No. of attendants	Average contribution per day £	Attendant(s) cost per day £	Average no. of vehicles/day driving on
1	312	40	81
2	520	80	29
3	576	120	16
4	600	160	2

Results of Simulation. Table 3

From the table it will be seen that there is little difference in net profit per day between 2, 3 and 4 attendants, although there is of course a substantial difference in the average number of vehicles driving on. The results of a simulation do not necessarily indicate an optimal solution but provide more information upon which a reasoned decision can be taken.

Computers and simulation

19. To carry out any sort of realistic simulation, the use of a computer becomes a necessity. This is not because of any great complexity, but merely because of the large number of times that one has to work through the model. In the above example a computer was used to establish the table of results, but it did nothing more than could have been achieved using the model flowchart and the worksheets and a lot of patience. Because of the importance of computers in simulation, special simulation languages have been developed to facilitate the program writing and the use of the model. The GPSS (General Purpose Systems Simulator) exists for most computer systems. In addition there is the universal availability of spreadsheet programs that are able to carry out many types of practical simulations.

Advantages of simulation

20. a) Can be applied in areas where analytical techniques are not available or would be too complex.

    b) Constructing the model inevitably must involve management and this may enable a deeper insight to be obtained into a problem.

    c) A well constructed model does enable the results of various policies and decisions to be examined without any irreversible commitments being made.

    d) Simulation is cheaper and less risky than altering the real system.

Disadvantages of simulation

21. a) Although all models are simplifications of reality, they may still be complex and require a substantial amount of managerial and technical time.

    b) Practical simulation inevitably requires the use of computers. Although there is near universal ownership of machines, a considerable amount of additional expertise is required to obtain worthwhile results from simulation exercises. This expertise is not always available.

    c) Simulations do not produce optimal results. The manager makes the decision after testing a number of alternative policies. There is always the possibility that the optimum policy is not selected.

Summary

22. a) Simulation is the process by which a model is experimented upon and the results of various policies examined.

    b) At the heart of simulation is the concept of a model. A model is any representation of reality and business models can take the forms of flowcharts, formulae, equations, etc.

    c) A model must reflect reality and reality inevitably involves variable or probabilistic elements. Practical simulations must include these variations and this is sometimes known as Monte Carlo simulation.

    d) Because of the iterative nature of simulation the use of a computer is a necessity for all practical problems.

Points to note

23. a) Simulation has many potential business applications; these include: queueing problems, capital rationing problems in investment analysis, corporate planning.

    b) Because it is very time consuming to carry out a full simulation, it is thought unlikely that students would be required to do this. However, the model construction or some general question on the applicability of simulation is well within the scope of most syllabuses.

Self review questions *Numbers in brackets refer to paragraph numbers*

1.   Why is the concept of a model basic to the technique of simulation? (2 to 4)

2.   What are the major factors to be considered in constructing a model? (5)

3.   What is the essential feature of Monte Carlo Simulations? (7)

4.   What types of variables are found in simulation models? (9 to 13)

5.   What are the steps in constructing a simulation model? (14)

6.   What is the role of computers in carrying out simulations (18)

7.   What are the advantages and disadvantages of simulation? (19 and 20).

Assessment and revision
Chapters 10 to 15

Examination questions with answers

Answers commence on page 536

A1. The total revenue (TR) and total cost (TC) functions for a business (for output between 1,000 and 1,300 units per period) are as follows:

$$TR = \frac{-x^2}{2} + 1,500x$$

$$TC = 330x + 415,000$$

(where x is the level of output in units and TR and TC are in pounds £)

Required:

a) Give an expression for total profit

b) Using differential calculus:

  i) determine the expressions for marginal revenue and marginal profit

  ii) find the profit maximising output, and confirm that it is a maximum value.

c) Draw a graph showing marginal revenue and marginal cost for output between 1,000 and 1,300 units per period, clearly indicating the profit maximising output.

ACCA Management Information

A2. Mayfair Electrics plc, open for 300 days each year, is to start the manufacture of a new product for which the company estimates an annual demand of 25,000 units; this demand being at a constant rate throughout the year. The product contains a component that they can purchase from either or both of two wholesalers, Apex Ltd and Bemax Ltd. Assume that in variable cost per component. Apex Ltd are able to supply orders in any quantities at a cost of £8 per component plus a fixed delivery charge of £50 per order. Bemax Ltd will supply the components at a cost of £7.95 each and will not charge anything for delivery but will only deliver in lots of 5,000 or more.

Required:

a) i) Determine the minimum total annual cost associated with each of the two suppliers, Apex Ltd and Bemax Ltd.

  ii) Advise Mayfair Electrics as to the best source of supply out of these two suppliers if they wish to minimise the total annual cost, stating the order quantity, the number of deliveries per year, and the time between deliveries.

      iii) State the assumptions of the simple stock control model and assess their likely validity for this situation.

  b) State any changes that will be necessary if Apex Ltd were to restrict the number of components supplied to any one customer to 15,000 per year and determine the additional annual cost to Mayfair Electrics.

ACCA Decision Making Techniques

A3. Orion Products Ltd is a retailer of materials used in the building trade, operating for 50 weeks each year. One of its fastest selling products is bought in from an outside manufacture at a cost of £8.50 per unit. Forecasted weekly demand for this item is 4,800 units. A three week lead time is required to obtain the product from the manufacturer and Orion's current practice is to order 24,000 units at a time. This order is placed when the stock level falls to 20,000 units.

Orion's financial analysts have established a cost of capital of 15% per annum for the use of inventory decisions within the company. In addition, an analysis of the purchasing operation shows that approximately 15 hours are required to process and co-ordinate an order for the item regardless of the quantity ordered. Purchasing salaries average £10 per hour, including employee benefits. In addition, a detailed analysis of 40 orders showed that £1,425 was spent on paper, postage and telephone related to the ordering process. Also the cost of receiving delivery of an order is estimated to be £80.

Required:

a) Show that C, the cost per order, is £265.625.

b) Calculate the annual total cost of purchasing, ordering and storing the product using the current practice.

c) Calculate the order quantity that minimises annual total cost. Determine the annual saving in using this optimum quantity assuming that Orion Products still requires the same level of protection as currently given by the use of safety stock.

d) The manufacturer offers a discount of 5% if Orion Products order 100,000 or more units at a time. If they wanted to change to these terms it would require extra storage space which would cost £25,000 per annum. Orion Products would still maintain the same level of safety stock.

    Write a short report to the finance director of Orion Products, giving advice regarding the decision of whether or not to accept the discount. Include both financial information and any further factors that might influence the decision.

ACCA Decision Making Techniques

A4. a) Describe the advantages and disadvantages of using simulation to investigate business problems compared with the use of mathematical formulae.

    The North Star Company runs a large machine that contains three identical vacuum tubes which are a major cause of down time. The current practice is to replace these tubes when they fail. A proposal has been made that all three tubes be replaced whenever any one of them fails in order to reduce the frequency with which the equipment must be shut down. The objective is to compare these alternatives on a cost basis.

The vacuum tubes themselves cost £400 each and installation costs are £100 for replacing one tube, £160 for two and £200 for three. Down time costs are estimated at £900 per hour and the time to replace one, two or three tubes is 30 mins, 50 mins and 60 mins respectively.

Historical data indicates that the time to failure of this type of tube has a distribution as shown below:

Time to failure – weeks	7	8	9	10	11	12
Probability of failing	0.1	0.1	0.2	0.3	0.2	0.1

b) Using the information above and by simulating the operation of this machine over a period of one year, estimate the costs of maintenance under each of the following proposals:

   i)   replace each tube as it fails;

   ii) replace all three after any one fails.

   Use the following random digits

Random digits									
Tube 1	8	0	0	1	7	2	1	9	9	7
Tube 2	8	8	8	4	8	5	2	3	7	0
Tube 3	5	7	2	0	2	8	6	2	8	5

c) Why are estimates, found in (b), not very reliable and how would you improve them?

d) The production manager proposes that each tube should be replaced after it fails, but that each tube with eight or more weeks of operating time should be replaced at the same time.

   Use simulation techniques to test this proposal.

   ACCA Decision Making Techniques

A5. a) Describe the features of a continuous stocktaking and perpetual inventory system.

   b) A retailer sells a large variety of products in its chain of stores. The number of weeks forward stock cover for a sample of 40 products has been calculated. The sample mean and standard deviation are 3.64 and 0.68 weeks respectively.

Required:

a) Calculate the 99% confidence interval for the mean value of stockholding weeks for all products.

279

b) Calculate the sample size required in order to establish the mean stock-holding value to within ± 0.15 weeks with 95% confidence.

c) One of the retailer's products has:

maximum daily sales	150 units
average daily sales	120 units
maximum lead time	10 days
average lead time	8 days
economic order quantity	2,900 units

   Note: The re-order level is established so as to ensure no stock-outs.

d) Calculate the average stockholding (in units)

ACCA Management Information

A6. A company uses Material Z (cost £3.50 per kg) in the manufacture of Products A and B. The following forecast information is provided for the year ahead:

	Product A	Product B
Sales (units)	24,600	9,720
Finished goods stock increase by year end (units)	447	178
Post-production rejection rate(%)	1	2
Material Z usage (kg per completed unit, net of wastage)	1.8	3.0
Material Z wastage (%)	5	11

Additional information:

- Average purchasing lead time for Material Z is two weeks.
- Usage of Material Z is expected to be even over the year.
- Annual stock holding costs are 18% of the material costs.
- The cost of placing orders in £30 per order.
- The re-order level for Material Z is set at the average usage in average lead time plus 1,000 kg of safety (buffer) stock.

Required:

a) State two items that would be regarded as 'stock holding costs' and explain how they may be controlled effectively.

b) Calculate for the year ahead:

   (i) the required production of Products A and B (in units);

   ii) the total requirements for Material Z (in kg);

   iii) the Economic Order Quantity for Material Z (in kg).

c) Calculate the average stock investment (£) and the annual stock holding costs (£) for Material Z.

ACCA Management Information

Examination questions without answers

B1.   FRN Ltd is a small company which produces a single product – a light aircraft called the Rustler. The Rustler is similar in capability and performance to aircraft produced by a number of other manufacturers, although the Rustler does have certain features which make it particularly useful for certain special purposes. In a given three-month period, all customers buying Rustlers are charged a single uniform price.

The construction of the Rustler by FRN Ltd is mainly an assembly operation using bought-in components. Most of FRN Ltd's factory staff are engaged on terms which include a fixed basic wage, an hourly rate and overtime bonuses. It is not possible to increase factory size of equipment levels within a three-month period.

You have been engaged as a consultant to advise FRN Ltd's management on output and product pricing during the coming three-month period. During this period the maximum output is 30 Rustlers. Your initial research indicates that the revenues and variable costs associated with output of 1 Rustler and 30 Rustlers during the period are:

Output of Rustlers	1	30
	£	£
Revenues	100,000	2,134,000
Variable costs	30,000	2,497,000

You may assume that the relationships of revenues to output and of variable costs to output are curvilinear ones that may be described by formulae based on the equation

$$y = ax^n$$

Required:

a)   Draw a diagram in order to illustrate the behaviour of revenue and variable costs over the whole range of possible levels of output.

b)   Determine two mathematical formulae to describe the relationship between

  – output and revenues,

  – output and variable costs.

c)   Determine the level of output and the price per Rustler that will maximise profit during the period.

d)   Give likely reasons why revenues and variable costs behave in relation to output in the way they do.

CIMA Management Accounting Applications

B2.   The Aquarius Automobile Company purchases, from a supplier, a component used in the manufacture of generators. Aquarius's generator production operation requires 400 components each month. The costs associated with ordering and delivery (carriage, invoicing etc.) amount to £200 per order. Each unit costs £20 and the company's cost of capital is believed to be 15% p.a. The present policy of Aquarius is to order in lots of 600, eight times per year.

Required:

a) Determine the total annual cost to the firm of the present ordering policy for this component.

b) Evaluate the frequency and size of an ordering policy so that total costs are minimised. What savings are made by switching to this policy? State the assumptions necessary for these calculations.

c) The supplier of this component will offer a 5% discount off the standard price if the components are ordered in lots of 1,600 or more.

i) Would you advise the firm to amend their current ordering policy to take advantage of the discount?

What reservations might they have about taking up this discount?

ii) Aquarius Automobile Company are unsure of their current cost of capital.

How large would this need to be for the firm to be indifferent between taking advantage of the quantity discount and maintaining the order policy found in (b)? Comment on the sensitivity of the company's cost of capital in respect of this decision.

iii) The warehouse manager estimates that lots of 1,600 or more would require structural extensions costing £25,000.

How long would it be before any savings made by taking up the discount offer covered the cost of these extensions? Comment on your answer.

ACCA Decision Making Techniques

B3. a) What are the advantages of the simulation approach? Are there any disadvantages?

b) A department store, open for 300 days each year, has noticed that it has been unsuccessful in trying to predict the customer buying pattern of one of its products, ladies gloves, costing £10 per pair. They have also discovered that there is some variability in the amount of lead time of each new order they place with the manufacturer. These circumstances have often resulted in high inventory levels at certain times and stockouts at other times. Customer demand data is collected and is shown in the following table.

Daily demand (pairs of gloves)	Probability
0–2	0.2
3–7	0.3
8–12	0.4
13–19	0.1

The cost of placing each order for gloves is estimated to be £15. For each pair of gloves it is estimated that the annual storage cost is equal to 24% of its cost.

Required:

i) Determine the expected annual demand.

ii) Using the simple EOQ formula show that the optimum order quantity is approximately 165 pairs of gloves.

c) The lead time distribution is recorded in the table below.

Lead time (days)	Probability
1	0.1
2	0.3
3	0.4
4	0.2

Use the random digits below to simulate the inventory operation for a period of 10 days in order to estimate the total weekly stockholding costs. Start with an initial stock of 30, use the order quantity found in (b), a reorder level of 20 and assume that the goodwill cost of being out of stock is £1 per pair of gloves demanded but not available.

Random digits:

Daily demand:	3	8	7	1	6	0	9	1	4	9
Lead time:	5	2	8	3						

State any assumptions that you make.

d) Explain how this simulation method might be used to determine the optimum order quantity and reorder level.

ACCA Decision Making Techniques

16 Linear programming – introduction

Objectives

1. After studying this chapter you will

- have been introduced to the techniques of Linear Programming (LP);
- know how the express LP problems in the standardised manner;
- understand the meaning of the objective function and the constraints;
- know how to formulate both maximising and minimising problems;
- understand the distinction between constraints of the 'greater than or equal' or 'less than or equal' types.

LP and allocation

2. Allocation problems are concerned with the utilisation of limited resources to best advantage. It is clear that many management decisions are essentially resource allocation decisions and various techniques exist to help management in this area. LP is one technique and is closely related to Assignment and Transportation techniques covered in Chapters 20 and 21 which are also allocation techniques.

LP definition

3. LP is a mathematical technique concerned with the allocation of scarce resources. It is a procedure to optimise the value of some objective (for example, maximum profit or minimum cost) when the factors involved (for example, labour or machine hours) are subject to some constraints (for example, only 1000 labour hours are available in a week).

Thus LP can be used to solve problems which conform to the following.

a) The problem must be capable of being stated in numeric terms.

b) All factors involved in the problem must have linear relationships e.g. a doubling of output requires a doubling of labour hours; if one unit provides £10 contribution 10 units will produce £100 and so on.

c) The problem must permit a choice or choices between alternative courses of action.

d) There must be one or more restrictions on the factors involved. These may be restrictions on resources (labour hours, tons of material etc) but they may be on particular characteristics, for example, a fertiliser must contain a minimum of 15% phosphates and 30% nitrogen or a patent fuel must contain not more than 6% ash, 2% phosphorous and 1% sulphur.

Note: The 'linear' part of the term LP is explained above. The 'programming' part refers to the solution method. This is invariably carried out by an iterative process whereby one moves from one solution to a better solution progressively on until a solution is reached which cannot be improved upon, i.e. optimum. In this context therefore the term programming is not connected with computer programming.

Linearity requirement

5. A major factor in LP is the requirement that all relationships are linear. Obviously not all factors involved in business decisions are linear and indeed in some situations non-linear relationships are typical and desirable, e.g. a production process could be increased in size so that economies of scale are realised and unit costs reduced.

   Fortunately it is found that many factors (e.g. labour hours, machine utilisation, contributions) are reasonably linear or can be linearly approximated over the operational range of the values being considered. It will be realised that this is a similar approach to that adopted by accountants when considering marginal costing and P/V analysis. If it is considered that the factor is definitely non-linear then LP cannot be used.

Expressing LP problems

5. Before considering the detailed methods of solving LP problems it is necessary to be able to express a problem in a standardised manner, This not only helps the calculations required for a solution but also ensures that no important element of the problem is over-looked. The two major factors are

   the objectives and

   the limitations or constraints.

   Objectives: The first step in LP is to decide what result is required, i.e. the objective. This may be to maximise profit or contribution, or minimise cost or time or some other appropriate measure. Having decided upon the objective it is now necessary to state mathematically the elements involved in achieving this. This is called the *objective function*.

Example 1

A factory can produce two products, A and B. The contribution that can be obtained from these products are

A contributes £20 per unit, B contributes £30 per unit

and it is required to maximise contribution.

The objective function for this factory can be expressed as

$$Maximise\ 20x_1 + 30x_2$$

where x_1 = number of units of A produced

and x_2 = number of units of B produced

Notes: This problem has 2 unknowns, x_1 and x_2. These are sometimes known as the *decision variables*. Only a single objective at the time (in this case, to maximise contribution) can be dealt with in a basic LP problem.

Example 2

A farmer mixes three products to feed his pigs. Feedstuff **M** costs 20p per kilo, feedstuff **Y** costs 40p per kilo and feedstuff **Z** costs 55p per kilo. Each feedstuff contributes some essential part of the pigs diet and the farmer wishes to feed the pigs as cheaply as possible.

The objective function is

$$Minimise\ 20x_1 + 40x_2 + 55x_3$$

where x_1 = Number of kilos of **M**

x_2 = Number of kilos of **Y**

x_3 = Number of kilos of **Z**

Alternatively, if the costs were required in £'s, the objective function could be expressed as follows,

$$Minimise\ 0.2x_1 + 0.4x_2 + 0.55x_3$$

Note: This example has 3 decision variables. The number of decision variables can vary from 2 to many hundreds. For examination purposes 4 or 5 is the maximum that is likely to be encountered. Linearity has been assumed in both examples above and is assumed in all those that follow.

Limitations or constraints

6. Circumstances always exist which govern the achievement of the objective. These factors are known as *limitations* or *constraints*. The limitations in any given problem must be clearly identified, quantified, and expressed mathematically.

In a problem concerned with the allocation of scarce resources restrictions take the form:

Resources used ≤ Resources available

The resources used must be expressed in linear form and the resources available form part of the given data.

Limitation examples

7.

Example 3

A factory can produce four products and wishes to maximise contribution. It has an objective function as follows:

$$\textit{Maximise } 5.5x_1 + 2.7x_2 + 6.0x_3 + 4.1x_4$$

where x_1 = number of units of A produced

x_2 = number of units of B produced

x_3 = number of units of C produced

x_4 = number of units of D produced

and the coefficients of the objective function (i.e. 5.5, 2.7, 6.0 and 4.1) are the contributions per unit of the products.

The factory employs 200 skilled workers and 150 unskilled workers and works a 40 hour week. The times to produce 1 unit of each product by the two types of labour are given below

	Products			
	A	**B**	**C**	**D**
Skilled hours	5	3	1	8
Unskilled hours	5	7	4	11

The limitations as regards to labour can be stated as follows

Skilled:	$5x_1$	+	$3x_2$	+	x_3	+	$8x_4$	$\leq$	8,000
Unskilled:	$5x_1$	+	$7x_2$	+	$4x_3$	+	$11x_4$	$\leq$	6,000

In addition a general limitation applicable to all maximising problems is that it is not possible to make negative quantities of a product, i.e.

$$x_1 \geq 0, \quad x_2 \geq 0, \quad x_3 \geq 0, \quad x_4 \geq 0$$

Notes:

a) The resource limitations in this maximising problem follow a typical pattern being of the less than or equal to type ($\leq$).

b) The formal statement of the non-negativity constraints on the unknowns (x_1, x_2 etc) has to be made for computer solutions but is normally inferred when solving by manual means.

c) This above restriction applies to labour hours: Machine hour restrictions would be dealt with in a similar fashion.

16 Linear programming – introduction

Example 4

A company produces four products **P, Q, R, S,** which are made from two basic materials, Sludge and Slurry. Only 1,000 tons of Sludge and 800 tons of Slurry are available in a period. The usage of the materials in the products is as follows.

Product		P	Q	R	S	
Materials	Sludge	2.6	1	5	0	tons/unit
	Slurry	3.3	4	0	6.2	tons/unit

$\therefore$ Limitations are as follows:

$$\text{Sludge restriction} \quad 2.6x_1 + x_2 + 5x_3 \leq 1{,}000$$
$$\text{Slurry restriction} \quad 3.3x_1 + 4x_2 + 6.2x_4 \leq 800$$

$$x_1 \geq 0, \quad x_2 \geq 0, \quad x_3 \geq 0, \quad x_4 \geq 0$$

$$\text{where } x_1 = \text{number of units of } \mathbf{P}$$
$$x_2 = \text{number of units of } \mathbf{Q}$$
$$x_3 = \text{number of units of } \mathbf{R}$$
$$x_4 = \text{number of units of } \mathbf{S}$$

Example 5

Sales limitations, because of quota and/or contract requirements, may be of the 'greater than or equal to' variety ($\geq$) or 'less than or equal to' ($\leq$).

A company is in a trading federation and has agreed to accept sales quotas on certain of its products. It has also contracted with some regular customers to supply minimum quantities of certain products. The details are as follows:

Quotas (as agreed with the Sales Federation)

	Product					
	A	B	C	D	E	
Maximum sales	500	1,000	No limit	2,000		Units/period
Contracts with regular customers						
Minimum quantities that must be supplied	–	300	400	–		Units/period

$\therefore$ Limitations are as follows

Quota constraints	x_1				≤ 500
Quota constraints		x_2			$\leq 1{,}000$
Quota constraints				$x_4 + x_5$	$\leq 2{,}000$
Contract constraints		x_2			≥ 300
Contract constraints			x_3		≥ 400

$$x_1 \geq 0, \quad x_4 \geq 0, \quad x_5 \geq 0.$$

where x_1 to x_5 are the quantities of **A** to **E** respectively to be produced.

Notes:

a) The usual non-negativity constraints are not required for x_2 and x_3 because the contract restrictions are of the greater than or equal to type.

b) Example 3,4 and 5 above are maximising problems where the restrictions are generally (but not exclusively) or the 'less than or equal to' type.

Minimisation problems

8. Minimisation problems (i.e. as Example 2) are normally concerned with minimising the cost of fulfilling some objective, subject to limitations. The limitations are generally of the 'greater than or equal to' type ($\geq$) for the the logical reason that the best way to minimise costs would be to produce nothing, therefore the limitations must be set so as to produce the minimum possible that fulfills certain requirements.

Example 6

A farmer feeds his pigs a mixture of swill, vitamins and a proprietary brand of feedmix. He owns 100 pigs who eat at least 20 kilogrammes of food per day each. He wishes to minimise the cost of feeding the pigs whilst at the same time ensuring that the animals receive a balanced diet. The following dietary and cost factors have been obtained.

	Calories	Vitamins 1	2	3	Costs
Minimum Daily Dietary Requirements/Pig	40	20	10	30	
Contents of foodstuffs					**Costs**
Swill/kg (x_1)	1.5	0.5	–	–	5p/kg
Feed mix/kg (x_2)	2.0	0.5	–	1	10p/kg
Vitamins/bottle (x_3)	–	0.5	7	14	20p/bottle

The LP formulation can be stated as follows:

Objective function: Minimise $5x_1 + 10x_2 + 20x_3$

Constraints:

Total Weights (i.e. 100 × 20)	$1.0x_1$	+	$1.0x_2$			$\geq$	2,000
Calories (i.e. 100 × 40)	$1.5x_1$	+	$2.0x_2$			$\geq$	4,000
Vitamin 1 (i.e. 100 × 20)	$0.5x_1$	+	$0.5x_2$	+	$5x_3$	$\geq$	2,000
Vitamin 2 (i.e. 100 × 10)					$7x_3$	$\geq$	1,000
Vitamin 3 (i.e. 100 × 30)			$1.0x_2$	+	$14x_3$	$\geq$	3,000

$$x_1 \geq 0, \quad x_2 \geq 0$$

where $x_1 =$ kilogrammes of swill

$x_2 =$ kilogrammes of feedmix

$x_3 =$ bottles of vitamins

Notes:

a) The restrictions are all of the 'greater than or equal' type ($\geq$) which is typical of minimisation problems.

b) In the total weight restriction the weight of the vitamins has been assumed to be negligible and has been ignored.

c) The non-negativity constraint for x_3 is not necessary in this problem because the Vitamin 2 constraint is for x_3 only and is of the 'greater than or equal to' type.

Development of the LP model

9. The basic approach to LP outlined in this chapter can be developed further using more advanced techniques. These include, for example, models which can deal with multiple objectives, the inclusion of fixed costs in the formulation and so on.

However, at the level of the syllabuses at which this book are aimed, fixed costs can generally be ignored so would not be included in the formulation. Only one objective at a time will be considered and linearity is assumed in all aspects of the model. One consequence of this is the use of *contribution* (i.e. sales less marginal costs) instead of profit in the objective function. This avoids the inclusion of fixed costs in the formulation and maintains linear relationships.

Summary

10. a) LP is a solution method to problems where an objective has to be optimised subject to constraints.

b) All factors concerned must be numeric and there must be linear relationships.

c) Before attempting to solve any LP problem it should be formulated in a standardised manner. The steps in the process are:

i) Decide upon the objective.

ii) Calculate the contribution per unit for maximising or the cost per unit for minimising (ignoring fixed costs in both circumstances).

iii) State the objective function noting how many unknowns, or decision variables, appear.

iv) Consider what factors limit or constrain the quantities to be produced or purchased.

v) State these factors mathematically taking care to ensure that the inequalities (i.e. $\geq$ or $\leq$) are of the correct type.

vi) Before attempting to solve the problem ensure that all the relationships established are linear or can be reasonably approximated by a linear function.

vii) Solve the problem by either drawing a graph (dealt with in Chapter 17) or by the use of what is known as the Simplex method (dealt with in Chapter 18).

Points to note

11. a) LP is an important practical technique with a wide variety of applications.

b) A substantial proportion of marks in examination questions is given for the formulation of the problem so this chapter will repay careful study.

c) In practice the sensitivity of a solution is a critical factor. An optimal solution may hold for only a very narrow range of constraint values and would thus be termed *sensitive*, alternatively it may hold good over a wide range of values and would thus be termed *robust*. The examination over which range of values the solution remains valid is termed sensitivity analysis. This is dealt with in later chapters.

Self review questions *Numbers in brackets refer to paragraph numbers*

1. What are the essential characteristics of problems than can be solved by LP methods? (3 and 4)

2. What is the 'objective function'? (5)

3. How many objectives can be dealt with at one time in an LP problem? (5)

4. What are limitations or constraints? (6)

5. Distinguish minimising from maximising problems. (8)

Exercises with answers

1. A factory produces four products **A, B, C** and **D** which earn contributions of £20, £25, £12 and £30 per unit respectively. The factory employs 500 workers who work a 40 hour week. The hours required for each product and the material requirements are set out below:

	Products			
	A	**B**	**C**	**D**
Hours per unit	6	4	2	5
Kgs Material **X** per unit	2	8.3	5	9
Kgs Material **Y** per unit	10	4	8	2
Kgs Material **Z** per unit	1.5	–	2	8

The total availability of materials per week is:

X	100,000 kgs
Y	65,000 kgs
Z	220,000 kgs

The company wish to maximise contribution:

Formulate the LP problem in the standard manner.

2. Green Bakeries Ltd have received a rush order from Camfam for high protein biscuits for famine relief. Costs must be minimised and the mix must meet minimum nutrition requirements.

The order will require 1,000 kgs of biscuits mix which is made from four ingredients R, S, T, U which cost £8, £2, £3 and £1 per kilogram respectively. The batch must contain a minimum of 400 kilos of protein, 250 kilos of fat, 300 kilos of carbohydrate and 50 kilos of sugar. The ingredients contain the following percentages by weight:

	Protein	Fat	Carbohydrate	Sugar	Filler
R	50%	30%	15%	5%	0
S	10%	15%	50%	15%	10%
T	30%	5%	30%	30%	5%
U	0	5%	5%	30%	60%

Only 150 kilos of **S** and 200 kilos of **T** are immediately available.

Formulate the LP problem in the standard manner.

Answers to exercises

1. Maximise $\quad$ **20A + 25B + 12C + 30D**

Subject to:	6A	+	4B	+	2C	+	5D	$\leq$	20,000
	2A	+	8.3B	+	5C	+	9D	$\leq$	100,000
	10A	+	4B	+	8C	+	2D	$\leq$	65,000
	1.5A	+			2C	+	8D	$\leq$	220,000

$$A, B, C \text{ and } D \geq 0$$

Where **A, B, C** and **D** represent the number of units of the Products A, B, C and D.

2. Minimise $\quad$ **8R + 2S + 3T + U**

Subject to:	R	+	S	+	T	+	U	$\geq$	1,000
	0.5R	+	0.1S	+	0.3T			$\geq$	400
	0.3R	+	0.15S	+	0.05T	+	0.05U	$\geq$	250
	0.15R	+	0.5S	+	0.3T	+	0.05U	$\geq$	300
	0.05R	+	0.15S	+	0.3T	+	0.3U	$\geq$	50

$$R, S, T, U \geq 0$$

Where **R, S, T** and **U** represent the number of kilograms of the ingredients R, S, T and U.

17 Linear programming – graphical solutions

Objectives

1. After studying this chapter you will
   - know that LP problems with 2 unknowns can be solved graphically;
   - be able to graph constraints of both types;
   - understand what is meant by the feasible region;
   - know how to identify the optimum solution;
   - understand the way that minimising problems can be solved graphically;
   - be able to find the shadow or dual prices of the binding constraints;
   - know how to invert an LP problem;
   - understand the principles of sensitivity analysis.

Graphical LP solution

2. Graphical methods of solving LP problems can only be used for problems with *two* unknowns or decision variables. Problems with *three or more* unknowns must be solved by techniques such as the simplex method (dealt with in Chapter 18). Graphical methods are the simplest to use and should be used wherever possible.

   a) Limitations. Graphical methods can deal with any number of limitations but as each limitation is shown as a line on a graph a large number of lines may make the graph difficult to read. This is rarely a problem in examination questions.

   b) Types of problems and limitations. Both maximisation and minimisation problems can be dealt with graphically and the method can also deal with limitations of the 'greater than or equal to' ($\geq$) type and the 'less than or equal to' ($\leq$) type.

   c) Graphical example. The method of solving LP problems graphically will be described step by step using the following maximising example as a basis.

Example 1

A manufacturer produces two products, Klunk and Klick. Klunk has a contribution of £3 per unit and Klick £4 per unit. The manufacturer wishes to establish the weekly production plan which maximises contribution.

Production data are as follows:

	Per unit		
	Machining (Hours)	Labour (Hours)	Material (kgs)
Klunk	4	4	1
Klick	2	6	1
Total available per week	100	180	40

Because of a trade agreement, sales of Klunk are limited to a weekly maximum of 20 units and to honour an agreement with an old established customer at least 10 units of Klick must be sold per week.

Step 1: Formulate the LP model in the standardised manner described in Chapter 16 thus:

$$\text{Maximise } 3x_1 + 4x_2$$

Subject to constraint

A	$4x_1$	$+$	$2x_2$	$\leq$	100	(Machining hours constraint)
B	$4x_1$	$+$	$6x_2$	$\leq$	180	(Labour hours constraint)
C	x_1	$+$	x_2	$\leq$	40	(Materials constraint)
D	x_1			$\leq$	20	(Klunk sales constraint)
E			x_2	$\geq$	10	(Klick sales constraint)

$$x_1 \geq 0$$

where $x_1 =$ number of units of Klunk

$x_2 =$ number of units of Klick

Notes:

As it is impossible to make negative quantities of the products it is necessary formally to state the non-negativity constraint (i.e. $x_1 \geq 0$).

The resource and sales constraints include both types of restrictions (i.e. $\geq$ and $\leq$).

As this is a problem with only 2 unknowns (i.e. x_1 and x_2) it can be solved graphically. The number of limitations does not exclude a graphical solution.

Step 2: Draw the axes of the graph which represent the unknowns, x_1 and x_2 thus

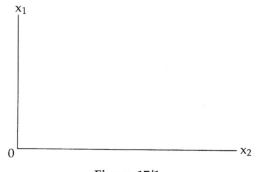

Figure 17/1

Note: The scales of the axes are best determined when the lines for the limitations are drawn. Each axis must start at zero and the scale must be constant (i.e. linear) along the axis but it is not necessary for the scales on both axes to be the same.

Step 3: Draw each limitation as a separate line on the graph.

Sales Limitations. These normally only affect one of the products at a time and in Example 1 the sales restrictions were

$$x_1 \leq 20 \text{ and } x_2 \geq 10$$

These limitations are drawn on the graph as follows:

The Klunk sales constraint (i.e. $x_1 \leq 20$).

Constraint D

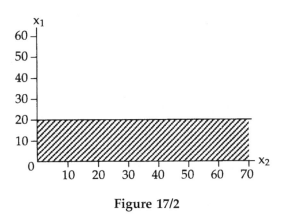

Figure 17/2

Note: The horizontal line represents $x_1 = 20$ and the hatched area below the line represents the area containing all the values less than 20.

The Klick sales constraint (constraint E) (i.e. $x_2 \geq 10$) is now entered thus:

295

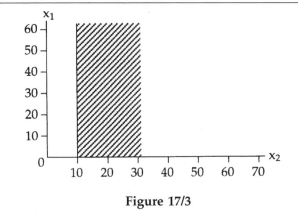

Figure 17/3

Note: The vertical line represents $x_2 = 10$ and the hatched area to the right of the line represents the area containing all values greater than 10. These two sales limitations can be shown on the same graph thus

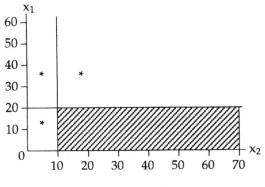

Figure 17/4

Note: The hatched area represents the area of possible production (i.e. which does not violate the constraints drawn) and is called the *feasible region*. The areas on the graph marked * violate one or both of the constraints.

Production and material limitations

3. In a similar fashion to above, the other restrictions should be drawn on the graph. Because these restrictions involve BOTH unknowns they will be sloping lines on the graph and not horizontal or vertical lines like the sales restrictions.

The three remaining restrictions are all of the same type and are dealt with as follows:

The machining constraint, constraint A, $4x_1 + 2x_2 \leq 100$ is drawn on the graph as $4x_1 + 2x_2 = 100$.

$$\text{Therefore when } x_1 = 0, x_2 = 50 \quad \left(\text{i.e.} \frac{100}{2}\right)$$

$$\text{and} \quad \text{when } x_2 = 0, x_1 = 25 \quad \left(\text{i.e.} \frac{100}{4}\right)$$

and so a line can be drawn from 25 on the x_1 axis to 50 on the x_2 axis thus

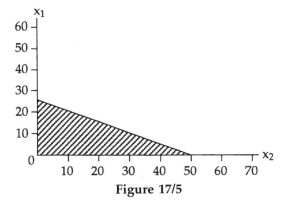

Figure 17/5

Note: As previously, the hatched area represents the area containing the 'less than' values. The other constraints are dealt with the same manner.

i.e. The labour constraint B $4x_1 + 6x_2 \leq 180$

is drawn on the graph as $4x_1 + 6x_2 = 180$

Therefore when $x_1 = 0, x_2 = 30$ $\left(\text{i.e.} \dfrac{180}{6}\right)$

and when $x_2 = 0, x_1 = 45$ $\left(\text{i.e.} \dfrac{180}{4}\right)$

and so a line can be drawn from 45 on the x_1 axis to 30 on the x_2 axis.

The materials constraint, (c), $x_1 + x_2 \leq 40$, is drawn on the graph as $x_1 + x_2 = 40$

Therefore when $x_1 = 0, x_2 = 40$ and when $x_2 = 0, x_2 = 40$ and so a line can be drawn from 40 on the x_1 axis to 40 on the x_2 axis.

All of the constraints (sales, production and material) can now be drawn on a single graph and the resulting *feasible region* defined.

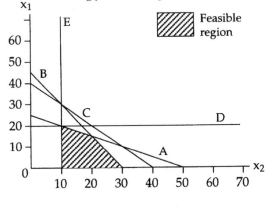

Figure 17/6

Notes:

a) The *feasible region* is the area which does not contravene any of the restrictions and is therefore the area containing all possible production plans.

b) The non-negativity restrictions (i.e. $x_1 \geq 0$, $x_2 \geq 0$) are automatically included in the graph because the graph quadrant used Figure 17/1 only shows positive values. It should be noted that as more restrictions are plotted the feasible region usually becomes smaller.

c) It will be noted that the material constraint C (line 40, 40) does not touch the feasible region. This is an example of a **redundant** constraint, i.e. it is non binding.

Step 4: Now that the feasible region has been defined it is necessary to find the point in or on the edge of the feasible region that gives the maximum contribution which, it will be recalled, is the specific objective.

This is done by plotting lines representing the *objective function* and thereby identifying the point in the feasible region which lies on the maximum value objective function line that can be drawn. These objectives function or contribution lines are straight lines representing different combinations of Klunk and Klick which yield the same contribution. For example,

> 20 units of Klunk and zero units of Klick yield £60 contribution.
> 12 units of Klunk and 6 units of Klick yield £60 contribution.
> 8 units of Klunk and 9 units of Klick yield £60 contribution.
> zero units of Klunk and 15 units of Klick yield £60 contribution, etc.

Very many other contribution lines could be drawn and if a number of these lines were drawn on the graph it would be noticed that:

a) They are parallel to each other with the same slope, which is determined by the relative contribution of the products.

b) The further to the right they are drawn the higher value of contribution they represent.

It therefore follows that the contribution line **furthest to the right but still touching the feasible region** shows the optimum production plan to provide the maximum possible contribution, thus

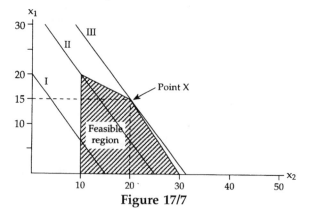

Figure 17/7

Optimum solution at point X, i.e. **15 units of x_1 and 20 units of x_2** yielding a contribution of **£125** (i.e. £3 × 15 + 4 × 20).

Notes:

a) The lines marked I to III are three of the many contribution lines that could be drawn and represent the following contributions.

$$I \ \ = 3x_1 + 4x_2 = \ \ 60$$
$$II \ = 3x_1 + 4x_2 = \ \ 90$$
$$III = 3x_1 + 4x_2 = 125$$

The contribution line has a slope of $\frac{4}{3}$ which is the ratio of the coefficients of x_1 and x_2. The intercept on, say, the x_1 axis is found by dividing the contribution by the x_1 coefficient, and vice versa. The intercepts for line II in Figure 7, for example, are as follows:

$$\text{Intercept on } x_1 \text{ axis} = \frac{\text{contribution}}{x_1 \text{ coefficient}} = \frac{90}{3} = \underline{\underline{30}}$$

$$\text{Intercept on } x_2 \text{ axis} = \frac{90}{4} = \underline{\underline{22.5}}$$

b) Only parts of lines I, II and III are feasible but as we require to maximise contribution we are only interested in Point X where line III touches the feasible region. It will be noted that the optimum is at a vertex or corner of the feasible region. This is ALWAYS the case.

c) Various contribution lines have been drawn on Figure 7 for instructional purposes. For examinations it is sufficient to draw only the contribution line representing the optimum position, i.e. in the example above, line III.

d) The contribution lines are sometimes termed **iso-profit** lines.

e) A simple way to check your answer is actually possible is to insert the values of the unknowns in the constraints and check whether the constraints are satisfied, e.g. the optimum solution of Example 1 found from Figure 17/7 is,

$$x_1 = \textbf{15 units}$$

$$x_2 = \textbf{20 units}$$

These values can be inserted into the constraints thus,

Constraint A	(4 × 15) + (2 × 20)	=	100	Constraint satisfied, no spare
Constraint B	(4 × 15) + (6 × 20)	=	180	Constraint satisfied, no spare
Constraint C	(1 × 15) + (1 × 20)	=	35	Constraint satisfied, 5 below maximum
Constraint D	Sales of x_1	=	15	Constraint satisfied, 5 below maximum
Constraint E	Sales of x_2	=	20	Constraint satisfied, 10 above minimum

It will be noted that the two constraints which intersect at the optimum vertex (see Figures 17/6, 17/7) are constraints A and B. These are the only constraints fully satisfied with no spare values. This is a general rule. They are known as **binding constraints**.

Minimisation example

4. Provided that they only have two unknowns, minimisation problems can also be dealt with by graphical means. The general approach of drawing the axes with appropriate scales and inserting lines representing the limitations is the same as for maximising problems but the following differences between maximising and minimising problems will be found.

    a) Normally in a minimising problem the limitations are of the greater than or equal to type ($\geq$) so that the feasible region will be *above* all or most of the limitations.

    b) The normal objective is to minimise cost so that the objective function line(s) represent cost and because the objective is to *minimise* cost the optimum point will be found from the **cost line furthest to the left which still touches the feasible region**, i.e. the converse of the method used for maximising problems.

Example 2

A manufacturer is to market a new fertiliser which is to be a mixture of two ingredients A and B. The properties of the two ingredients are:

	Ingredients analysis				
	Bone meal	**Nitrogen**	**Lime**	**Phosphates**	**Cost/kg**
Ingredient A	20%	30%	40%	10%	1.2p
Ingredient B	40%	10%	45%	5%	0.8p

It has been decided that

    a) the fertiliser will be sold in bags containing 100 kgs

    b) it must contain at least 15% nitrogen

    c) it must contain at least 8% phosphates

    d) it must contain at least 25% Bone Meal

The manufacturer wishes to meet the above requirements at the minimum cost possible.

Solution

Because of the basic similarity between graphing minimising and maximising problems (except for the two differences mentioned above) the detailed intermediate steps given in Example 1 will not be repeated.

The problem in standardised format is as follows

Objective function

$$\textit{Minimise } 12x_1 + 8x_2 \text{ (cost expressed in tenths of a penny)}$$

subject to

Constraint							
A	x_1	+	x_2	=	100	(Weight constraint)
B	$0.3x_1$	+	$0.1x_2$	$\geq$	15	(Nitrogen constraint)
C	$0.1x_1$	+	$0.05x_2$	$\geq$	8	(Phosphates constraint)
D	$0.2x_1$	+	$0.4x_2$	$\geq$	25	(Bone Meal constraint)

$$x_1 \geq 0 \text{ and } x_2 \geq 0$$

where x_1 = kgs of ingredient A

x_2 = kgs of ingredient B

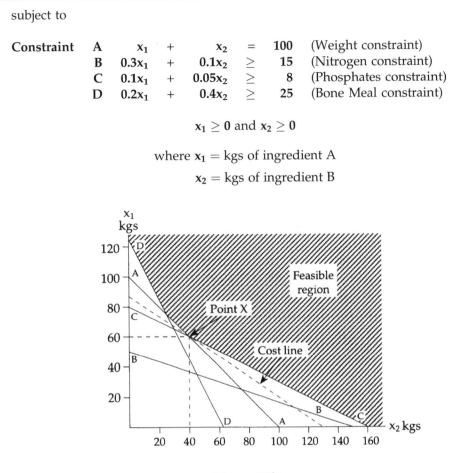

Figure 17/8

Notes on Figure 17/8:

Optimum solution at point X i.e. 60 kgs of x_1 and 40 kgs of x_2 giving a cost of

$$1.2 \times 60 + .8 \times 40 = \text{£1.04 per 100 kilo bag}$$

a) Only one cost line ($12x_1 + 8x_2 = 1{,}040$ in tenths of a penny) has been drawn so as to show the optimum position.

b) Each restriction on the graph is labelled A, B, C, D and can be cross referenced to those stated in the standardised format above.

c) The optimum position in the feasible region is the furthest *point to the left* touched by the cost line. This is because it is a *minimising* problem. Note again that the optimum is at a corner or vertex of the feasible region.

Using simultaneous equations

5. An alternative method of finding the solution values of the constraints at any

intersection of the graph, including the optimal vertex, is to solve the simultaneous equations of the relevant binding constraints. To illustrate this, the answers to the maximising problem in Example 1 and the minimising problem in Example 2 will be recalculated using simultaneous equations.

Reworking Example 1

The two constraints which intersected at optimum were constraints A and B (see Figures 17/6, and 17/7) thus:

$$\text{Constraint A} \quad 4x_1 + 2x_2 = 100$$
$$\text{Constraint B} \quad 4x_1 + 6x_2 = 180$$

Solving by deducting A from B gives

$$4x_2 = 80$$
$$\therefore x_2 = 20$$

and by substitution, $x_1 = 15$, which confirms the answer found by reading off the intercepts of the graph.

Reworking example 2

The intersecting constraints were A and C (see Figure 17/8) thus

$$\text{Constraint A} \qquad x_1 + x_2 = 100$$
$$\text{Constraint C} \qquad 0.1x_1 + 0.05x_2 = 8$$
$$\therefore A - 10C = \qquad 0.5x_2 = 20$$
$$\therefore x_2 = \underline{\underline{40}}$$

and substituting, $x_1 = 60$, again confirming the results from the graph.

If it was known what constraints appeared at the optimum intersection before the graph was drawn, then the problem could be solved by simultaneous equations without having to draw a graph. However this is unlikely to occur but it is sometimes useful to calculate the exact solution values using the equations, rather than use the approximate values obtainable from a graph.

The valuation of scarce resources

6. The graphing of an LP problem not only provides the optimal answer but also identifies the binding constraints, alternatively called the limiting factors. For example, in example 1 machining hours and labour hours are the binding constraints, the shortage of which limit further production and profit. Not all factors in a given problem are limiting factors. For instance in example 1, Constraints C, D and E are not at their maxima and therefore are not scarce or limiting.

It is important management information to value the scarce resources. These valuations are known as the *dual prices* or *shadow prices* and are derived from the amount of the increase (or decrease) in contribution that would arise if one more (or one less) unit of the scarce resource was available. Only a scarce resource can have a positive dual price and the calculated price assumes that there is only a marginal increase or decrease in the availability of the scarce resource and that all other factors are held constant.

Finding the shadow or dual prices

7.   There are two methods of calculating these prices – the arithmetic method and the dual formulation – and both are illustrated using the data and results from Example 1, reproduced below:

Original problem

$$\text{maximise} \quad 3x_1 \quad + \quad 4x_2$$

subject to

Constraint A	$4x_1$	+	$2x_2$	$\leq$	100	machining hours
Constraint B	$4x_1$	+	$6x_2$	$\leq$	180	labour hours
Constraint C	x_1	+	x_2	$\leq$	40	materials
Constraint D	x_1			$\leq$	20	sales
Constraint E			x_2	$\geq$	10	sales

where x_1 = units of klunk

x_2 = units of klick

The solution was: Produce $15x_1$ and $20x_2$ giving a contribution of £125. Constraints A and B are binding.

The problem now is to find the shadow prices of the two binding constraints, machine hours and labour hours, i.e. what is the valuation of one more (or less) machine hour and one more (or less) labour hour?

Arithmetic method of finding shadow prices

8.   Dealing first with machine hours, we assume that 1 more machine hour is available (but labour hours are constant at 180) and calculate the resulting difference in contribution, thus:

The binding constraints become

Machine hours	$4x_1 + 2x_2 = 101$	(i.e. original $100 + 1$)
Labour hours	$4x_1 + 6x_2 = 180$	(unchanged)

Solving these simultaneous equations, new values for x_1 and x_2 are obtained:

$x_1 = 15.375$ and $x_2 = 19.75$ and substituting into the objective function gives a new contribution.

$$3(15.375) + 4(19.75) = \text{£125.125}$$
$$\text{Original contribution} = \underline{\text{£125}}$$
$$\text{Difference} = \underline{\underline{\text{£0.125}}}$$

Thus 1 extra machine hour has resulted in an increase in contribution of £0.125 which is the shadow price per machining hour.

A similar process for labour hours is shown below:

New constraints with extra labour hour (but machine hours constant at 100).

Machine hours	$4x_1 + 2x_2 = 100$	
Labour hours	$4x_1 + 6x_2 = 181$	

and solving gives, $x_1 = \textbf{14.875}$ and $x_2 = \textbf{20.25}$

$$\text{New contribution} = 3(14.875) + 4(20.25) = \text{£125.625}$$
$$\text{Original contribution} = \underline{\text{£125}}$$
$$\text{Difference} = \underline{\underline{\text{£0.625}}}$$

$\therefore$ shadow price per labour hour = **£0.625**

The two values obtained can be verified by showing that the quantities of the binding constraints at their shadow price valuations do produce the contribution of £125 shown in the optimum solution thus:

$$100(\text{£0.125}) + 180(\text{£0.625}) = \underline{\underline{\text{£125}}}$$

Notes:

a) Similar results would be obtained in each case if 1 less hour had been used in the calculations. Verify this yourself.

b) The shadow prices calculated above only apply whilst the constraint is binding. If, for example, more and more machining hours became available there would eventually be so many machining hours that they would no longer be scarce and some other constraint would become binding. This point is developed further in Para 12.

Dual formulation method for shadow prices

9. Every LP problem has an *inverse* or *dual* formulation. If the original problem, known as the *primal problem*, is a maximising one then the dual formulation is a minimising one and vice versa. Thus, as the Example 1 primal formulation is a maximising problem, its dual is the minimising problem. Solving the dual problem gives the shadow prices of the binding constraints, hence the alternative term, dual prices.

The stages in finding and solving the dual for the solution to Example 1 are shown below:

The relevant parts of the original full formulation which appear in the solution are:

maximise	$3x_1$	+	$4x_2$			(objective function)
subject to	$4x_1$	+	$2x_2$	$\leq$	100	(machine hours)
	$4x_1$	+	$6x_2$	$\leq$	180	(labour hours)

(Constraints C, D, E are non-binding so do not appear in the solution.)

The minimising dual problem is formed by inverting the above formulation, i.e. making the columns into rows and the rows into columns thus:

Dual formulation

minimise	100M	+	180L			(i.e. originally the constraint column)
subject to	4M	+	4L	$\geq$	3	(originally the x_1 column)
	2M	+	6L	$\geq$	4	(originally the x_2 column)

What are now the constraints can be solved by simultaneous equations:

$$4M + 4L = 3$$

$$2M + 6L = 4$$

Solving gives **M = 0.125 and L = 0.625**, which will be recognised as the valuations already calculated in para. 8.

If these dual prices are inserted into the objective function of the dual, exactly the same value is obtained as in the primal problem.

$$\text{i.e. } 100\ (0.125) + 180\ (0.625) = \underline{\underline{£125}}$$

This result is identical to the Primal problem and this will always be so.

Notes:

a) The dual formulation above is a contraction of the full dual formulation. This is possible here because the problem had already been solved and it was thus known that three of the constraints (C, D and E) were non-binding.

b) The dual formulation is developed further in Chapter 19.

Interpretation of shadow prices

10. The shadow price of a binding constraint provides valuable guidance because it indicates to management the extra contribution they would gain from increasing by one unit the amount of the scarce resource. As an example, the shadow price of labour hours calculated above is £0.625 per hour. This means that management would be prepared to pay up to £0.625 per hour extra in order to gain more labour hours. If the current labour cost was £8 per hour then management would be prepared to pay up to £8.625 per hour for extra labour hours; perhaps from overtime working.

It is important to realise that only binding constraints have shadow or dual prices. Those that are not binding have zero shadow prices. This accords with common sense for there would be little point in paying more to increase the supply of a resource of which you already have a surplus. In example 1, constraints C, D and E have zero shadow prices.

Sensitivity analysis

11. This is the process of changing the values and relationships within the problem and observing the effect on the solution. The general aim is to discover how sensitive is the optimal solution to the changes made. With two variable problems i.e. those that can be solved graphically, sensitivity analysis can be visualised as the process of altering the angle, or distance from the origin, of the the lines on the graph.

    As an example, it was mentioned earlier that if more and more machine hours became available there would come a point where machine hours would cease to be a restriction. This is, incidentally, the range over which the shadow price of £0.125 applies. Assume that management wish to know how many extra machine hours are necessary before machine hours cease to be a binding constraint and also the same information relating to labour hours.

Number of extra machine hours

If Figure 17/9 is examined it will be seen that the original solution to Example 1 is reproduced together with new constraints labelled (A) and (B) shown by dotted lines.

Dealing first with A, the machine hour constraint, the dotted line is drawn parallel to the original constraint and up to the intersection of constraints B and D, where (A) ceases to be binding as it is on the edge of the new feasible area.

The BD intersection Point M on the graph is then solved using simultaneous equations in the normal way.

$$4x_1 + 6x_2 = 180 \quad \text{(Constraint B)}$$

$$x_1 = 20 \quad \text{(Constraint D)}$$

Solving gives $x_1 = 20$ and $x_2 = 16\frac{2}{3}$, which is the new production plan assuming that machine hours are increased. At this point $113\frac{1}{3}$ machine hours are required, i.e. $4(20) + 2(16\frac{2}{3}) = 113\frac{1}{3}$.

The original number of machine hours was 100 so that only $13\frac{1}{3}$ extra machine hours are necessary for the machine hour constraint to be non-binding. The $13\frac{1}{3}$ hours is also the range over which the calculated shadow price of £0.125 applies.

Number of extra labour hours

This follows a similar pattern with the intersection being where constraint C joins the x_2 axis, Point L. Equations are not required here as the new optimum output is clearly 40 units of x_2 and nil x_1.

Substituting in the labour hour constraint gives

$$4(0) + 6(40) = 240$$

Thus the extra labour hours are 60 (i.e. $240 - 180$) and the shadow price of £0.625 would apply over the range from 180 hours to 240 hours.

Note: Although not shown on Figure 17/9 for clarity, changing the constraints as done above will cause the feasible region to change.

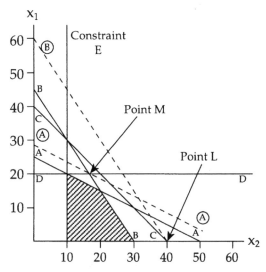

Sensitivity analysis on Example 1. Figure 17/9

Further aspects of sensitivity analysis

12. If the *amount* of a constraint is altered, then the new line on the graph is parallel to the original constraint line. This is the case for the machine hours and labour hours constraints dealt with in Para 11.

On the other hand, if the *relationship changes* between the variables in the constraints or objective function, then the *slope* of the line changes.

As an illustration of the way the slope of the line changes when relationships change, the following example deals with changing the relationships in the objective function from Example 1. This was originally:

$$\text{maximise } 3x_1 + 4x_2$$

Example 3

Assume that management are able to adjust the contribution of x_1 in Example 1, with all other factors remaining the same. Management wish to know

a) What would be new *reduced* contribution per unit of x_1 to cause the optimum solution to change from the existing one of 15 units x_1, and 20 units x_2?

b) Conversely, what would be *increased* unit contribution to cause a change?

Solution

The original objective function, $3x_1 + 4x_2$, had a slope of

$$\frac{\text{coefficient of } x_2}{\text{coefficient of } x_1} \text{ i.e. } \frac{4}{3} \text{ or } \frac{1\frac{1}{3}}{1}$$

This slope produced an optimum solution at the intersection of constraints A and B. Reducing the coefficient (the contribution) of x_1 causes the slope of the objective function to steepen.

At present the objective function has a less steep slope than constraint B, which has a

$$\frac{6}{4} \text{ or } \frac{1\frac{1}{2}}{1} \text{ slope.}$$

If follows, therefore that when the objective functions slope steepens to the slope of constraint B (in fact, just marginally steeper) the optimum solution will change.

Thus, the new objective function's minimum slope must be $\frac{x_2}{x_1} = \frac{1\frac{1}{2}}{1}$.

As the coefficient of x_2 is unchanged at **4**, x_1 can be deduced as $\frac{4}{1\frac{1}{2}} = 2^2/_3$.

At just marginally lower than this value the objective function shows that the optimum point is Point B on Figure 17/9. This gives a solution of

$$\text{Nil units } x_1, \text{ 30 units } x_2$$
$$\text{Contribution } 30 \times \pounds 4 = \underline{\underline{\pounds 120}}$$

Conversely when the contribution of x_1 *increases*, the slope of the objective function becomes *less steep*. The maximum slope it can be is that of Constraint A.

$$\therefore \text{maximum slope} = \text{constraint A slope} = \frac{2}{4} = \frac{1}{2}$$

$$\therefore \text{as } x_2 = 4 \text{ and } \frac{x_2}{x_1} = \frac{1}{2}, \ x_1 = \underline{\underline{8}}$$

Thus, when the contribution of x_1 is marginally more than £8 the new optimum point is the intersection of constraints D, A and E. This gives a solution of

$$20x_1, 10x_2$$

Contribution: 20 (£8+) + 10 (£4) = £200 + a marginal amount.

Summary

13. a) LP is a resource allocation technique where some objective, for example to maximise contribution, is required to be optimised subject to resource constraints.

   b) All factors involved – objective, constraints – must have linear relationships.

c) LP problems should be expressed initially in a standardised format consisting of an objective function followed by the various constraints taking care to identify whether the constraints are of the 'less than or equal to' ($\leq$) type or of the 'greater than or equal to' ($\geq$) type.

d) LP problems can be maximising or minimising problems.

e) If an LP problem has 2 unknowns it may be solved by graphical means regardless of the number of constraints and whether it is a maximising or minimising problem.

f) To solve graphically it is necessary to draw two axes representing the 2 unknowns and then plot lines representing the constraints.

g) When the constraints have been drawn the feasible region can be identified. The feasible region is the area on the graph which does not contravene any of the constraints. Any point inside or on the edge of the feasible region is a possible solution although not necessarily the optimum solution.

h) To establish the optimal solution it is necessary to draw a line representing the objective function. For maximising problems typically this is a contribution line and for minimising problems a cost line.

i) In a maximising problem the optimum position in the feasible region is the furthest point *to the right* touched by the contribution line, conversely in a minimising problem it is the furthest point *to the left* touched by the cost line.

j) The exact values at any intersection can be found using simultaneous equations.

k) Shadow or dual prices are the valuations of scarce resources.

l) Shadow prices can be calculated arithmetically or by solving the dual.

m) Shadow prices show the *extra* contribution per unit of a binding constraint.

n) Sensitivity analysis changes values and relationships in the problem to see the effect on the original solution.

Points to note

14. a) In examination questions it is often difficult to formulate the LP problem in the standardised format. If you do have difficulty concentrate on these basic questions – What are the unknowns? What is the objective? What factors govern the achievement of the objective?

b) Guidance on the scales for graphs can be obtained if the calculations are done for the constraints as described in the early part of the chapter. Remember the scales do not have to be the same for each axis but the chosen scale must be constant and continuous along each axis.

c) Although the examples used in this chapter have produced whole number solutions, fractional solutions are quite possible and often occur.

d) Although the identification of the optimum point for simple LP problems may appear obvious without drawing the objective line this is not recommended. Students should **always** draw the objective function line to identify the correct point that gives the optimum solution.

e) A simple way to check your answer is actually possible is to insert the values of the unknowns in the constraints and check whether the constraints are satisfied.

f) The solution is *always found at a vertex on the graph,* i.e. the junction of two or more of the constraint lines. Having located the optimum vertex it is useful, on occasions, to evaluate the coordinates by solving simultaneous equations. This can be important when it is difficult to choose between two points, both sets of coordinates can be evaluated and the values of the resulting objective functions compared.

g) Every primal problem has a *dual* problem which is minimised if the primal is maximised or maximised if the primal is minimised.

Self review questions *Numbers in brackets refer to paragraph numbers*

1. What types of LP problems can be solved by graphical methods? (2)
2. How many constraints can be handled by graphical methods? (2)
3. What is the feasible region? (3)
4. How is optimum point on the feasible region determined? (3)
5. What are the differences in dealing with minimising and maximising problems by graphical means? (4)
6. How can simultaneous equations be used in solving LP problems? (5)
7. What are shadow or dual prices? (6)
8. What methods can be used to calculate shadow prices? (7)
9. What is the primal problem and what is the dual problem? (9)
10. What do shadow prices tell management? (10)
11. What is sensitivity analysis and how is it carried out? (11 & 12)

Exercises with answers

1. A firm produces two products, X and Y with a contribution of £8 and £10 per unit respectively.

   Production data are: (per unit)

	Labour hours	Material A	Material B
X	3	4	6
Y	5	2	8
Total available	500	350	800

   Formulate the LP model in the standardised manner.

2. Solve the model in question 1 using the graphical method.
3. Calculate the shadow prices of the binding constraints and interpret.

Answers to exercises

1.   Maximise **8X + 10Y**

    Subject to   Labour $3X + 5Y \leq 500$ (L)

                            Material A $4X + 2Y \leq 350$ (A)

                            Material B $6X + 8Y \leq 800$ (B)

                                $X, Y \geq O$

            Where

                        X = number of units of product X

                        Y = number of units of product Y

2.

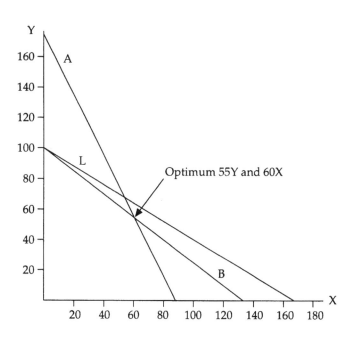

    Optimum Solution 55Y and 60X
    giving $60(8) + 55(10) = \underline{\textbf{£1,030}}$

3.   **Shadow price calculations**

    Optimum solution   Objective **8X + 10Y**

                        Subject to $4X + 2Y \leq 350$

                                        $6X + 8Y \leq 800$

    Dual formulation   Minimise **350S + 800T**

$$\text{Subject to} \qquad 4S + 6Y \geq 8$$

$$2S + 8T \geq 10$$

$$\text{Solving} \qquad 4S + 6T = 8$$

$$2S + 8T = 10$$

Gives

$$T = 1.2 = \text{value of one unit of material B}$$

and

$$S = 0.2 = \text{value of one unit of material A}$$

Proof

$$350(0.2) + 800(1.2) = \underline{\mathbf{£1,030}}$$

18 Linear programming – Simplex method for maximising

Objectives

1.   After studying this chapter you will
     - be able to use the Simplex method for solving maximising LP problems;
     - know how to set up the Initial Simplex Tableau;
     - be able to identify the pivot element and form the next tableau;
     - understand how to identify the optimum solution;
     - be able to interpret the final Simplex Tableau;
     - know how sensitivity analysis is used in the simplex method.

Simplex method definition

2.   A step by step arithmetic method of solving LP problems whereby one moves progressively from a position of, say, zero production, and therefore zero contribution, until no further contribution can be made. Each step produces a feasible solution and each step produces an answer better than the one before, i.e. either greater contribution in maximising problems, or less cost in minimising problems. The mathematics behind the Simplex method are complex and this chapter does not try to explain why the method works but it does describe how to use the technique.

Formulating the Simplex model

3.   To use the Simplex method it is first necessary to state the problem in the standardised manner previously described in Chapter 16. It will be recalled that this results in an objective function and a number of constraints which are inequalities either of the $\geq$ or $\leq$ type. Having stated the problem in the standardised format, the inequalities must be converted to equations. For example, if a manufacturer made two products A and B, which took 3 and 5 hours respectively to machine on a drilling machine which was available for up to 320 hours per period the constraint would be written in the standardised format as follows:

$$3x_1 + 5x_2 \leq 320$$

$$\text{where } x_1 = \text{units of A}$$

$$x_2 = \text{units of B}$$

This is converted into an equation by adding an extra variable called a *slack variable* thus

$$3x_1 + 5x_2 + x_3 = 320$$

The slack variable, x_3 in this case, represents any unused capacity in the constraint and can thus take any value from **320 hours** (i.e. the position of zero production and therefore maximum unused capacity), to **0 hours** (i.e. the position of the machine being fully utilised and therefore, zero unused capacity).

Notes:

a) Each constraint will have its own slack variable.

b) Once the slack variable has been incorporated into the constraint the Simplex method automatically assigns it an appropriate value at each iteration.

A Simplex maximising example

4. The following maximising example will be used to provide a step by step approach to the Simplex method.

Example 1

A company can produce three products, A, B and C. The products yield a contribution of £8, £5 and £10 respectively.

The products use a machine which has 400 hours capacity in the next period.

Each unit of the products uses 2, 3 and 1 hour respectively of the machine's capacity.

There are only 150 units available in the period of a special component which is used singly in products A and C.

200 kgs only of a special alloy is available in the period. Product A uses 2 kgs per unit and Product C uses 4 kgs per units.

There is an agreement with a trade association to produce no more than 50 units of Product B in the period.

The company wishes to find out the production plan which maximises contribution.

Step 1: Express the problem in the standardised format thus:

Objective function:

maximise	$8x_1$	$+$	$5x_2$	$+$	$10x_3$		
subject to	$2x_1$	$+$	$3x_2$	$+$	x_3	$\leq$ 400	(machine hours constraint)
	x_1			$+$	x_3	$\leq$ 150	(component constraint)
	$2x_1$			$+$	$4x_3$	$\leq$ 200	(alloy constraint)
			x_2			$\leq$ 50	(sales constraint)

$$x_1 \geq 0, \ x_2 \geq 0, \ x_3 \geq 0$$

where x_1 = no. of units of Product A

x_2 = no. of units of Product B

x_3 = no. of units of Product C

Step 2: Make the inequalities in the constraints into equalities by adding a 'slack variable' in each constraint, thus:

maximise	$8x_1$	+	$5x_2$	+	$10x_3$				
subject to	$2x_1$	+	$3x_2$	+	x_3	+	x_4	=	400
	x_1			+	x_3	+	x_5	=	150
	$2x_1$			+	$4x_3$	+	x_6	=	200
			x_2			+	x_7	=	50

Note: x_4, x_5, x_6, x_7 are the slack variables and represent the spare capacity in the limitations.

Step 3: Set up the initial Simplex Tableau by arranging the objective function and equalised constraints from Step 2 in the following form.

Initial Simplex tableaux. Table 1

Solution variable	Products			Slack variables				Solution quantity
	x_1	x_2	x_3	x_4	x_5	x_6	x_7	
x_4	2	3	1	1	0	0	0	400
x_5	1	0	1	0	1	0	0	150
x_6	2	0	4	0	0	1	0	200
x_7	0	1	0	0	0	0	1	50
Z	8	5	10	0	0	0	0	0

Notes:

a) It will be seen that the values in the body of the table are the values from the objective function and constraints in Step 2.

b) The variable 'Z' has been used for the objective function and represents total contribution.

c) The tableau shows that x_4 = 400, x_5 = 150, x_6 = 200, x_7 = 50 and Z = 0.

d) The tableau shows a feasible solution, that of nil production, nil contribution, and maximum unused capacity as represented by the values of the slack variables x_4, x_5, x_6 and x_7.

e) Although feasible, this plan can obviously be improved and this is done as follows:

Step 4: Improve the previous feasible solution by making as many as possible of the product with the most contribution, i.e. the highest figure in the Z row. The number that can be made will be limited by one or more of the constraints becoming operative thus:

Select highest contribution in Z row – i.e. 10 under x_3.

Divide the positive numbers in the x_3 column into the solution quantity column.

$$i.e.\ 400 \div 1 = 400$$
$$150 \div 1 = 150$$
$$200 \div 4 = 50$$
$$50 \div 0 = ignore$$

Select the row that gives the *lowest* answer (in this case the row identified x_6). Ring the element which appears in both the identified column (x_3) and the identified row (x_6), this element is known as the *pivot element* thus

Initial Simplex tableau reproduced. Table 2

Solution variable	Products			Slack variables				Solution quantity
	x_1	x_2	x_3	x_4	x_5	x_6	x_7	
x_4	2	3	1	1	0	0	0	400
x_5	1	0	1	0	1	0	0	150
→ x_6	. 2	0	4	0	0	1	0	200
x_7	0	1	0	0	0	0	1	50
Z	8	5	10	0	0	0	0	0

Step 5: Divide all the elements in the identified row (x_6) by the value of the pivot element (4) and change the solution variable to the heading of the identified column (x_3) thus

Old row

| x_6 | 2 | 0 | 4 | 0 | 0 | 1 | 0 | 200 |

Divide by 4 and retitling the new row becomes

New row

| x_3 | $\frac{1}{2}$ | 0 | 1 | 0 | 0 | $\frac{1}{4}$ | 0 | 50 |

Enter this row in a new tableau

Second Simplex tableau. Table 3

Rows no.	Solution variable	Products			Slack variables				Solution quantity
		x_1	x_2	x_3	x_4	x_5	x_6	x_7	
1	x_4	2	3	1	1	0	0	0	400
2	x_5	1	0	1	0	1	0	0	150
3	x_3	$\frac{1}{2}$	0	1	0	0	$\frac{1}{4}$	0	50
4	x_7	0	1	0	0	0	0	1	50
5	Z	8	5	10	0	0	0	0	0

Notes:

a) The row numbers have been included to aid the explanatory material which follows and form no part of the Simplex Tableaux.

b) It will be seen that this second tableau is identical to the first except for row 3 which was calculated above.

c) Row 3 means that 50 units of x_3 are to be produced.

Step 6: As a consequence of producing 50 units of x_3 it is necessary to adjust the other row so as to take up the appropriate number of hours, components etc. used and to show the contribution produced for the 50 units of x_3. This is done by repetitive row by row operations using Row 3 which makes all the other elements in the pivot elements column into zeros. To maintain each row as an equality it is, of course, necessary to alter each element along the row on both sides of the equality sign. Using Row 1 in the Second Tableau as an example the process is as follows.

	Row 1	x_4	2	3	1	1	0	0	0	=	400
Minus	Row 3	x_3	$\frac{1}{2}$	0	1	0	0	$\frac{1}{4}$	0	=	50
Produces a new row		x_4	$1\frac{1}{2}$	3	0*	1	0	$-\frac{1}{4}$	0	=	350

Notes:

a) This new row will be inserted into a third tableau along with all the other altered rows and Row 3 from the second tableau.

b) The aim of this row operation was to produce the zero (marked *). In this case a simple subtraction was all that was necessary but to make a zero in other cases may require further operations using Row 3 as a basis.

The other rows in the second tableau are operated on a similar fashion.

	Row 2	x_5	1	0	1	0	1	0	0	=	150
Minus	Row 3	x_3	$\frac{1}{2}$	0	1	0	0	$\frac{1}{4}$	0	=	50
Produces a new row		x_5	$\frac{1}{2}$	0	0*	0	1	$-\frac{1}{4}$	0	=	100

Row 4 needs no operation because the element in column x_3 is already zero.

Row 5	**Z**	8	5	10	0	0	0	0	=	0
Minus 10 × Row 3	x_3	5	0	10	0	0	$2\frac{1}{2}$	0	=	500
Produces a new row	**Z**	3	5	0*	0	0	$-2\frac{1}{2}$	0	=	−500

Notes:

a) To produce the required zero (0 *) it was necessary to multiply the Row 3 by 10 and then subtract from Row 5.

b) The '−500' at the end of the new Z row is the contribution earned by 50 units of x_3 at £10 i.e. £500. The negative sign is merely a result of the Simplex method and the fact that the contribution is shown as a negative figure can be disregarded.

Step 7: When all the row operations have been done a third tableau can be produced thus.

Third Simplex tableau. Table 4

Rows no.	Solution variable	Products			Slack variables				Solution quantity
		x_1	x_2	x_3	x_4	x_5	x_6	x_7	
6 (Row 1 – Row 3)	x_4	$1\frac{1}{2}$	3	0	1	0	$-\frac{1}{4}$	0	350
7 (Row 2 – Row 3)	x_5	$\frac{1}{2}$	0	0	0	1	$-\frac{1}{4}$	0	100
8 (i.e. Row 3)	x_3	$\frac{1}{2}$	0	1	0	0	$\frac{1}{4}$	0	50
9 (i.e. Row 4)	x_7	0	1	0	0	0	0	1	50
10 (Row 5 – 10× Row 3)	**Z**	3	5	0	0	0	$-2\frac{1}{2}$	0	−500

Notes:

a) All the new rows produced by the row operations in Step 6 have been inserted into the third tableau.

b) The rows have been consecutively numbered again and a summary of the operations carried out in Step 6 to produce the new lines has been given against the new row numbers.

e.g. Row 10 was produced by multiplying Row 3 by 10 and subtracting it from Row 5.

Step 8: To produce subsequent tableaux and eventually an optimum solution, steps 4 to 7 are repeated until no positive numbers can be found in the Z row.

From Row 10 it will be seen that the maximum contribution is 5.

∴ x_2 column is chosen.

The positive numbers in the x_2 column are divided into the solution quantities and the lowest result selected.

i.e. Row 6 $350 \div 3 = 116\frac{2}{3}$

Row 7 $100 \div 0$ ignore

Row 8 $50 \div 0$ ignore

Row 9 $50 \div 1 = 50$

$\therefore$ Row 9 is selected and the pivot element identified and the solution variable altered to x_2 thus

Row 9 x_2 0 $\boxed{1}$ 0   0   0   0   1 $=$ 50

As the pivot element is already 1 no further action is necessary on it but the other elements in the pivot element column (x_2) must be made into zeros by using row operations based on Row 9 thus

	Row 6	x_4	$1\frac{1}{2}$	3	0	1	0	$-\frac{1}{4}$	0	$=$ 350
Minus 3 ×	Row 9	x_2	0	3	0	0	0	0	3	$=$ 150
Produces a new row		x_4	$1\frac{1}{2}$	0^*	0	1	0	$-\frac{1}{4}$	-3	$=$ 200

Note:

It was necessary to multiply Row 9 by 3 to produce the zero in column x_2 (0^*)

Rows 7 and 8 need no operation because the elements in column x_2 are already zero.

	Row 10	Z	3	5	0	0	0	$-2\frac{1}{2}$	0	$=$ -500
Minus 5 ×	Row 9	x_2	0	5	0	0	0	0	5	$=$ 250
Produces a new row		Z	3	0^*	0	0	0	$-2\frac{1}{2}$	-5	$=$ -750

The new row produced can now be entered into a fourth tableau as follows

Fourth Simplex tableau. Table 5

Row no.	Solution variable	Products			Slack variables				Solution quantity
		x_1	x_2	x_3	x_4	x_5	x_6	x_7	
11 (Row 6 – 3× Row 9)	x_4	$1\frac{1}{2}$	0	0	1	0	$-\frac{1}{4}$	-3	200
12 (as Row 7)	x_5	$\frac{1}{2}$	0	0	0	1	$-\frac{1}{4}$	0	100
13 (as Row 8)	x_3	$\frac{1}{2}$	0	1	0	0	$\frac{1}{4}$	0	50
14 (Pivot Row as Row 9)	x_2	0	1	0	0	0	0	1	50
15 (Row 10 – 5× Row 9)	Z	3	0	0	0	0	$-2\frac{1}{2}$	-5	-750

Notes:

a) The above tableau shows that 50 units of Products B and C could be made (x_2 and $x_3 = 50$).

b) As a result of this amount of production £750 contribution would be gained ($Z = -750$)

Step 9: Because there is still a positive number in the Z row (3 under column x_1) the iterative process is repeated in precisely the same manner. Column x_1 is chosen and the positive numbers in the x_1 column are divided into the solution quantities and the lowest number selected.

Row 11	$200 \div$	$1\frac{1}{2}$	$=$	$133\frac{1}{3}$
Row 12	$100 \div$	$\frac{1}{2}$	$=$	200
Row 13	$50 \div$	$\frac{1}{2}$	$=$	100
Row 14	$50 \div$	0	$=$	ignore

$\therefore$ Row 13 is selected and the pivot element identified and the solution variable altered to x_1 thus

| Row 13 | x_1 | $\boxed{\frac{1}{2}}$ | 0 | 1 | 0 | 0 | $\frac{1}{4}$ | 0 | $=$ | 50 |

The pivot element must be made into a 1 so the whole row is multiplied by 2 thus

| Row 13 | x_1 | $\boxed{1}$ | 0 | 2 | 0 | 0 | $\frac{1}{2}$ | 0 | $=$ | 100 |

The rest of the elements in column x_1 must now be made into zeros by the usual row operations

	Row 11	x_4	$1\frac{1}{2}$	0	0	1	0	$-\frac{1}{4}$	-3	$=$	200
Minus $1\frac{1}{2} \times$	Row 13	x_1	$1\frac{1}{2}$	0	3	0	0	$\frac{3}{4}$	0	$=$	150
Produces a new row		x_4	0^*	0	-3	1	0	-1	-3	$=$	50

	Row 12	x_5	$\frac{1}{2}$	0	0	0	1	$-\frac{1}{4}$	0	$=$	100
Minus $\frac{1}{2} \times$	Row 13	x_1	$\frac{1}{2}$	0	1	0	0	$\frac{1}{4}$	0	$=$	50
Produces a new row		x_5	0^*	0	-1	0	1	$-\frac{1}{2}$	0	$=$	50

Row 14 needs no operation because the element in column x_1 is already zero

	Row 15	Z	3	0	0	0	0	$-2\frac{1}{2}$	-5	$=$	-750
Minus $3 \times$	Row 13	x_1	3	0	6	0	0	$-1\frac{1}{2}$	0	$=$	300
Produces a new row		Z	0^*	0	-6	0	0	-4	-5	$=$	$-1,050$

The new rows produced can now be entered into a fifth tableau thus

Fifth Simplex tableau. Table 6

Row no.	Solution variable	Products			Slack variables				Solution quantity
		x_1	x_2	x_3	x_4	x_5	x_6	x_7	
16 (Row 11 – $1\frac{1}{2}$× Row 18)	x_4	0	0	–3	1	0	–1	–3	50
17 (Row 12 –$\frac{1}{2}$× Row 18)	x_5	0	0	–1	0	1	$-\frac{1}{2}$	0	50
18 (Pivot Row 2× Row 13)	x_1	1	0	2	0	0	$\frac{1}{2}$	0	100
19 (as Row 14)	x_2	0	1	0	0	0	0	1	50
20 (Row 15 – 3× Row 13)	Z	0	0	–6	0	0	–4	–5	–1,050

As there are no positive values in the Z row the optimum solution has been reached.

Step 10: All that remains is to obtain the maximum information from the fifth tableau.

Dealing first with the solution variables: x_1, x_2, x_4 and x_5

Optimum product mix

$$x_1 = 100 \text{ i.e. produce 100 units of Product A}$$

$$x_2 = 50 \text{ i.e. produce 50 units of Product B}$$

Value of Slack Variables

$$x_4 = 50 \text{ i.e. there are 50 machine hours unused at Optimum}$$

$$x_5 = 50 \text{ i.e. there are 50 components unused at Optimum}$$

Note:

It will be seen that the other two slack variables, x_6 and x_7, do not have values in the Solution Quantity column. Their values are both zero, which means x_6 and x_7, representing the alloy and sales constraints respectively, have no unused capacity at optimum and the constraints they represent are fully utilised.

Contribution and resource valuations

It will be seen from Row 20 in Table 6 that Z has the Solution Quantity of –1,050. This means, at optimum, the maximum contribution is £1,050 (this can be confirmed by calculating the contributions of the optimum product mix, i.e. 100A and 50B = (100 × 8) + (50 × 5) = 1050).

The value of –6 under Product x_3, the product that was not in the optimum plan, means that if any unit of x_3 was produced then overall contribution would fall by £6.

The values of Row 20 for the slack variables are of great importance. These are the valuations of resources and are known as **shadow prices**. These have the following meanings:

$x_4 = 0$ i.e. there are no value to be gained from increasing machine hours

$x_5 = 0$ i.e. there are no value to be gained from increasing the number of components

$x_6 = -4$ i.e. for every extra kilo of alloy available £4 extra overall contribution would be gained

$x_7 = -5$ i.e. for every extra unit of B that was allowed to be produced overall contribution would increase by £5

It will be seen that **only the constraints x_6 and x_7 that are binding, i.e. are fully utilised, have non-zero shadow prices**. This is a general rule, for there would be no value in increasing the availability of a resource already in surplus. Thus, in this example, machine hours and the supply of components have shadow prices of zero.

The shadow prices of the binding constraints can be used to confirm again the overall contribution thus:

Alloy availability	200 kg × £4 =	£800
Sales constraint	50 unit × £5 =	£250
		= **£1,050**

When solving LP problems by graphical means the shadow prices have to be calculated separately. When using the Simplex process they are an automatic by product.

Notes:

a) Each variable in the final solution variable column has a specific meaning which is detailed above.

b) Using step 10 as a guide interpret the meanings of the solution variables and the valuations in the Z row of the intermediate tableaux.

c) Alternative names for shadow prices are: shadow costs, dual prices or simplex multipliers.

d) Take heart. To work through a normal Simplex problem with 3 unknowns is a very quick process once the foregoing steps are mastered.

Mixed limitations

5.   The maximising example given above had constraints all of which were of the 'less than or equal to' type ($\leq$). This is a common situation but on occasions the constraints contain a mixture of $\leq$ and $\geq$ varieties. The usual cause of one or more 'greater than or equal to ($\geq$) constraints is the requirement to produce at least a given number of certain products.

In such circumstances the simplest approach is to reduce the capacity of the other limitations by the amounts required to make the required number of the product(s) specified. Then maximise in the normal way and add back the quantities which were required to be produced, to the optimum solution found by the normal Simplex method. The method is shown in the following example.

Example containing mixed constraints

6. Assume that an LP problem had been set up in the usual standardised format as follows:

Example 2

Objective function

maximise $\quad 5x_1 + 3x_2 + 4x_3$

subject to a. $\quad 3x_1 + 12x_2 + 6x_3 \leq 660 \quad$ (machine hours constraint)

b. $\quad 6x_1 + 6x_2 + 3x_3 \leq 1{,}230 \quad$ (labour hours constraint)

c. $\quad 6x_1 + 9x_2 + 9x_3 \leq 990 \quad$ (component onstraint)

d. $\qquad\qquad\qquad x_3 \geq 10 \quad$ (sales constraint)

where x_1, x_2 and x_3 represent units of Products A, B and C.

Only one of the constraints d. is of the $\geq$ variety so we decide to make the minimum quantity possible to satisfy constraint d. i.e. 10 units of x_3. The resource requirements for 10 units of x_3 must be subtracted from the total available in constraints a, b. and c thus

Constraint a. $\quad 3x_1 + 12x_2 + 6x_3 \quad \leq \quad 660 \quad - \quad (6 \times 10)$

b. $\quad 6x_1 + 6x_2 + 3x_3 \quad \leq \quad 1{,}230 \quad - \quad (3 \times 10)$

c. $\quad 6x_1 + 9x_2 + 9x_3 \quad \leq \quad 990 \quad - \quad (9 \times 10)$

Notes:

a) In each case the expression in the bracket represents the coefficient of x_3 in each constraint multiplied by the 10 units being made.

b) An alternative way of dealing with the $\geq$ constraint is to multiply both sides by -1 and change the inequality sign

$$(x_3 \geq 10) \times -1$$

$$\text{becomes} \quad -x_3 \leq -10$$

This constraint can then be used in the normal manner.

Having eliminated constraint d. and made the appropriate reductions in the other constraints the initial Simplex Tableau can be set up.

Initial Simplex tableau Example 2. Table 7

Modified constraint	Solution variable	Products			Slack variables			Solution quantity
		x_1	x_2	x_3	x_4	x_5	x_6	
a)	x_4	3	12	6	1	0	0	600
b)	x_5	6	6	3	0	1	0	1,200
c)	x_6	6	9	9	0	0	1	900
	Z	5	3	4	0	0	0	0

Note: This is now a normal maximising Simplex model with a slack variable for each constraint and is solved by exactly the same process covered in the earlier part of the chapter.

After carrying out the usual Simplex procedure the final tableau becomes:

Final Simplex tableau Example 2. Table 8

Solution variable	Products			Slack variables			Solution quantity
	x_1	x_2	x_3	x_4	x_5	x_6	
x_4	0	$7\frac{1}{2}$	$1\frac{1}{2}$	1	0	$-\frac{1}{2}$	150
x_5	0	-3	-6	0	1	-1	300
x_1	1	$1\frac{1}{2}$	$1\frac{1}{2}$	0	0	$\frac{1}{6}$	150
Z	0	$-4\frac{1}{2}$	$-3\frac{1}{2}$	0	0	$-\frac{5}{6}$	-750

Solution from final tableau: -150 units of x_1 producing £750 contribution.

Plus **production to satisfy constraint d.** 20 units of x_3 producing £40 contribution.

∴ Total solution is 150 units of x_1 and 10 units of x_3 giving £790 contribution.

As previously described in Step 10 the tableau can be interpreted as follows:

Spare capacity $x_4 = 150$ means that there are 150 spare machining hours

$x_5 = 300$ means that there are 300 spare labour hours

Products not being made

From the final Tableau Table it will be seen that x_2 and x_3 do not appear in the Simplex solution and they have valuations in the Z row of $-4\frac{1}{2}$ and $-3\frac{1}{2}$ respectively. We already know that 10 units of x_3 will be made and the tableau informs us that the 10 units of x_3 which we have had to make have cost £35 in reduced contribution. Any units of x_2 that were made would similarly reduce contribution by £4.50 per unit.

Shadow prices

Only one constraint (constraint c, components) is fully utilised and thus has a non-zero shadow price. The tableau shows that for every extra component that could be obtained contribution would be increased by $£\frac{5}{6}$.

As previously the shadow price of the binding constraint can be used to confirm the overall contribution as follows:

$$\text{Modified component constraint} = 900 \times £\tfrac{5}{6} = £750$$

An alternative method of dealing with mixed limitations uses what are termed *artificial variables* as well as slack variables. This method is outside the scope of this book.

Comparing Simplex and graphical solutions

7. The Simplex method can be used for problems with any number of unknowns – even those with only 2 unknowns that can also be solved graphically. To illustrate both solution methods Example 1 used and solved graphically in Chapter 17 is solved below using the Simplex method.

    Example 1 reproduced from Chapter 17

    Maximise $3x_1 + 4x_2$

$4x_1$	$+$	$2x_2$	$\leq$	100	A
$4x_1$	$+$	$6x_2$	$\leq$	180	B
x_1	$+$	x_2	$\leq$	40	C
x_1			$\leq$	20	D
		x_2	$\geq$	10	E

    This problem is inserted into the initial tableau with a slack variable for each of the five constraints and with constraint E multiplied by -1 to reverse the inequality sign.

    Initial tableau: equivalent to point 0 on Figure 18/1. Table 9

Solution variable	Products		Slack variables					Solution quantity
	x_1	x_2	x_3	x_4	x_5	x_6	x_7	
x_3	4	2	1	0	0	0	0	100
x_4	4	6	0	1	0	0	0	180
x_5	1	1	0	0	1	0	0	40
x_6	1	0	0	0	0	1	0	20
x_7	0	-1	0	0	0	0	1	-10
Z	3	4	0	0	0	0	0	0

The problem is then solved by the usual Simplex iterations. Each iteration improves on the one before and the process continues until optimum is reached. The following tables show the position after each iteration and each one is cross referenced to Figure 18/1 which is a reproduction of the graphical solution.

First iteration: equivalent to point 1 on Figure 18/1. Table 10

Solution variable	Products		Slack variables					Solution quantity
	x_1	x_2	x_3	x_4	x_5	x_6	x_7	
x_2	0	1	0	0	0	0	–1	10
x_3	4	0	1	0	0	0	2	80
x_4	4	0	0	1	0	0	6	120
x_5	1	0	0	0	1	0	1	30
x_6	1	0	0	0	0	1	0	20
Z	3	0	0	0	0	0	4	–40

This shows $10x_2$ being produced and £40 contribution. The first four constraints have surpluses of 80, 120, 30 and 20 respectively. *Not* optimum, as there are still positive values in Z row.

Second iteration: equivalent to point 2 on Figure 18/1. Table 11

Solution variable	Products		Slack variables					Solution quantity
	x_1	x_2	x_3	x_4	x_5	x_6	x_7	
x_2	0.667	1	0	0.167	0	0	0	30
x_3	2.667	0	1	–0.333	0	0	0	40
x_4	–0.333	0	0	–0.167	0	0	0	10
x_5	1	0	0	0	1	1	0	20
x_6	0.333	0	0	0.167	0	0	1	20
Z	0.333	0	0	–0.667	0	0	0	–120

This shows $30x_2$ being produced and £120 contribution. All constraints have surpluses except labour hours. Not optimum as there is a positive value in Z row.

Third iteration: equivalent to point 3 on Figure 18/1. Table 12

Solution variable	Products		Slack variables					Solution quantity
	x_1	x_2	x_3	x_4	x_5	x_6	x_7	
x_1	1	0	0.375	0.125	0	0	0	15
x_2	0	1	–0.25	0.25	0	0	0	20
x_5	0	0	–0.125	–0.125	1	0	0	5
x_6	0	0	–0.375	0.125	0	1	0	5
x_7	0	0	–0.25	0.25	0	0	1	10
Z	0	0	–0.125	–0.625	0	0	0	–125

This is the optimum and shows all the information obtained previously.

Solution

Optimum production:

$$x_1 = 15 \text{ i.e. produce 15 units of Klunk}$$

$$x_2 = 20 \text{ i.e. produce 20 units of Klick giving an overall contribution of £125}$$

Shadow prices of binding constraints:

$$x_3 = £0.125 \text{ i.e. each additional machine hour would produce £0.125 extra contribution labour}$$

$$x_4 = £0.625 \text{ i.e. each additional hour would produce £0.625 extra contribution}$$

The non-binding constraints represented by x_5, x_6 and x_7 have 5, 5 and 10 spare, respectively.

Note: By seeing how the Simplex interactions move from one vertex to an improved vertex on the feasible region it is possible to gain a better understanding of the process.

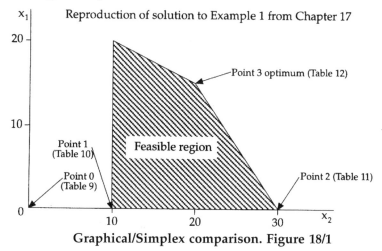

Graphical/Simplex comparison. **Figure 18/1**

Summary of Simplex method

8. Flowchart 1 (on p. 329) provides a concise summary of the Simplex method and is cross referenced to key points in the chapter.

Sensitivity analysis and the Simplex

9. In practice virtually all LP problems are solved by computer packages which incorporate numerous facilities to test the sensitivity of the solution to changes in the problem. Typically these facilities include showing the effects of changes in:

    a) the amounts of the resources and constraints

    b) the coefficients of the objective function and the constraints.

Most of these forms of sensitivity analysis are outside the scope of the book and the syllabuses at which it is aimed. One form of sensitivity analysis is however covered, as it is a typical examination topic. That is, the aspect concerned with variations in the total amount of the resources/constraints.

This is demonstrated by again using example 1 from Chapter 17 which was solved by the Simplex method in Para 7 above.

The procedure is to use the *Simplex multipliers* which are the + and − values found in the slack variable columns in the final tableau, Table 12. Multiplication of the variation in the constraint by the multipliers gives the change in the solution quantities.

Assume that it is required to calculate the effect on the solution of having 5 extra labour hours (represented by slack variable x_3).

Multipliers for x_3	×	Change in constraint	=	Change in solution	+	Original solution	=	New solution
0.375	×	5	=	1.875	+	15	=	16.875
−0.25	×	5	=	−1.25	+	20	=	18.75
−0.125	×	5	=	−0.625	+	5	=	4.375
−0.375	×	5	=	−1.875	+	5	=	3.125
−0.25	×	5	=	−1.25	+	10	=	8.75
−0.125	×	5	=	−0.625	+	(−125)	=	−125.625

Thus if 5 extra labour hours were available the solution is 16.875 units of x_1, 18.75 units of x_2 giving a contribution of £125.625. The other values in the solution column are, as before, the unused amounts of constraints C, D and E represented by slack variables x_5, x_6 and x_7.

Points to note

10. a) The Simplex method is not only capable of dealing with more unknowns than the graphical method but more information is automatically produced.

b) In every tableau it will be noted that the solution variable column always contains a single 1 with the rest of the column elements being zeros.

c) The valuations of the Z row of the final tableau are known as shadow prices or shadow costs. These valuations can be of considerable value in valuing the contributions made by scarce resources.

d) On occasions examination questions merely ask for the initial Simplex Tableau to be set up. Make sure that you obtain as much practice as possible in defining problems in the standardised format and setting up the Simplex Tableau.

e) At other times, the final Simplex Tableau is provided and you have to interpret the results. Using the examples given in the chapter, make sure you understand all the figures in the Optimum Tableau.

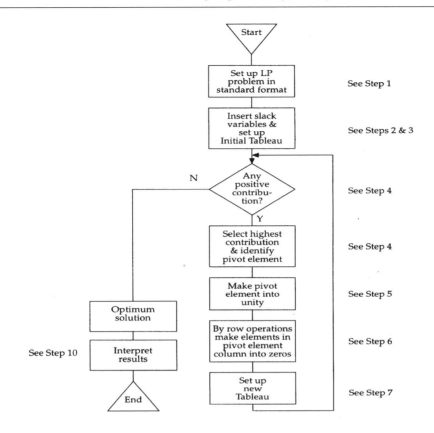

Simplex method. Flowchart 1

Self review questions *Numbers in brackets refer to paragraph numbers*

1. Can the Simplex method deal with more than 2 unknowns? (1)

2. Define the Simplex method. (2)

3. Why are slack variables necessary? (3)

4. What does the Initial Simplex Tableau show? (4, Step 3)

5. What is the pivot element? (4, Step 4)

6. How is the optimum solution identified in Simplex table? (Table 6)

7. What information can be derived from the final Simplex Tableau? (4, Step 10)

8. How are mixed limitations dealt with? (5)

9. Can you relate the various Simplex Tableau to vertices on a graph? (7)

9. What are the key steps in the Simplex method? (8)

10. What is one method of sensitivity analysis with the Simplex method? (9)

Exercises with answers

1. A firm produces three products X, Y, Z with a contribution of £20, £18 and £16 respectively. Production data are as follows:

Products	Machine hours	Labour hours	Materials (kg)
		Per unit	
X	5	2	8
Y	3	5	10
Z	6	3	3
Availability	3,000 hours	2,500 hours	10,000 kg

Set up the initial Simplex Tableau including the necessary slack variables.

2. Carry out the first iteration of the tableau in question 1.

3. Interpret the final tableau of a Simplex solution shown below:

	A	B	C	S1	S2	S3	S4	
S2	0	0	0	$1\frac{1}{4}$	1	2	−1.5	520
A	1	0	0	$\frac{3}{4}$	0	$1\frac{3}{4}$	3.0	276
B	0	1	0	0	0	3	1	68
C	0	0	1	−1.5	0	$−1\frac{1}{4}$	1.5	127
	0	0	0	−6	0	−12.5	−25	−48,600

A = Units of Product A

B = Units of Product B

C = Units of Product C

S1 = Slack variable labour hours

S2 = Slack variable material kilogrammes

S3 = Slack variable machine hours

S4 = Slack variable sales restriction A

Answers to exercises

1.

Solution variable	X	Y	Z	S_1	S_2	S_3	
S_1	5	3	6	1	0	0	3,000
S_2	2	5	3	0	1	0	2,500
S_3	8	10	3	0	0	1	10,000
	20	18	16	0	0	0	0

2.

First iteration

Solution variable	X	Y	Z	S_1	S_2	S_3	
X	1	$\frac{3}{5}$	$\frac{6}{5}$	$\frac{1}{5}$	0	0	600
S_2	0	$3\frac{4}{5}$	$\frac{3}{5}$	$-\frac{2}{5}$	1	0	1,300
S_3	0	$5\frac{1}{5}$	$-6\frac{3}{5}$	$-1\frac{3}{5}$	0	1	5,200
	0	6	–8	–4	0	0	–12,000

3. Optimum production

Produce   276 units of Product A
68 units of Product B
127 units of Product C

giving a contribution of £48,600

There are 520 kg of materials spare and there are the following shadow prices of the binding constraints for labour hours (S_1), machine hours (S_3) and the sales restriction (S_4)

Shadow price **S1** = £6 per labour hour

Shadow price **S3** = £12.50 per machine hour

Shadow price **S4** = £25 per unit

19 Linear programming – Simplex method for minimising

Objectives

1. After studying this chapter you will
   - know how to use the Simplex method to solve minimising LP problems;
   - understand how to invert a standard LP formulation;
   - be able to interpret the final tableau of a minimising problem.

Minimisation example

2. The following simple example will be used as a basis for explaining the methods to be used.

Example 1

A chemical manufacturer processes two chemicals, Arkon and Zenon, in varying proportions to produce three products, A, B and C. He wishes to produce at least 150 units of A, 200 units of B and 60 units of C. Each ton of Arkon yields 3 of A, 5 of B and 3 of C. Each ton of Zenon yields 5 of A, 5 of B and 1 of C.

If Arkon costs £40 per ton and Zenon £50 per ton advise the manufacturer how to minimise his cost.

Step 1. In exactly the same manner as previously described formulate the problem in the standardised way, thus:

Objective function:

Minimise	$40x_1$	$+$	$50x_2$			
Subject to:	$3x_1$	$+$	$5x_2$	$\geq$	150	Product A constraint
	$5x_1$	$+$	$5x_2$	$\geq$	200	Product B constraint
	$3x_1$	$+$	x_2	$\geq$	60	Product C constraint

$$x_1 \geq 0, \quad x_2 \geq 0$$

where x_1 = number of tons of Arkon

x_2 = number of tons of Zenon

Note: This will be seen as a simple minimisation problem with all constraints being of the $\geq$ type.

Step 2. Formulate the *inverse* or *dual* of the formulation above. That is, make the formulation above into a maximising problem by making a *column* for each *limitation* and a *constraint row* for each *element* in the objective function thus:

Minimise	150A	+	200B	+	60C		
Subject to:	3A	+	5B	+	3C	≤	40
	5A	+	5B	+	C	≤	50

Notes:

a) In step 1 there were 2 columns (x_1 and x_2) and 3 constraints (Products A, B, C). There are now 3 columns (one for each of original constraints) and 2 constraint rows (one for each of the original objective function elements, x_1 and x_2).

b) There is a dual or inverse of every LP problem, i.e. for every maximising problem there is an equal but opposite minimising problem, and for every minimising problem there is an equal but opposite maximising problem. Because solving Simplex maximising problems is quite straightforward it is normal to convert minimising problems into maximising problems.

c) It will be noted that the original quantity column (150, 200 and 60) has become the objective function and the original costs (40 and 50) have become the amounts of the constraints.

d) What are now the constraints are of the ≤ type, i.e. as a normal maximising problem.

e) The original problem is known as the *primal* and the inverse is known as the *dual*.

Step 3. Now that a standard maximising problem has been obtained, the initial Simplex tableau, complete with slack variables, can be set up in the normal manner thus:

Initial Simplex tableau. Table 1

Solution variable	A	B	C	Slack variables S_1	S_2	Cost
S_1	3	5	3	1	0	40
S_2	5	5	1	0	1	50
Quantity	150	200	60	0	0	0

Step 4. Proceed in the step by step manner previously described until the tableau is reached showing optimum i.e. when there are no positive values in the bottom row thus:

Final Simplex tableau. Table 2

Solution variable	A	B	C	Slack variables S_1	S_2	Cost
B	0	1	$\frac{6}{5}$	$-\frac{3}{10}$	$-\frac{3}{10}$	5
A	1	0	-1	$-\frac{1}{2}$	$+\frac{1}{2}$	5
Quantity	0	0	-30	-25	-15	$-1{,}750$

Step 5 The tableau above is the normal result of a Simplex maximising routine. However, to obtain the solutions to the original minimising problem some differences in the usual interpretation are necessary as follows.

Optimum solution

Quantities to be purchased

The figures under the slack variable columns represent the quantities to be purchased.

The -25 in the S_1 column means   Purchase **25** tons of **Arkon**

The -15 in the S_2 column means   Purchase **15** tons of **Zenon**

Total cost

The figure at the bottom of the cost column represents total cost.

i.e. $-$ **1750 represents a cost of £1750**

$$(25 \times £40 + 15 \times £50 = £1750)$$

Valuation of constraints

Rows titled B and A indicate the cost of changing the limitation of **200 B** and **150 A**. If either of these limitations are changed by one unit the total cost will change by £5, i.e. these are the shadow prices.

Proof : $(200 \times 5 + 150 \times 5) = $ **£1750** total cost

Over-production. The -30 under column C indicates an over-production of 30 units.

Proof:

	A	B	C
25 tons **Arkon** yields	$25 \times 3 = \quad 75$	$25 \times 5 = \quad 125$	$25 \times 3 = \quad 75$
15 tons **Zenon** yields	$15 \times 5 = \quad \underline{75}$	$15 \times 5 = \quad \underline{75}$	$15 \times 1 = \quad \underline{15}$
	$\underline{150}$	$\underline{200}$	$\underline{90}$
	= Minimum required	= Minimum required	= Over production of 30 units

Notes:

a) Once the inversion process and the method of interpretation is understood, minimising problems are no more difficult to solve than maximising problems.

b) The particular example used here had only 2 unknowns (tons of Arkon and Zenon) so it could also have been solved by graphical means. The Simplex method is, of course, equally suitable for any number of unknowns.

Summary

3. a) The simple transformation given above may be applied to all minimisation problems with all constraints of the greater than or equal type. The Simplex method can be applied to all combinations but only a restricted range of problems is covered here.

b) After formulating the problem in the standardised manner, the formulation must be *inverted*, i.e. the columns made into rows and the constraint rows into columns. This converts it into a maximising problem.

c) The maximising problem is solved by the usual Simplex steps.

d) Optimum is recognised in the usual manner, i.e. no positive values in base row.

e) The solutions to the minimising problem can be read directly from the final tableau. The quantities required are found at the bottom of the slack variable columns and the total cost at the bottom of the cost column.

Points to note

4. a) Most past examination questions appear to be maximising problems but students should be capable of dealing with minimising problems of the type covered in this chapter.

b) If you have the option and there are only 2 unknowns, solve minimising problems by the graphical method already described.

c) The process of converting a primal problem into its dual can, on occasions, be used to simplify the solution method. For example, if a primal problem contained three decision variables and two constraints, this could not be solved graphically. However, if its dual was formed, this would be a two-decision variable, three constraint problem which could be solved graphically. If this technique is used take care with the interpretation of the graphical solution. Remember, it is the solution to the original *primal* problem that is required.

Self review questions *Numbers in brackets refer to paragraph numbers*

1. What is the dual of a minimising LP problem? (2, Step 2)

2. Why is it usual to convert a minimising problem into a maximising? (2)

3. How is the final Simplex Tableau interpreted? (2, Step 5)

Exercises with answers

1. Form the dual of the following problem

Minimise	30A	+	60B	+	20C		
subject to	5A	+	10B	+	15C	$\geq$	2,000
	2A	+	3B	+	1C	$\geq$	300
	8A	+	6B	+	4C	$\geq$	650

2. Form the dual of the following problem:

Minimise	200X	+	250Y	–	100Z	+	50W		
subject to	6X	+	10Y	–	3Z			$\geq$	55
	9X	+	7Y			+	3W	$\geq$	40

$$X, Y, Z, W \geq 0$$

Answers to exercises

1.

Maximise	2,000L	+	300M	+	650N		
Subject to	5L	+	2M	+	8N	$\leq$	30
	10L	+	3M	+	6N	$\leq$	60
	15L	+	1M	+	4N	$\leq$	20

$$L, M, N \geq 0$$

2.

Maximise	55R	+	40S		
Subject to	6R	+	9S	$\leq$	200
	10R	+	7S	$\leq$	250
	3R			$\geq$	100
(or –3R				$\leq$	–100)
			3S	$\leq$	50
			R, S	$\geq$	0

20 Transportation

Objectives

1.   After studying this chapter you will

   • be able to recognise Transportation problems;

   • know how to set up the initial Transportation table;

   • be able to make the initial feasible allocations;

   • understand how to improve the initial allocation by calculating the shadow costs;

   • know when to include a dummy destination;

   • be able to use the Transportation technique for maximising problems.

Transportation problems defined

2.   The typical transportation problem deals with a number of sources of supply (e.g. warehouses) and a number of destinations (e.g. retail shops). The usual objective is to minimise the transportation cost of supplying quantities of a commodity from the warehouses to the shops. The major requirement is that there must be a constant transportation cost per unit, i.e. if one unit costs £10 to transport from Warehouse A to Shop X, five units will cost £50. This will be recognised as the linearity requirement fundamental to all forms of LP.

The transportation technique

3.   Although the method of solving transportation problems described below differs in appearance from the Simplex method it has some basic similarities, as follows:

   a)   It is an iterative, step by step, process.

   b)   It starts with a feasible solution and each succeeding solution is also feasible.

   c)   At each stage a test is made to see whether transportation costs can be reduced.

   d)   Optimum is reached when no further cost reductions are possible.

Transportation example

4.   The following simple example will be used as a basis for the step-by-step explanation of the transportation technique.

> **Example 1**
>
> A firm of office equipment suppliers has three depots located in various towns. It receives orders for a total of 15 special filing cabinets from four customers. In total in the three depots there are 15 of the correct filing cabinets available and the management wish to minimise delivery costs by despatching the filing cabinets from the appropriate depot for each customer.
>
> Details of the availabilities, requirements, and transport costs per filing cabinet are given in the following table.

Note: The body of the table contains the transportation costs per cabinet from the depots to the customer. For example, it cost £14 to send 1 cabinet from Depot Y to customer B (and £28 for 2, £42 for 3, etc.).

Table 1

		cabinets	Customer A	Customer B	Customer C	Customer D	Total
			3	3	4	5	15
	Depot X	2	£13	11	15	20	transport-
Available	**Depot Y**	6	£17	14	12	13	ation cost
	Depot Z	7	£18	18	15	12	per unit
	Total	15					

Step 1. Make an initial feasible allocation of deliveries. The method used for this initial solution does not affect the value of the optimum but a careful initial choice may reduce the number of iterations that have to be made. The method to be used in this book is to select the cheapest route first, and allocate as many as possible then the next cheapest and so on. The result of such an allocation is as follows.

Table 2

			A 3	B 3	C 4	D 5
	X	2 units		2(1)		
Available	Y	6 units	1(4)	1(3)	4(2)	
	Z	7 units	2(5)			5(2)

Requirements (column header over A B C D)

Note: The numbers in the table represent deliveries of cabinets and the numbers in the brackets (1), (2), etc., represent the sequence in which they are inserted, lowest cost first, i.e.

£

1.	2 units X → B £11/unit	Total cost	22
2.	4 units Y → C £12/unit	Total cost	48
	5 units Z → D £12/unit	Total cost	60
3.	The next lowest cost move which is feasible i.e. does not		14
	exceed row or column totals is 1 unit Y → B £14/unit		
4.	Similarly the next lowest feasible allocation		17
	1 unit Y → A £17/unit		
5.	Finally to fulfil the row/column totals 2 units		
	Z → A £18/unit		36

197

Step 2. Check solution obtained to see if it represents the minimum cost possible. This is done by calculating what are known as 'shadow costs' (i.e. an imputed cost of not using a particular route) and comparing these with the real transport costs to see whether a change of allocation is desirable.

This is done as follows:

Calculate a nominal 'despatch' and 'reception' cost for each occupied cell by making the assumption that the transport cost per unit is capable of being split between despatch and reception costs thus:

$$
\begin{aligned}
D(X) + R(B) &= 11 \\
D(Y) + R(A) &= 17 \\
D(Y) + R(B) &= 14 \\
D(Y) + R(C) &= 12 \\
D(Z) + R(A) &= 18 \\
D(Z) + R(D) &= 12
\end{aligned}
$$

where $D(X)$, $D(Y)$ and $D(Z)$ represent Despatch cost from depots **X, Y** and **Z**, and $R(A)$, $R(B)$, $R(C)$ and $R(D)$ represent Reception costs at customers **A, B, C, D**.

By convention the first depot is assigned the value of zero, i.e. $D(X) = 0$ and this value is substituted in the first equation and then all the other values can be obtained thus

$$
\begin{aligned}
R(A) &= 14 & D(X) &= 0 \\
R(B) &= 11 & D(Y) &= 3 \\
R(C) &= 9 & D(Z) &= 4 \\
R(D) &= 8 &
\end{aligned}
$$

Using these values the shadow costs of the unoccupied cells can be calculated. The unoccupied cells are **X : A, X : C, X : D, Y : D, Z : B, Z : C**.

								Shadow costs
∴	$D(X)$	+	$R(A)$	=	0	+	14	=	14
	$D(X)$	+	$R(C)$	=	0	+	9	=	9
	$D(X)$	+	$R(D)$	=	0	+	8	=	8
	$D(Y)$	+	$R(D)$	=	3	+	8	=	11
	$D(Z)$	+	$R(B)$	=	4	+	11	=	15
	$D(Z)$	+	$R(C)$	=	4	+	9	=	13

339

These computed 'shadow costs' are compared with the actual transport costs (from Table 1), Where the *actual* costs are *less than shadow costs*, overall costs can be reduced by allocating units into that cell.

	Actual cost	−	Shadow cost	=	+ Cost increase − Cost reduction
Cell X : A	13	−	14	=	−1
X : C	15	−	9	=	+6
X : D	20	−	8	=	+12
Y : D	13	−	11	=	+2
Z : B	18	−	15	=	+3
Z : C	15	−	13	=	+2

The meaning of this is that total costs could be reduced by £1 for every unit that can be transferred into cell **X : A**. As there is a cost reduction that can be made the solution in Table 2 is not optimum.

Step 3. Make the maximum possible allocation of deliveries into the cell where actual costs are less than shadow costs using occupied cells, i.e. Cell **X : A** from Step 2, The number that can be allocated is governed by the need to keep within the row and column totals. This is done as follows:

Table 3

			Requirements			
			A 3	B 3	C 4	D 5
	X	2 units	+	2−		
Available	Y	6 units	1−	1+	4	
	Z	7 units	2			5

Table 3 is a reproduction of Table 2 with a number of + and − inserted. These were inserted for the following reasons.

Cell **X : A** + indicates a transfer *in* as indicated in Step 2.

Cell **X : B** − indicates a transfer *out* to maintain Row X total.

Cell **Y : B** + indicates a transfer *in* to maintain Column B total.

Cell **Y : A** − indicates a transfer *out* to maintain Row Y and Column A totals.

The maximum number than can be transferred into Cell **X : A** is the lowest number in the minus cells i.e. cells **Y : A**, and **X : B** which is 1 unit.

∴ 1 unit is transferred in the + and − sequence described above resulting in the following table.

Table 4

			Requirements			
			A	B	C	D
			3	3	4	5
	X	2 units	1	1		
Available	Y	6 units		2	4	
	Z	7 units	2			5

The total cost of this solution is

			£
Cell **X : A**	1 unit @ £13	=	13
Cell **X : B**	1 unit @ £11	=	11
Cell **Y : B**	2 units @ £14	=	28
Cell **Y : C**	4 units @ £12	=	48
Cell **Z : A**	2 units @ £18	=	36
Cell **Z : D**	5 units @ £12	=	60
			£196

The new total cost is £1 less than the total cost established in Step 1. This is the result expected because it was calculated in Step 2 that £1 would be saved for every unit we were able to transfer to Cell **X : A** and we were able to transfer 1 unit only.

Notes: Always commence the + and − sequence with a + in the cell indicated by the (actual cost − shadow cost) calculation. Then put a − in the occupied cell in the same row which has an occupied cell in its column. Proceed until a − appears in the same column as the original +.

Step 4. Repeat Step 2, i.e. check that solution represents minimum cost. Each of the processes in Step 2 are repeated using the latest solution (Table 4) as a basis, thus:

Nominal despatch and reception costs for each occupied cell.

$$\begin{array}{ccccc}
\mathbf{D(X)} & + & \mathbf{R(A)} & = & 13 \\
\mathbf{D(X)} & + & \mathbf{R(B)} & = & 11 \\
\mathbf{D(Y)} & + & \mathbf{R(B)} & = & 14 \\
\mathbf{D(Y)} & + & \mathbf{R(C)} & = & 12 \\
\mathbf{D(Z)} & + & \mathbf{R(A)} & = & 18 \\
\mathbf{D(Z)} & + & \mathbf{R(D)} & = & 12 \\
\end{array}$$

Setting **D(X)** at zero the following values are obtained

$$\begin{array}{llll}
\mathbf{R(A)} & = & 13 & \quad \mathbf{D(X)} & = & 0 \\
\mathbf{R(B)} & = & 11 & \quad \mathbf{D(Y)} & = & 3 \\
\mathbf{R(C)} & = & 9 & \quad \mathbf{D(Z)} & = & 5 \\
\mathbf{R(D)} & = & 7 & \\
\end{array}$$

Using these values the shadow costs of the *unoccupied* cells are calculated. The unoccupied cells are **X : C, X : D, Y : A, Y : D, Z : B**, and **Z : C**.

$$
\begin{aligned}
\therefore D(X) + R(C) &= 9 \\
D(X) + R(D) &= 7 \\
D(Y) + R(A) &= 16 \\
D(Y) + R(D) &= 10 \\
D(Z) + R(B) &= 16 \\
D(Z) + R(C) &= 14
\end{aligned}
$$

The computed shadow costs are compared with actual costs to see if any reduction in cost is possible.

	Actual cost	−	Shadow cost		+ Cost increase − Cost reduction
Cell X : C	15	−	9	=	+6
X : D	20	−	7	=	+13
Y : A	17	−	16	=	+1
Y : D	13	−	10	=	+3
Z : B	18	−	16	=	+2
Z : C	15	−	14	=	+1

It will be seen that all the answers are positive, therefore no further cost reduction is possible and optimum has been reached.

Optimum solution

1 unit	X → A
1 unit	X → B
2 units	Y → B
4 units	Y → C
2 units	Z → A
5 units	Z → D

with a total cost of £196

This solution is shown in the following tableau.

	A	B	C	D
X	1	1		
Y		2	4	
Z	2			5

Note: In this example only one iteration was necessary to produce an optimum solution mainly because a good initial solution was chosen. The principles explained above would, of course, be equally suitable for many iterations.

Unequal availability and requirement quantities

5. Example 1 above had equal quantities of units available and required. Obviously

this is not always the case and the most common situation is that there are more units available to be despatched than are required. The transportation technique can be used in such circumstances with only a slight adjustment to the initial table. A dummy destination, with zero transport costs, is inserted in the table to absorb the surplus available. Thereafter the transportation technique is followed. The following example explains the procedure.

Transportation example with a dummy destination

6. **Example 2**

A firm of wholesale domestic equipment suppliers, with 3 warehouses, received orders for a total of 100 deep freezers from 4 retail shops. In total in the 3 warehouses there are 110 freezers available and the management wish to minimise transport costs by despatching the freezers required from the appropriate warehouses, Details of availabilities, requirements, and transport costs are given in the following table.

Table 5

			Required				
			Shop A	Shop B	Shop C	Shop D	Total
		freezers	25	25	42	8	100
Available	Warehouse I	40	£3	16	9	2	transport costs per freezer
	Warehouse II	20	£1	9	3	8	
	Warehouse III	50	£4	5	2	5	
	Total	110					

Step 1 Add a ***dummy*** destination to Table 5 with a zero transport costs and a requirements equal to the surplus availability.

∴ Dummy requirement = 110 – 100 = **10 Freezers**

Table 6

			Required					
			Shop A	Shop B	Shop C	Shop D	Dummy	Total
		freezers	25	25	42	8	10	110
Available	Warehouse I	40	£3	16	9	2	0	transport costs per freezer
	Warehouse II	20	£1	9	3	8	0	
	Warehouse III	50	£4	5	2	5	0	
	Total	110						

Step 2. Now that the quantity available equals the quantity required (because of the insertion of the dummy) the solution can proceed in exactly the same manner described in Para. 4 Example 7. First set up an initial feasible solution.

Table 7

		A 25	B 25	C 42	D 8	Dummy 10
				Requirements		
	I 40	5(4)	17(6)		8(3)	10(7)
Available	II 20	20(1)				
	III 50		8(5)	42(2)		

The numbers in the table represent the allocations made and the numbers in brackets represent the sequence they were inserted based on lowest cost and the necessity to maintain row/column totals. The residue of 10 was allocated to the dummy.

The cost of this allocation is

		£		£
I → A	5 units @ 3		=	15
I → B	17 units @ 16		=	272
I → D	8 units @ 2		=	16
I → Dummy	10 units @ zero cost			
II → A	20 units @ 1		=	20
III → B	8 units @ 5		=	40
III → C	42 units @ 2		=	84
				£447

Step 3. Check solution to see if it represents the minimum cost possible in the same manner as previously described, i.e.

Despatch & Reception Costs of used routes:

				£
D(I)	+	R(A)	=	3
D(I)	+	R(B)	=	16
D(I)	+	R(D)	=	2
D(I)	+	R(Dummy)	=	0
D(II)	+	R(A)	=	1
D(III)	+	R(B)	=	5
D(III)	+	R(C)	=	2

Setting D (I) at zero the following values can be obtained:

R(A)	=	3	D(I)	=	0
R(B)	=	16	D(II)	=	−2
R(C)	=	13	D(III)	=	−11
R(D)	=	2			
R(Dummy)	=	0			

Using these values the shadow costs of the unused routes can be calculated. The unused routes are **I : C, II : B, II : C, II : D, II : Dummy, III : A, III : D,** and **III : Dummy**.

								Shadow costs £
∴ D(I)	+	R(C)	=	0	+	13	=	13
D(II)	+	R(B)	=	−2	+	16	=	14
D(II)	+	R(C)	=	−2	+	13	=	11
D(II)	+	R(D)	=	−2	+	2	=	0
D(II)	+	R(Dummy)	=	−2	+	0	=	−2
D(III)	+	R(A)	=	−11	+	3	=	−8
D(III)	+	R(D)	=	−11	+	2	=	−9
D(III)	+	R(Dummy)	=	−11	+	0	=	−11

The shadow costs are then deducted from actual costs.

	Actual cost	−	Shadow cost		+Cost increase −Cost reduction
Cell I : C	9	−	13	=	−4
II : B	9	−	14	=	−5
II : C	3	−	11	=	−8
II : D	8	−	0	=	+8
II : Dummy	0	−	(−2)	=	+2
III : A	4	−	(−8)	=	+12
III : D	5	−	(−9)	=	+14
III : Dummy	0	−	(−11)	=	+11

It will be seen that total costs can be reduced by £8 per unit for every unit that can be transferred into Cell **II : C**.

Step 4. Make the maximum possible allocation of deliveries into Cell **II : C**. This is done by inserting a sequence of + and −, maintaining row and column totals.

Table 8

			Requirements				
			A 25	B 25	C 42	D 8	Dummy 10
	I	40	5+	17−		8	10
Available	II	20	20−		+		
	III	50		8+	42−		

Note: This is a reproduction of Table 7 with a sequence of + and − added starting with a + in Cell **II : C** and then − and + as necessary to maintain row and column balances. The maximum number that can be transferred is the lowest number in the minus cells, i.e. 17 units in cell **1 : B**.

This transfer is made and the following table results.

Table 9

			A 25	B 25	C 42	D 8	Dummy 10
	I	40	22	17−		8	10
Available	II	20	3		17		
	III	50		25	25		

Requirements (column headers A, B, C, D, Dummy)

The cost of this allocation is

$$(22 \times 3) + (8 \times 2) + (10 \times 0) + (3 \times 1) + (17 \times 3) + (25 \times 5) + (25 \times 2) = £311$$

This can be verified by deducting from the cost of the original allocation (Table 7) the savings of £8 per unit for the 17 units transferred, i.e.

$$£447 - (17 \times 8) = £311$$

Step 5. Repeat Step 3 to check if cost is at a minimum. After setting **D(I) = 0**, the following values can be obtained.

R(A)	=	3		D(I)	=	0
R(B)	=	8		D(II)	=	−2
R(C)	=	5		D(III)	=	−3
R(D)	=	2				
R(Dummy)	=	0				

These values are used to calculate the shadow costs of the unused routes.

Shadow costs

£

Cell I : B	=	8
I : C	=	5
II : B	=	6
II : D	=	0
II : Dummy	=	−2
III : A	=	0
III : D	=	−1
III : Dummy	=	−3

When these shadow costs are deducted from the actual costs no negative values result: the allocation shown in Table 9 is optimum with a minimum transportation cost of £311.

Notes:

a)   It will be seen that, apart from setting up the initial table with a Dummy destination, the solution method for Example 2 is identical to Example 1.

b) A frequently advocated method of making the initial feasible allocation is what is called the North-West corner method. This means filling the requirements starting from the top left hand cell regardless of the costings. The same optimum solution is reached but more iterations are usually required. Example 2, for instance, would take 5 iterations to reach optimum starting from a North-West corner initial allocation as opposed to 2 iterations when a more rational choice is made.

c) Frequently it is found that alternative solutions exist, i.e. different series of allocations can be found with the same overall cost.

Maximisation and the transportation technique

7. Although transportation problems are usually minimising problems, on occasions problems are framed so that the objective is to make the allocations from sources to destinations in a manner which maximises contribution or profit. These problems are dealt with relatively easily as follows.

Steps involved in maximisation

8. Using Example 1, paragraph 4, for comparison the following procedures should be followed.

See	Minimising technique	Maximising technique
Step 1	Make initial feasible allocation lowest cost first and so on.	Make initial feasible allocation on basis of maximum contribution first, then next highest and so on.
Step 2	Test differences between actual and shadow costs on unused routes for optimality, i.e. all positive. If not, make allocation into cell with largest negative difference.	For optimum, the differences between actual and shadow contributions for the unused routes should be *all negative*. If not, make allocation into cell with the *largest positive* difference. Apart from the differences above the transportation technique can be followed as usual.

Maximising and dummies

9. It will be recalled from Para 5 that where availability and requirements are unequal a dummy destination with zero transport costs was introduced and thereafter the normal technique was followed. In a maximising problem where there are more items available than are required, a dummy destination with zero contribution should also be introduced and the maximising procedure in Para 8 followed.

Degeneracy and the transportation technique

10. If Tables 2, 3, 4, 5, 7, 8 and 9 are examined it will be seen in each case that the number of allocations made is 1 less than the number of rows added to the number of columns. For example, Table 9 (3 rows and 5 columns) has 7 allocations. To be able to calculate the shadow despatch and receiving costs this condition is essential but on occasions the number of allocations turns out to be less than *rows* + *columns* − *1*. This condition is known as degeneracy.

Dealing with degeneracy

11. If the number of allocations is less than number of *rows* + *columns* − *1*, then it is necessary to make one or more zero allocations to routes to bring up the number of allocations to *rows* + *columns* − *1*. This means that, for shadow cost calculation purposes, one or more cells with zero allocation are treated as occupied. For example in a 3 row, 5 column table if 6 allocations had been made which satisfied the requirements, any 1 cell would be given a zero allocations and treated as occupied. For moves following a zero allocation the cell with zero allocation is treated as a positive allocation and can therefore be moved to and from as with other allocations. The allocation must be into an independent cell.

Summary

12. a) Figure 20/1 opposite gives an outline of the transportation method for minimising problems and is cross referenced to appropriate parts of the chapter.

    b) Maximising problems follow the same general pattern except that the initial feasible solution is made on the basis of maximum contribution first and optimum is recognised when all the values are negative after deducting shadow from actual contribution.

    c) The number of allocations made must be one less than the number of rows plus the number of columns. If this does not happen after making the necessary actual allocations one or more cells, with zero allocation, must be treated as occupied.

Points to note

13. a) The transportation method is simple but tedious therefore reduce the number of iterations necessary by making the best possible choice of initial solution.

    b) The figure obtained by deducting the shadow costs from actual costs is the netted result of the + and − sequence. For example in Table 8 the + and − sequence and costs were:

    +**Costs in cells, I : A, II : C and III : B** i.e. £3, £3 and £5 = **£11**

    −**Cost in cells, I : B, I : A and III : C** i.e. £16, £1 and £2 = **£19**

    making a net cost reduction of £8 which was the value indicated by the shadow cost calculations.

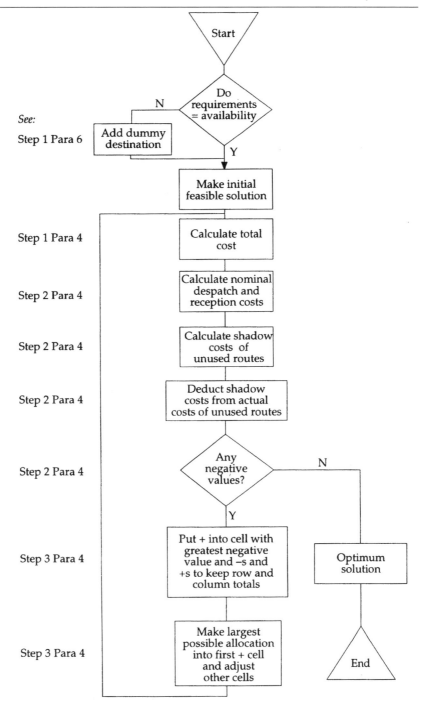

Outline of transportation method flowchart. Figure 20/1

c) An Appendix to this chapter describes a shorter approach to solving Transportation problems using the principles developed in the Chapter.

Self review questions Numbers in brackets refer to paragraph numbers

1.   What is a typical transportation problem? (2)
2.   What similarities to the Simplex method has the usual transportation solution technique? (3)
3.   What are the shadow costs? (4, Step 2)
4.   How is the basic transportation technique modified to deal with the situation where there are more items available than are required? (5)
5.   How is the transportation technique modified to solve maximising problems? (7 and 8)
6.   What is degeneracy? (10)
7.   How is degeneracy dealt with? (11)

Appendix

The methodology developed and summarised in Flowchart 20/1 provides a detailed description of the principles involved in the Transportation Technique. However in the time constrained examination a shorter approach using the same principles can be employed.

Example 2 is re-worked using such a technique.

The initial *Step 1* is repeated below.

→	A	B	C	D	Du	Total
	3	16	13	2	0	
W1	5 + x	17 – x	[−4] 8		10	
0	3	16	9 13 2		0	40
W2	20 – x	[−5]	x [−8] [8]		[2]	
−2	1	9 14 3	11 8 0	0(−2)		20
W3	[12]	8 + x	42 – x [−4]		[11]	
−11	4(−8)	5	2 5 9	0(−11)		50
Total	25	25	42	8	10	110

The initial feasible solution is inserted as in *Step 2*.

Step 3 proceeds with the insertion of Despatch and Reception Costs for cells used. Put 0 in the box marked W1 to represent the despatch cost from Warehouse 1. The receipt cost in A is 3. This is put in the box labelled A. Repeat the process for B, D and Du. Row W1 can be linked to row W2 through column A. The despatch cost for W2 is −2. Row W1 can be linked to row W3 through Column B. The despatch cost for W3 is −11. The receipt cost for Column C is obtained from W3 through (W3 to C) and is 13.

Add together the receipt and despatch costs for unused cells and write them by the actual cost. These are the shadow costs to compare with the actual costs. The difference is noted in each cell. The only ones of interest are those which are circled. The minus values show savings and the maximum saving is through cell W2 to C. Let x units be put through this cell. This is exactly the same as *Step 4* in the main text. A closed loop of reallocation is made. As before the maximum value of x is 17.

At this point the diagram is redrawn below:

	A 3	B 8	C 5	D 2	Du 0	Total
W1 0	22 3	8 16 8	4 9 5	8 2	10 0	40
W2 −2	3 1	3 9 9	17 3	8 8 0	2 0(−2)	20
W3 −3	4 4 0	25 5	25 2	1 5(−1)	3 0(−3)	50
Total	25	25	42	8	10	110

The process described above is now repeated.

There are no more savings, i.e. negative values, in the circles and so the minimum cost allocation has been reached.

This technique is the same as that developed in the main text, except that all the calculations are performed on one diagram, thus saving valuable minutes in the examination.

Exercises with answers

1. A firm has three shops with a total of 80 televisions. An order is received from the Local Authority for 70 sets to be delivered to 4 schools. The transport costs from shops to schools are shown below together with the availabilities and requirements.

			A	B	C	D	
		Sets	20	30	15	5	Requirements
Available	Shop I	40	2	4	1	6	
	Shop II	20	4	3	3	3	Costs
	Shop III	20	1	2	5	2	

Schools (header over A B C D)

It is required to make the most economic deliveries.

Set up the initial tableau and make the initial feasible deliveries.

2. Work out the shadow prices for the tableau in question 1.

3. Solve the problem in question 1.

Answers to exercises

1.

	A 20	B 30	C 15	D 5	Dummy 10
I 40		15(5)	15(1)		10(6)
II 20		15(4)		5(3)	
III 20	20(2)				

NB solution is degenerate

∴ Treat DIII B as occupied

2.

DI	→	B	=	4
DI	→	C	=	1
DI	→	Dummy	=	0
DII	→	B	=	3
DII	→	D	=	3
DIII	→	A	=	1
DIII	→	B	=	2	(zero allocation)

∴ if

DI	=	0
B	=	4
DII	=	−1
D	=	4
DIII	=	−2
A	=	3
C	=	1
Dummy	=	0

352

Shadow costs of unused routes

I	+	A	=	4
I	+	D	=	3
II	+	A	=	2
II	+	C	=	0
II	+	Dummy	=	−1
III	+	C	=	−1
III	+	D	=	3
III	+	Dummy	=	−2

Actual − shadow

I	+	A	=	2–4	=	−4
I	+	D	=	6–3	=	3
II	+	A	=	4–2	=	2
II	+	C	=	3–0	=	3
II	+	Dummy	=	0–(−1)	=	1
III	+	C	=	5 − (−1)	=	6
III	+	D	=	2–3	=	−1
III	+	Dummy	=	0–(−2)	=	2

3.

New allocation

	A	B	C	D	Dummy
I	15		15		10
II		15		5	
III	5	15			

DI	→	A	=	2
DI	→	C	=	1
DI	→	Dummy	=	0
DII	→	B	=	3
DII	→	D	=	3
DIII	→	A	=	1
DIII	→	B	=	2
∴ if		DI	=	0
		A	=	2
		B	=	3
		C	=	1
		DII	=	0
		DIII	=	−1
		Dummy	=	0
		D	=	3

Unused routes

	Actual	−	Shadow		
I + B	4	−	3	=	1
I + D	6	−	3	=	3
II + A	4	−	2	=	2
II + C	3	−	1	=	2
II + Dummy	0	−	0	=	0
III + C	5	−	0	=	0
III + D	2	−	2	=	0
III + Dummy	0	−	0	=	0

∴ optimum.

21 Assignment

Objectives

1. After studying this chapter you will

    - understand when the Assignment technique should be used;
    - know how to set up the initial Assignment table;
    - be able to move to the next improved table;
    - know how to recognise the optimum solution;
    - understand how the Assignment technique can be applied to maximising problems;
    - be able to deal with unequal sources and destinations.

The assignment technique for minimising

2. The following example will be used as a basis of the step-by-step explanation.

Example 1

A company employs service engineers based at various locations throughout the country to service and repair their equipment installed in customers' premises. Four requests for service have been received and the company finds that four engineers are available. The distances each of the engineers is from the various customers is given in the following table and the company wishes to assign engineers to customers to minimise the total distance to be travelled.

Table 1

Customers

		W	X	Y	Z
	Alf	25	18	23	14
Service engineers	Bill	38	15	53	23
	Charlie	15	17	41	30
	Dave	26	28	36	29

Distances in miles from engineers to customers

Setting up the Initial Table

Step 1. Reduce each column by the smallest figure in that column. The smallest figures are 15, 15, 23 and 14 and deducting these values from each element in the columns produces the following table.

Table 2

	W	X	Y	Z
A	10	3	0	0
B	23	0	30	9
C	0	2	18	16
D	11	13	13	15

Step 2. Reduce each row by the smallest figure in that row.

The smallest figures are 0, 0, 0 and 11 and deducting these values gives the following table.

Table 3

	W	X	Y	Z
A	10	3	0	0
B	23	0	30	9
C	0	2	18	16
D	0	2	2	4

Note: Where the smallest value in a row is zero (i.e. as in rows A, B and C above) the row is, of course, unchanged.

Step 3. Cover all the zeros in Table 3 by the *minimum possible* number of lines. The lines may be horizontal or vertical.

Table 4

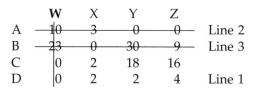

Note: Line 3, covering Row B, could equally well have been drawn covering column X.

Moving to an Improved Table

Step 4. Compare the number of lines with the number of assignments to be made (in this example there are 3 lines and 4 assignments). If the number of lines *equals* the number of assignments to be made go to Step 6.

If the number of lines is *less* than the number of assignments to be made (i.e. as in this example which has three lines and four assignments) then

a) Find the smallest *uncovered* element from Step 3, called *X* (in Table 4 this value is 2).

b) Subtract *X* from every element in the matrix.

c) Add back X to every element covered by a line. If an element is covered by two lines, for example, cell **A : W** in Table 4, X is added twice.

Note: The effect of these steps is that X is subtracted from all uncovered elements, elements covered by one line remain unchanged, and elements covered by two lines are increased by X.

Carrying out this procedure on Table 4 produces the following result:

In Table 4 the smallest element is 2. New table is

Table 5

	W	X	Y	Z
A	12	3	0	0
B	25	0	30	9
C	0	0	16	14
D	0	0	0	2

Note: It will be seen that cells **A : W** and **B : W** have been increased by 2; cells **A : X, A : Y, A : Z, B : X, B : Y, B : Z, C : W** and **D : W** are unchanged, and all other cells have been reduced by 2.

Step 5. Repeat Steps 3 and 4 until the number of lines covering the zeros equals the number of assignments to be made.

In this example covering the zeros in Table 5 by the minimum number of lines equals the number of assignments without any further repetition, thus:

Table 6

	W	X	Y	Z	
A	12	3	0	0	Line 1
B	25	0	30	9	Line 2
C	0	0	16	14	Line 3
D	0	0	0	2	Line 4

Recognising the Optimum Solution

Step 6. When the number of lines *equals* the number of assignments to be made this is optimum and the actual assignments can be made using the following rules:

a) Assign to any zero which is unique to *both* a column and a row.

b) Assign to any zero which is unique to a column or a row.

c) Ignoring assignments already made repeat rule (b) until all assignments are made.

Carrying out this procedure for our example results in the following:

a) (Zero unique to *both* a column and a row). None in this example.

b) (Zero unique to column or row). Assign **B** to **X** and **A** to **Z**. The position is now as follows.

Table 7

	W		X		Y		Z	
A	Row		Satisfied ↔			↕	Column Satisfied ↕	
B	Row		Satisfied ↔			↕	Column Satisfied ↕	
C	0	↕	Column Satisfied	↕	16	↕	Column Satisfied ↕	
D	0	↕	Column Satisfied	↕	0	↕	Column Satisfied ↕	

c) Repeating rule (b) results in assigning **D** to **Y** and **C** to **W**.

Notes:

a) Should the final assignment not be to a zero, then more lines than necessary were used in Step 3.

b) If a block of 4 or more zero's is left for the final assignment, then a choice of assignment exists with the same mileage.

Step 7 Calculate the total mileage of the final assignment.

A to Z	Mileage	14
B to X		15
C to W		15
D to Y		36
		80 miles

The assignment technique for maximising

3. A maximising assignment problem typically involves making assignments so as to maximise contribution. To maximise only Step 1 from above differs – the columns are reduced by the *largest* number in each column. From then on the same rules apply that are used for minimising.

Maximising example

4.

> **Example 2**
>
> The previous example No. 1 will be used with the changed assumption that the figures relate to contribution and not mileage and that it is required to maximise contribution. The solution would be reached as follows. (In each case the Step number corresponds to the solution given for Example No 1.)

Original data

Table 8

	W	X	Y	Z	
A	25	18	23	14	
B	38	15	53	23	Contributions
C	15	17	41	30	to be gained
D	26	28	36	29	

Step 1. Reduce each column by *largest* figure in that column and ignore the resulting minus signs.

Table 9

	W	X	Y	Z
A	13	10	30	16
B	0	13	0	7
C	23	11	12	0
D	12	0	17	1

Step 2. Reduce each row by *smallest* figure in that row.

Table 10

	W	X	Y	Z
A	3	0	20	6
B	0	13	0	7
C	23	11	12	0
D	12	0	17	1

Step 3. Cover zeros by minimum possible number of lines.

Table 11

	W	X	Y	Z
A	3	0	20	6
B	0	13	0	7
C	23	11	12	0
D	12	0	17	1

Step 4. If number of lines equals the number of assignments to be made go to Step 6. If less, (as in this example), carry out the 'uncovered element' procedure previously described. This results in the following table:

Table 12

	W	X	Y	Z
A	0	0	17	6
B	0	16	0	10
C	20	11	9	0
D	9	0	14	1

Step 5. Repeat steps 3 and 4 until number of lines equals the number of assignments to be made. In this example this occurs without further repetition, thus:

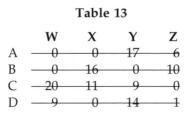

Table 13

	W	X	Y	Z
A	0̶	0̶	17̶	6̶
B	0̶	16̶	0̶	10̶
C	20̶	11̶	9̶	0̶
D	9̶	0̶	14̶	1̶

Step 6. Make assignments in accordance with the rules previously described which result in the following assignments:

C to Z
D to X
S to W
B to Y

Step 7. Calculate contribution to be gained from the assignments.

	£
C to Z	30
D to X	28
A to W	25
B to Y	53
	£136

Notes:

a) It will be apparent that maximising assignment problems can be solved in virtually the same manner as minimising problems.

b) The solution methods given are suitable for any size of matrix. If a problem is as small as the illustration used in this chapter, it can probably be solved merely by inspection.

Unequal sources and destinations

5. To solve assignment problems in the manner described the matrix must be square, i.e. the supply must equal the requirements. Where the supply and requirements are not equal, an artificial source or destination must be created to square the matrix. The costs/mileage/contributions, etc., for the fictitious column or row will be zero throughout.

Solution method

6. Having made the sources equal the destinations, the solution method will be as normal, treating the fictitious elements as though they were real. The solution method will automatically assign a source or destination to the fictitious row or column and the resulting assignment will incur zero cost or gain zero contribution.

Summary

7. a) Flowchart 21/1 gives an outline of the assignment method and is cross referenced to appropriate parts of the chapter.

    b) Because of the similarity in approach both *minimising* and *maximising* problems can be dealt with in the same flowchart.

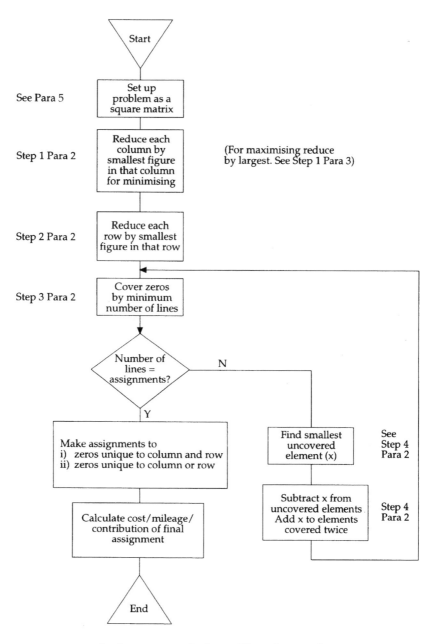

Assignment technique. Flowchart 21/1

Points to note

8. a) The assignment technique can be used for any pairing type of problem, e.g. taxis to customers, jobs to personnel.

   b) Most practical problems of the size illustrated could be solved fairly readily using nothing more than commonsense. However, the technique illustrated can be used to solve much larger problems.

Self review questions *Numbers in brackets refer to paragraph numbers*

1. What is the assignment problem? (1)
2. What is the basis of the solution method? (2)
3. How does the solution method for maximising problems differ from minimising problems? (3)
4. How are unequal numbers of sources and destinations dealt with? (5)

Exercises with answers

1. A foremen has four fitters and has been asked to deal with five jobs. The times for each job are estimated as follows:

	Alf	Bill	Charlie	Dave
			Fitters	
Job 1	6	12	20	12
2	22	18	15	20
3	12	16	18	15
4	16	8	12	20
5	18	14	10	17

   Allocate the men to the jobs so as to minimise the total time taken and identify the job which will not be dealt with.

2. A company has four salesmen who have to visit four clients. The profit records from previous visits are shown in the table and it is required to maximise profits by the best assignments.

	A	B	C	D
Customer 1	£6	12	20	12
2	22	18	15	20
3	12	16	18	15
4	16	8	12	20

Answers to exercises

1. Dummy fitter inserted to Square matrix

	A	B	C	D	Dummy
1	6	12	20	12	0
2	22	18	15	20	0
3	12	16	18	15	0
4	16	8	12	20	0
5	18	14	10	17	0

    Reduce columns by smallest element and cover by lines

0	4	10	0	0
16	10	5	8	0
6	8	8	3	0
10	0	2	8	0
12	6	0	5	0

    4 lines so not optimum, smallest element 3

    ∴ reduce uncovered elements by 3 and increase elements crossed by 2 lines by 3

		A	B	C	D	Dummy
Jobs	1	0	4	10	0	3
	2	13	7	2	5	0
	3	3	5	5	0	0
	4	10	0	2	8	3
	5	12	6	0	5	3

    5 lines so optimum.

    > Assignments
    > B to 4
    > C to 5
    > A to 1
    > Dummy to 2
    > D to 3
    > Job 2 not done

2. Deducting each value from 22, the largest value gives

	A	B	C	D
Customer 1	16	10	2	10
2	0	4	7	2
3	10	6	4	7
4	6	14	10	2

    Step 1

	A	B	C	D
Customer 1	14	8	0	8
2	0	4	7	2
3	6	2	0	3
4	4	12	8	0

Step 2

Customer				
1	14	6	0	8
2	0	2	7	2
3	6	0	0	3
4	4	10	8	0

At least 4 lines are required, so this is optimum.

Assignment	Profit
C to Customer 1	20
A to Customer 2	22
B to Customer 3	16
D to Customer 4	20
	£78

Assessment and revision
Chapters 16 to 21

Examination questions with answers

Answers commence on page 542

A1. A factory is about to buy some machines to produce boxes and has a choice of Type X or Type Y machine. £160,000 has been budgeted for the purchase of machines. Type X machines cost £5,000 each, require 25 hours of maintenance a week and produce 1,500 units a week. Type Y machines cost £10,000 each, require 10 hours of maintenance a week and produce 2,000 units a week.

Each machine, X or Y, needs 50 square metres of floor area. There are available 1,000 square metres of floor area and 400 hours of maintenance time each week. Since all production can be sold, the factory management wishes to maximise output.

You are required

a) to list the objective function and constraints;

b) to graph the constraints, shading the feasible region;

c) to state the optimum mix of machines to buy, with reasons;

d) to add any comments that would be useful to management.

CIMA Quantitative Methods

A2. A company manufactures two products, X and Y. Each product comprises three materials, in the following quantities per unit of product:

Product	Raw materials (kg)		
	Material A	Material B	Material C
X	0.4	0.2	0.6
Y	0.3	0.5	0.2

Supplies of Materials A and B in the following period are limited to a maximum of 2,200 kgs and 2,500 kgs respectively. These quantities are insufficient to fully satisfy demand for the two products. There is no limit on the quantity of Material C available.

Selling prices and variable costs of the products are as follows:

Product	Selling price £ per unit	Variable costs £ per unit
X	10.00	4.50
Y	13.50	7.00

No finished goods stocks are held.

You are required to:

a) Formulate a linear programming model that could be used to determine the production quantities of each product in the following period so as to maximise profit.

b) Use simultaneous equations to determine the optimal production plan in the following period.

ACCA Management Information

A3. Fertily plc is a small firm which produces a variety of chemical products for agricultural use. In a particular production process, four raw materials are mixed to produce three products, Xtragrow (X), Youngrow (Y), and Zupergrow (Z). Each tonne of Xtragrow is a mixture of 0.3 tonnes of material A, 0.2 tonnes of material B, and 0.5 tonnes of material C. Each tonne of Youngrow is a mixture of 0.5 tonnes of material B, 0.1 tonnes of material C and 0.4 tonnes of material D. Each tonne of Zupergrow is a mixture of 0.2 tonnes of material A, 0.4 tonnes of material B, 0.1 tonnes of material C, and 0.3 tonnes of material D.

The maximum quantity available for each of the four materials (in tonnes per week) and their production costs are as follows.

Material	Max available (tonne)	Cost (£/tonne)
A	500	10
B	1,000	12
C	800	8
D	600	10

The current market prices of the three chemical products are as follows:

Xtragrow	£25 per tonne
Youngrow	£20 per tonne
Zupergrow	£24 per tonne

Required:

a) Formulate this problem to maximise the weekly contribution to profit.

State any assumptions relevant to your model.

b) The final tableau of the Simplex solution is as follows:

Basis	X	Y	Z	S_1	S_2	S_3	S_4	Value
Z	0	0	1	5.35	0.70	−3.49	0	581.4
Y	0	1	0	−4.19	1.63	1.86	0	1,023.3
X	1	0	0	−0.23	−0.47	2.33	0	1,279.1
S_4	0	0	0	0.07	−0.86	0.30	1	16.3
	0	0	0	29.54	17.07	6.65	0	37,158.1

where X, Y and Z are the weekly production levels for the three chemical products, and S_1, S_2, S_3 and S_4 are the amounts (in tonnes) by which the usage of the four materials A, B, C and D respectively falls short of the maximum weekly availability.

i) State the optimum weekly production levels for the three chemical products.

ii) Determine the weekly contribution, and the amounts of any unused materials.

iii) State the dual value for each of the four materials, explaining clearly their meaning.

iv) Fertily plc have an opportunity to sell material A alone at a price of £15 per tonne. Should the company take up this opportunity?

ACCA Decision Making Techniques (part question)

A4. Explain the value of sensitivity analysis in linear programming problems and show how dual values are useful in identifying the price worth paying to relax constraints.

Venus Computers is a small manufacturer of personal computers. It concentrates production on three models – a Desktop-386, a Desktop-286, and a Laptop-286, each containing one CPU chip. Due to its limited assembly facilities, Venus Computers are unable to produce more than 500 desktop models or more than 250 laptop models per month. It has 120 80386 chips) these are used in Desktop-386) and 400 80286 chips (used in Desktop-286 and Laptop-286) for the month. The Desktop-386 model requires five hours of production time, the Desktop-286 model requires four hours of production time, and the Laptop-286 requires three hours of production time. Venus Computers have 2000 hours of production time available for the coming month. The company estimates that the profit on a Desktop-386 is £250, for a Desktop-286 the estimated profit is £170 and £150 estimated profit for a Laptop-286.

Required:

a) Formulate this problem as a profit maximisation problem.

b) An extract of the output from a computer package for this problem is given below:

Output

  Solution:

  $X1 = 120$, $X2 = 200$, $X3 = 200$

Dual values:

  Constraint 3   150, Constraint 4   90, Constraint 5   20

---

where variables X1, X2, X3 are the monthly production levels for Desktop-386, Desktop-286, and Laptop-286 respectively and constraints 1–5 describe the limitation on the numbers of desktops, laptops, CPU 80386 chips, CPU 80286 chips, and production hours.

i) Interpret the output clearly, including optimum product mix, monthly profit, unused resources, and dual values.

ii) The computer package also printed the following output:

Sensitivity analysis of objective function coefficients			
Variable	Lower limit	Original value	Upper limit
X1	100	250	no limit
X2	150	170	200
X3	127.5	150	170
Sensitivity analysis of right-hand side ranges			
Constraint	Lower limit	Original value	Upper limit
1	320	500	no limit
2	200	250	no limit
3	80	120	130
4	350	400	412.5
5	1950	2000	2180

Give a full description of this further information. Calculate the increase in profit if the company is able to obtain a further 10 CPU 80386 chips.

ACCA Decision Making Techniques

A5. Computico Limited, currently operating in the United Kingdom, assembles electronic components at its two factories, located at Manchester and London, and sells these components to three major customers. Next month the customers' orders are (in units) 3,000 from customer A, 4,200 from customer B, and 5,300 from customer C.

As Computico's three customers are in different industries it allows them to charge different prices to the different customers (slight alterations are made to the components for each customer at insignificant extra cost). For each component customer A pays £108, customer B pays £105 and customer C pays £116. The variable costs of assembling the components at the two factories vary because of differing labour and power costs; these are £28 per unit at Manchester and £36 per unit at London.

The transportation costs per unit between each factory and customer are:

From	To customer	Cost (£)
Manchester	A	45
Manchester	B	40
Manchester	C	55
London	A	42
London	B	43
London	C	44

The financial director at Computico must decide how to allocate next month's orders between the two factories.

Required:

a) Set up a table showing the contribution to profit of supplying one unit to each customer from each factory.

b) Assuming no restriction on capacity at the two factories calculate the maximum total contribution from next month's orders and the associated capacities for the two factories.

c) Next month's orders from its three customers does, in fact, exceed the combined capacity of the two factories. The maximum output for Manchester is 4,000 units and that for London is 6,000 units.

   i) Use the transportation algorithm to determine the number of units that Computico should supply from each factory to each customer if it wishes to maximise total contribution.

   ii) Calculate the total contribution and indicate which customers will have unsatisfied demand.

   iii) Determine the decrease in total contribution compared with (ii) if Computico wishes to completely satisfy two out of the three customer orders.

ACCA Decision Making Techniques

A6. RAB Consulting Ltd specialises in two types of consultancy project.

- Each Type A project requires twenty hours of work from qualified researchers and eight hours of work from junior researchers.

- Each Type B project requires twelve hours of work from qualified researchers and fifteen hours of work from junior researchers.

Researchers are paid on an hourly basis at the following rates:

Qualified researchers	£30/hour
Junior researchers	£14/hour

Other data relating to the projects:

Project type	A (£)	B (£)
Revenue per project	1,700	1,500
Direct project expenses	408	310
Administration*	280	270

* Administration costs are attributed to projects using a rate per project hour. Total administration costs are £28,000 per four-week period.

During the four-week period ending on 30 June 2000, because of holidays and other staffing difficulties, the number of working hours available are:

Qualified researchers	1,344
Junior researchers	1,120

An agreement has already been made for twenty type A projects with XYZ group. RAB Consulting Ltd must start and complete these projects in the four-week period ending 30 June 2000.

A maximum of 60 type B projects may be undertaken during the four-week period ending 30 June 2000.

RAB Consulting Ltd is preparing its detailed budget for the four-week period ending 30 June 2000 and needs to identify the most profitable use of the resources it has available.

Required:

a) i) Calculate the contribution from each type of project.

   ii) Formulate the linear programming model for the four-week period ending 30 June 2000.

   iii) Calculate, using a graph, the mix of projects that will maximise profit for RAB Consulting Ltd for the four-week period ending 30 June 2000. (*Note:* projects are not divisible.)

b) Calculate the profit that RAB Consulting Ltd would earn from the optimal plan.

c) Explain the importance of identifying scarce resources when preparing budgets and the use of linear programming to determine the optimum use of resources.

CIMA Management Accounting – Performance Management

Examination questions without answers

B1. A company owns two small factories, A and B, which make the same product.

A management decision has to be taken to find the number of hours for which each factory should be operated next week. Factory A can operate for a maximum of 75 hours per week, producing 45 kilograms of product an hour. Factory B can operate for a maximum of 50 hours per week, producing 30 kilograms an hour. To meet customer demand, at least 2,700 kilograms must be produced next week. The hourly cost of running Factory A is £2,000 an hour and for Factory B it is £1,000 an hour. Because of State grant regulations, the company must operate Factory A for at least as many hours as Factory B. The company's aim is to minimise running costs.

You are required

a) To write down the objective function and the constraints;

b) To draw a graph of the problem, shading the feasible region;

c) To recommend a production policy with reasons and comments.

CIMA Quantitative Methods

B2. John Frederickson has been offered early retirement and is currently considering how to invest the lump sum of £50,000 which he will receive from his company's pension scheme. He has decided that he will invest in two managed investment funds, the Harlequin, and the Prince, but that he will also put some of the money into his building society account which will earn 7% per annum interest. John's objective is to maximise his total return on the investment over a five-year period, at which time he intends making major changes to his lifestyle At the moment the investment with the Harlequin is forecast to earn 9% per annum over the five years and in addition has 2% per annum capital growth (making a total return of 11% per annum) whereas the Prince managed fund is forecast to earn 16% per annum (made up from 10% earnings and 6% capital growth). John has taken advice from his accountant, who believes that he should spread the investment over the two managed funds, leaving some in his building society account. The accountant advises that at least 70% of the total sum should be allocated to either the building society or the Harlequin. For safety, at least 60% of the money invested in the two managed funds should be invested in the Harlequin. In order to achieve good capital growth at least 30% of the money invested in the two managed funds should be in the Prince fund. Furthermore the accountant recommended that no more than £20,000 should be invested in the Harlequin managed fund and that no more than £25,000 should be invested in the building society account.

Required:

a) i) Given John Frederickson's objective, formulate a 3-variable linear programming model to show how the lump sum should be distributed between the various investments.

ii) Use the equality constraint to reduce the problem to a 2-variable model involving the investment sums for the Harlequin and Prince managed funds.

b) Solve the problem graphically. State the optimum investment amounts and the final value of the funds.

c) If the recommendation that no more than £20,000 should be invested in the Harlequin was ignored, calculate John Frederickson's possible increase in return.

ACCA Decision Making Techniques

B3. A company manufactures two types of aluminium piping for use in plumbing and other building work. It only manufactures for 48 weeks per year (averaging four weeks per month) to allow cleaning and other maintenance work to be carried out during the remainder of the year. Production of the piping involves three basic processes – heating, forming and rolling – and each process requires specialist machinery. The pipes are produced in 10 metre lengths and both types, X and Y, differ only in the alloys used but consequently require differing times in the three processes.

The number of machines, the number of eight-hour shifts per week for these machines, and the percentage of time the machines are under repair (downtime) is given for each process in the following table.

Process	Number of machines	Number of eight hour shifts per week per machine	Downtime (%)
Heating	5	15	4
Forming	6	5	5
Rolling	3	20	10

Information concerning the output rate per machine, monthly sales limits, and contribution per pipe is included in the table below.

	Output rate per machine (pipes)			Maximum monthly sales	Contribution per pipe (£)
	Heating	Forming	Rolling		
Alloy X	25 per hour	1 per minute	25 per hour	40,000	5
Alloy Y	20 per hour	1 per minute	40 per hour	30,000	7.5

The operations manager has to determine the weekly production of each alloy, within the limitations of sales and machine capacities, in order to maximise contribution to profit.

Required:

a) Show that the weekly machine capacity within the three processes is 576 hours for heating, 228 hours for forming, and 432 hours for rolling.

b) Formulate a linear programming model to determine the optimum weekly production level of each alloy.

c) Solve the problem graphically. State the optimum weekly production levels and the corresponding contribution to profit.

d) The operations manager is able to reduce downtime for each process separately by introducing new working practices. However, he would only do this if there was a substantial saving.

For which of the three processes might it be worth while reducing the downtime? For one of these processes calculate the weekly saving to the company if the downtime percentage is halved.

ACCA Decision Making Techniques

B4. A wholesale company has nine storage depots which it proposes to rationalise. Four depots, **Q, R, S**, and **T** are to be expanded and five depots, **A, B, C, D**, and **E** are to be closed. Thirty six of the mechanical loaders in the depots to be closed will be required for use in the enlarge depots.

The mechanical loaders in the five depots to be closed are:

A : 5, B : 7 , C : 11, D : 8, and E : 9

The additional loaders required at the depots to be expanded are:

Q : 8, R : 9, S : 11 and T : 8.

The cost of transporting one mechanical loader, in hundreds of pounds, between depots is given below:

Depots to be closed	Depots to be expanded			
	Q	R	S	T
A	3	3	7	9
B	6	5	3	3
C	6	4	8	7
D	5	4	5	4
E	4	3	6	5

You are required to:

a) show by calculation the minimum cost plan for meeting the rationalisation requirement by transfers between depots;

b) state at which depot there will be surplus loaders;

c) state with reasons whether or not your solution is unique.

CIMA Quantitative Techniques

22 Network analysis – introduction and terminology

Objectives

1. After studying this chapter you will
    - have been introduced to the control and planning technique called Network Analysis;
    - know what is meant by Activity, Event and Dummy Activity;
    - understand the key rules for drawing Networks;
    - know the various ways Activities are identified;
    - be able to use Dummy Activities correctly.

Definition

2. Network analysis is a generic term for a family of related techniques developed to aid management to plan and control projects. These techniques show the inter-relationship of the various jobs or tasks which make up the overall project and clearly identify the critical parts of the project. They can provide planning and control information on the time, cost and resource aspects of a project. Network analysis is likely to be of most value where projects are:

    a) complex, i.e. they contain many related and interdependent activities; and/or

    b) large, i.e. where many types of facilities, high capital investments, many personnel are involved; and/or

    c) where restrictions exists, i.e. where projects have to be completed within stipulated time or cost limits, or where some or all of the resources (material, labour) are limited.

Background

3. A basic form of network analysis was being used in the UK and USA in the mid-1950s in an attempt to reduce project times.

    In 1958 the US Naval Special Projects Office set up a team to devise a technique to control the planning of complex projects. The outcome of the team's efforts was the development of the network technique known as PERT (Programme Evaluation and Review Technique). Pert was used to plan and control the development of the Polaris missile and was credited with saving two years in the missile's development.

    Since 1958 the technique has been developed and nowadays many variants exist which handle, in addition to basic time factors, costs, resources, probabilities and combinations of all these factors. A variety of names exist and some of the more commonly used are:

Critical Path Planning   CPP
Critical Path Analysis   CPA
Critical Path Scheduling   CPS
Critical Path Method   CPM
Programme Evaluation and Review Technique etc. PERT, PERT/COST

Basic network terminology

4. Only the basic elements of networks are covered to start with, other more complex features are introduced as required in later chapters.

Activity

This is a task or job of work which takes time and resources, e.g. Build a Wall, Verify the debtors in a sales ledger, Dig foundations, etc. An activity is represented in a network by an arrow thus:

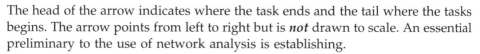

The head of the arrow indicates where the task ends and the tail where the tasks begins. The arrow points from left to right but is *not* drawn to scale. An essential preliminary to the use of network analysis is establishing.

a) what activities are involved in the project.

b) their logical relationship, e.g. the activity of Building a Wall must take place after the activity, Dig Foundations.

c) an estimate of the time the activity is expected to take. Note that the basic time estimate is always necessary but in addition other estimates of times, costs, resources, probabilities etc may also be required. These other factors are dealt with later.

Event

This is a point in time and indicates the start or finish of an activity, or activities, e.g. Wall built, Debtors verified, Foundations Dug, etc. An event is represented in a network by a *circle* or node thus:

It will be noted that the establishment of activities automatically determines events because they are the start and finish of activities and represent the achievement of a specific stage of a project.

Dummy activity

This is an activity which does not consume time or resources. It is used merely to show clear, logical dependencies between activities so as not to violate the rules for drawing networks. It is represented on a network by a *dotted arrow* thus:

- - - - - - - ►

Note that dummy activities are not usually listed with the real activities but become necessary as the network is drawn. Dummy activity examples are given after the rules for drawing networks have been discussed.

Network

This is the combination of activities, dummy activities and events in logical sequence according to the rules for drawing networks. Thus a small network might appear as follows:

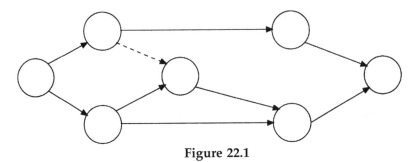

Figure 22.1

Rules for drawing networks

5. The following rules are all logically based and should be thoroughly learned before attempting to draw networks.

   a) A complete network should have only one point of entry – a *start* event and only one point of exit – a *finish* event.

   b) Every activity must have one preceding or 'tail' event and one succeeding or 'head' event. Note that many activities may use the same tail event and many may use the head event, e.g.

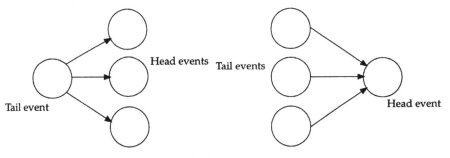

Figure 22.2

However an activity must not share the same tail event *and* the same head event with any other activities (this is dealt with in detail in Para 8 on Dummies).

   c) No activity can start until its tail event is reached.

   d) An event is not complete until all activities leading in to it are complete. This is an important rule and invariably has to be applied in examination questions.

   e) 'Loops', i.e. a series of activities which lead back to the same event, are not allowed because the essence of networks is a progression of activities always moving onwards in time.

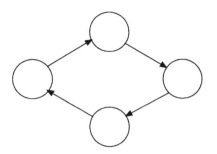

'Loops' not to be used. Figure 22.3

f)   All activities must be tied into the network, i.e. they must contribute to the progression or be discarded as irrelevant. Activities which do not link into the overall project are termed 'danglers'.

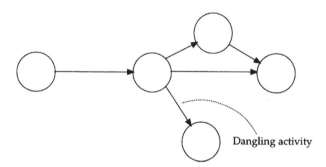

Dangling activity

Dangler. Not to be used. Figure 22.4

Conventions for drawing networks

6.   In addition to the Rules in Paragraph 5 above, which must not be violated, certain conventions are usually observed and for the sake of uniformity and easier communication students are recommended to follow the normal conventions.

a)   Networks proceed from left to right.

b)   Networks are not drawn to scale i.e. the length of the arrow does not represent time elapsed.

c)   Arrows need not be drawn in the horizontal plane but unless it is totally unavoidable they should proceed from left to right.

d)   If there are not already numbered, events or nodes should be progressively numbered from left to right. Simple networks may have events numbered in simple numeric progression, i.e. 0, 1, 2, 3, etc., but larger, more realistic networks may be numbered in 'fives', i.e. 0, 5, 10, 15, etc., or 'tens', i.e. 0, 10, 20, 30, etc. This enables additional activities to be inserted subsequently without affecting the numbering sequence of the whole project.

e) In addition to the normal conventions, the following hints will provide some help in drawing a network from a given precedence table:

- remember there is one starting event and one finishing event;
- enter first the activity(ies) with no pre-requisites;
- next enter the activity(ies) which follow on from activities already in the network.

In this way the network will develop logically from left to right.

Activity identification

7. Activities may be identified in several ways and students should familiarise themselves with the various methods so that unfamiliar presentation does not cause confusion. Typical of the methods to be found include:

a) Shortened description of the job e.g. plaster wall, order timber, etc.

b) Alphabetic or numeric code. e.g. A, B, C etc. or 100, 101, 108, etc.

c) Identification by the tail and head event numbers e.g. 1–2, 2–3, 2–5, etc.

Dummy activities

8. It will be recalled that a dummy activity is one that does not consume time or resources but merely shows a logical relationship. It is shown on a network by a dotted arrow.

Dummy example. Assume that part of the network involves a car arriving at a service station during which two independent activities take place, filling with petrol (A) and topping up with oil (B).

This could be shown thus (incorrectly):

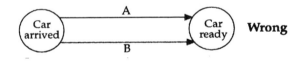

Figure 22.5

NB. This is wrong because it contravenes rule (b) para 5.

With the use of a dummy activity it could be shown thus (correctly):

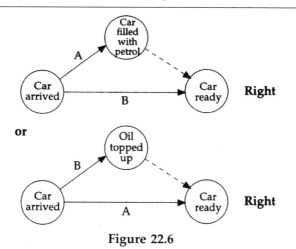

Figure 22.6

On occasions, dummies are necessary to preserve the logic contained in the precedence table. Consider the following:

Activity	Preceding Activity
A	–
B	–
C	A
D	C
E	B and C

The project is complete when both D and E are complete, but it will be noted that E follows both B and C whereas D follows only C. As a consequence, a dummy activity becomes necessary as shown in Figure 22.7.

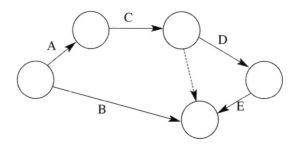

Figure 22.7

Network example

9. To summarise the material so far, a simple network example is given. Try to draw the network yourself and compare the logic of your network with the solution given.

   Using the hint for drawing networks given in Para 6, Activities A and B are entered first, then Activities C, D and E as they follow A, and so on progressively through the network.

Project XXX Building a Boat

Activity	Preceding activity	Activity description
A	–	Design Hull
B	–	Prepare Boat Shed
C	A	Design Mast and Mast mount
D	A	Obtain Hull
E	A	Design Sails
F	C	Obtain Mast Mount
G	C	Obtain Mast
H	C	Design Rigging
J	B,D	Prepare Hull
K	F,J	Fit Mast Mount to Hull
L	E,H,G,K	Step Mast
M	E,H	Obtain Sails and Rigging
N	L,M	Fit Sails and Rigging

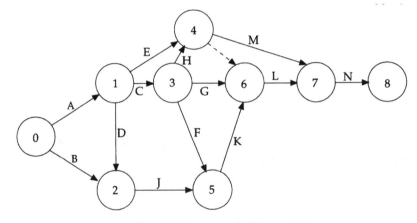

Project XXX network Figure 22.8

Notes on solution given.

a) The events have been numbered from 0, the start event, through to 8, the finish event.

b) A dummy (4–6) was necessary because of the preceding activity requirements of activity L. If activities E, H had not been specified as preceding activity L, the dummy would not have been necessary.

c) The shape of the network is unimportant but the logic must be correct.

Summary

10. a) Network analysis is used for the planning and control of large, complex projects.

b) Networks comprise activities (represented thus ⟶) and events (represented thus ◯). Activities consume time and resources, events are points in time.

c) Networks have one start event and one end event. An event is not complete until all activities leading into it are complete.

d) The length of the arrows representing activities is not important because networks are not drawn to scale.

e) Dummy activities (represented thus ------►) are necessary to show logical relationships. They do not consume time or resources. They become necessary as the network is drawn.

Points to note

11. a) Network analysis is a popular type of examination question.

b) Network analysis is an important management tool and in some industries notably civil engineering and construction, is used on a day by day basis.

c) Rarely does one draw a network in a neat, orderly fashion at the first attempt. Accordingly it is is a useful examination technique to do a draft and copy this out neatly for the official answer.

Self review questions *Numbers in brackets refer to paragraph numbers*

1. What is Network Analysis? (2)
2. What are activities and events? (4)
3. What are the basic rules for drawing networks? (5)
4. Does the length of an activity arrow represent the time taken? (6)
5. What is a dummy activity? (8)

Exercises with answers

1. Draw the network for the following problem.

Activity	Preceding activity
1	–
2,3,4	1
5	2
6	3
7	5
8	6
9	7,8
10	3
11	4
12	9,10,11

2. Draw the network in question 1 except that Activity 8 is preceded by 6 and 2.

Answers to exercises

1.

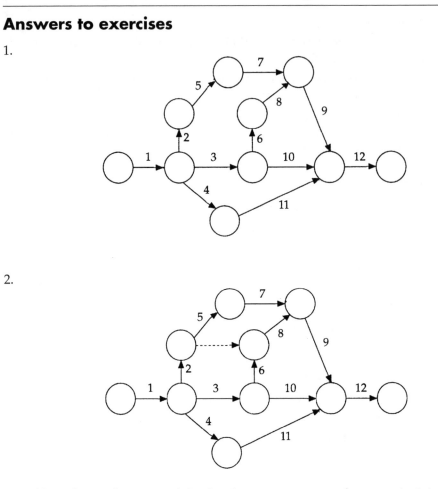

2.

Note that a dummy activity has become necessary because Activity 8 is now preceded by both Activities 6 and 2.

23 Network analysis – time analysis

Objectives

1. After studying this chapter you will
    - know that there may be single or multiple time estimates for each activity;
    - be able to define the Critical Path;
    - understand how to find the Critical path using the forward pass and the backward pass;
    - know what is meant by Float and how it is calculated;
    - understand how basic time analysis can be extended using multiple time estimates;
    - be able to estimate the Standard deviation of the project duration;
    - know how to use both Continuous and Discrete probabilities in Networks.

Assessing the time

2. Once the logic has been agreed and the outline network drawn it can be completed by inserting the activity duration times.

    a) Time estimates. The analysis of project times can be achieved by using:

       i) single time estimates for each activity. These estimates would be based on the judgement of the individual responsible or by technical calculations using data from similar projects;

       ii) multiple time estimates for each activity. The most usual multiple time estimates are three estimates for each activity, i.e. Optimistic (O), Most Likely (ML), and Pessimistic (P). These three estimates are combined to give an expected time and the accepted formula is:

$$\textbf{Expected time} = \frac{O + P + 4ML}{6}$$

For example assume that the three estimates for an activity are

Optimistic	11 days
Most likely	15 days
Pessimistic	18 days

$$\text{Expected time} \quad = \quad \frac{11 + 18 + 4(15)}{6}$$

$$= \quad \textbf{14.8 days}$$

    b) Use of time estimates. As the three time estimates are converted to a single time estimate there is no fundamental difference between the two methods as regards the basic time analysis of a network. However, on completion of the basic time analysis, projects with multiple time estimates can be further analysed to give an estimate of the probability of completing the project by a scheduled date. (This is dealt with in detail in para 8).

c) Time units. Time estimates may be given in any unit, i.e. minutes, hours, days, weeks, depending on the project. All times estimates within a project must be in the same units otherwise confusion is bound to occur.

Basic time analysis – critical path

3. However sophisticated the time analysis becomes, a basic feature is always the calculation of the project duration which is the duration of the critical path.

Critical path. The critical path of a network gives the **shortest time in which the whole project can be completed**. It is the chain of activities with the longest duration times. There may be more than one critical path in a network and it is possible for the critical path to run through a dummy. The following paragraphs give step by step, the procedure for establishing the critical path.

Earliest start times (EST), Once the activities have been timed it is possible to assess the total project time by calculating the ESTs for each activity. The EST is the earliest possible time at which a succeeding activity can start and the method of calculation will be apparent from the following example.

Assume the following network has been drawn and the activity times estimated in days.

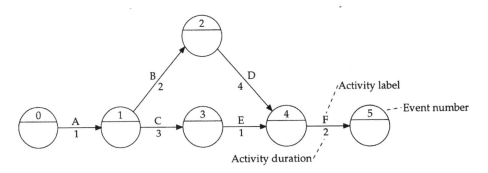

Figure 23/1

The ESTs can be inserted as follows.

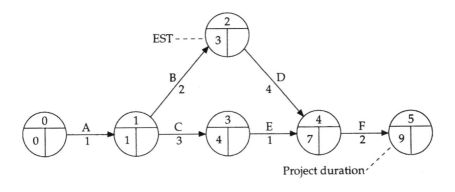

Figure 23/2

Notes on calculation of EST (termed the *forward pass*).

a) The EST of a head event is obtained by adding onto the EST of the tail event the linking activity duration starting from Event 0, time 0 and working forward through the network.

b) Where two or more routes arrive at an event the *longest* route time must be taken, e.g. Activity F depends on completion of D and E. E is completed by day 5 and D is not complete until day 7 ∴ F cannot start before day 7.

c) The EST in the finish event No. 5 is the project duration and is the shortest time in which the whole project can be completed.

Latest start times (LST). To enable the critical path to be isolated, the LST for each activity must be established. The LST is the latest possible time at which a preceding activity can finish without increasing the project duration. Using the example above the LSTs are as follows.

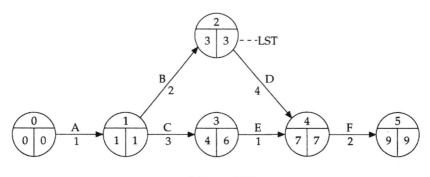

Figure 23/3

Notes on calculating LST (termed the *backward pass*).

a) Starting at the finish event No. 5, insert the LST (i.e. day 9) and work backwards through the network deducting each activity duration from the previously calculated LST.

b) Where the tails of activities B and C join event No. 1, the LST for C is day 3 and the LST for B is day 1. The lowest number is taken as the LST for Event No. 1 because if event No. 1 occurred at day 3 then activities B and D could not be completed by day 7 as required and the project would be delayed.

Critical path

4. Examination of Figure 23/3 shows that one path through the network (A, B, D, F) has EST's and LST's which are identical. This is the **critical path** which it should be noted is the chain of activities which has the **longest duration**.

An activity is critical when EST = LST of the tail event, and EST = LST of the head event, and the EST of the head event *minus* the EST of the tail event equals the activity duration.

The critical path can be indicated on the network either by a different colour or by two small transverse lines across the arrows on the critical path thus:

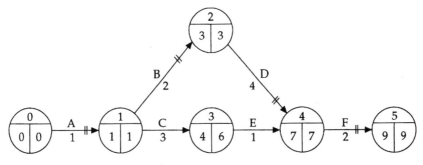

Figure 23/4

Critical path implications. The activities along the critical path are vital activities which must be completed by their ESTs/LSTs otherwise the project will be delayed. The non critical activities (in the example above, C and E) have spare time or *float* available, i.e. C and/or E could take up to an additional 2 days in total without delaying the project duration. If it is required to reduce the overall project duration then the time of one or more of the activities on the critical path must be reduced perhaps by using more labour, or more or better equipment or some other method of reducing job times.

Note that for simple networks the critical path can be found by inspection, i.e. looking for the longest route but the above procedure is necessary for larger projects and must be understood. The procedure is similar to that used by most computer programs dealing with network analysis.

Float

5. Float or spare time can only be associated with activities which are non-critical. By definition, activities on the critical path cannot have float. There are three types of float, Total Float, Free Float and Independent Float. To illustrate these types of float, part of a network will be used together with a bar diagram of the timings thus:

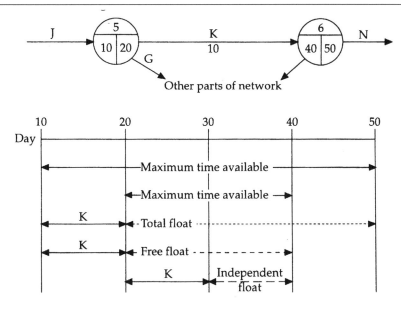

Figure 23/5

a) *Total float*

This is the amount of time a path of activities could be delayed without affecting the overall project duration. (For simplicity the path in this example consists of one activity only, i.e. Activity K).

Total Float = Latest Headtime − Earliest Tail time − Activity Duration
Total Float = 50 − 10 − 10
     = **30 days**

b) *Free float*

This is the amount of time an activity can be delayed without affecting the commencement of a subsequent activity at its earliest start time, but may affect float of a previous activity.

Free Float = Earliest Head time − Earliest Tail time − Activity Duration
Free Float = 40 − 10 − 10
     = **20 days**

c) *Independent float*

This is the amount of time an activity can be delayed when all preceding activities are completed as late as possible and all succeeding activities completed as early as possible. Independent float therefore does not affect the float of either preceding or subsequent activities.

Independent float = Earliest Head time − Latest Tail time − Activity Duration
Independent float = 40 − 20 − 10
     = **10 days**

Notes:

a) For examination purposes the most important type of float is Total Float because it is involved with the overall project duration. On occasions the term 'Float' is used without qualification. In such cases assume that Total Float is required.

b) The total float can be calculated separately for each activity but it is often useful to find the total float over chains of non-critical activities between critical events. For example in Figure 23/4 the only non-critical chain of activities is C, E for which the following calculation can be made:

Non-critical chain	Time required	Time available	Total float over chain
C, E	$3 + 1 = $ **4 days**	$7 - 1 = $ **6 days**	**= 2 days**

If some of the 'chain float' is used up on one of the activities in a chain it reduces the leeway available to other activities in the chain.

c) Alternative terms for Earliest Head Time and Latest Headtime are Earliest Finishing Time (EFT) and Latest Finishing Time (LFT), respectively.

Example of float calculations

6. The example used in the preceding chapter is reproduced below with the addition of activity durations. It is required to find the critical path and all floats.

Project XXX Building a Boat			
Activity	Preceding activity	Activity description	Activity duration (days)
A	–	Design Hull	9
B	–	Prepare Boat Shed	3
C	A	Design Mast and Mast Mount	8
D	A	Obtain Hull	2
E	A	Design Sails	3
F	C	Obtain Mast Mount	2
G	C	Obtain Mast	6
H	C	Design Rigging	1
J	B, D	Prepare Hull	4
K	F, J	Fit Mast Mount to Hull	1
L	E, H, G, K	Step Mast	2
M	E, H	Obtain Sails and Rigging	3
N	L M	Fit Sails and Rigging	4

Solution

The network is shown in the normal manner in Figure 23/6 from which it will be seen that the critical path is:

Activities A, C, G, L, N with a duration of 29 days.

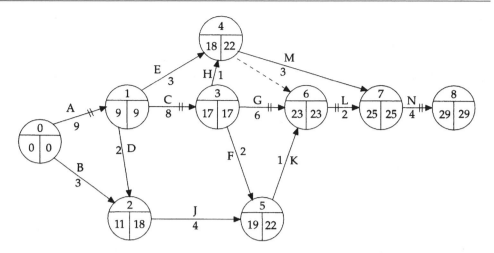

Project XXX network. Figure 23/6

The float calculations are shown in Table 1.

Activity	EST	LST	EFT	LFT	D	Total float (LFT-EST-D)	Free float (EFT-LST-D)	Independent float (EFT-LST-D)
A*	0	0	9	9	9	–	–	–
B	0	0	11	18	3	15	8	8
C*	9	9	17	17	8	–	–	–
D	9	9	11	18	2	7	–	–
E	9	9	18	22	3	10	6	6
F	17	17	19	22	2	3	–	–
G*	17	17	23	23	6	–	–	–
H	17	17	18	22	1	4	–	–
J	11	18	19	22	4	7	4	–
K	19	22	23	23	1	3	3	–
L*	23	23	25	25	2	–	–	–
M	18	22	25	25	3	4	4	–
N*	25	25	29	29	4	–	–	–

* Critical activities   Float calculations

Project XXX. Table 1

The total float on the non critical chains can also be calculated:

Non-critical chain	Time required	Time available	Total float over chain
B, J, K	8	23	15
D, J, K	7	14	7
F, K	3	6	3
E, M	6	16	10
H, M	4	8	4
E, dummy	3	14	11
H, dummy	1	6	5

Slack

7. This is the difference between the EST and LST for each event. Strictly it does not apply to activities but on occasions the terms are confused in examination questions and unless the context makes it abundantly clear that event slack is required, it is likely that some form of activity float is required. Events on the critical path have zero slack.

Further project time analysis

8. More sophisticated time analysis presupposes that some form of distribution is available for each activity time estimate, for example, the three time estimates described in Para 2. These can be used to make statements about the probability of achieving scheduled dates.

Probability example. Assume that a simple project has the following network shown in Fig 23/7. The activity times are in weeks and three estimates have been given for each activity. The expected durations can be found using what is known as the PERT formula:

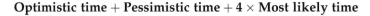

$$\frac{\text{Optimistic time} + \text{Pessimistic time} + 4 \times \text{Most likely time}}{6}$$

For example:

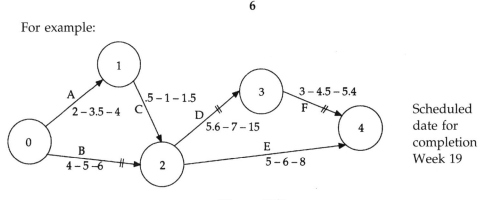

Scheduled date for completion Week 19

Figure 23/7

Critical path (B, D, F) expected duration

$$B = \frac{4 + 6 + 4(5)}{6} = 5 \text{ weeks}$$

$$D = \frac{5.6 + 15 + 4(7)}{6} = 8.1 \text{ weeks}$$

$$F = \frac{3 + 5.4 + 4(4.5)}{6} = \underline{4.4}$$

$$\underline{\underline{17.5}}$$

If the critical activities were to occur at their optimistic times, event 4 would be reached in 12.6 weeks but if the critical activities occurred at their pessimistic times, event 4 would be reached in 26.4 weeks. As these durations span the scheduled date of week 19 some estimate of the probability of achieving the schedule date must be calculated, as follows.

a) Make an estimate of the Standard Deviation for each of the critical activities. If no additional information is available the following PERT formula can be used.

$$\frac{\text{Pessimistic time} - \text{Optimistic time}}{6}$$

i.e. Standard Deviation   Activity B $= \dfrac{6-4}{6}$ $= 0.33$

Activity D $= \dfrac{15 - 5.6}{6}$ $= 1.57$

Activity F $= \dfrac{5.4 - 3}{6}$ $= 0.4$

b) Find the standard deviation of event 4 by calculating the statistical sum (the square root of the sum of the squares) of the standard deviations of all activities on the critical path.

i.e. Standard Deviation of Event $4 = \sqrt{0.33^2 + 1.57^2 + 0.4^2}$

$$= 1.65 \text{ weeks}$$

c) Using principles developed previously, the Z score is calculated thus:

$$x = \text{duration of project}$$

and $Z = \dfrac{x - \mu}{\sigma}$ where $\mu = 17.5$

and $\sigma = 1.65$

thus $Z = \dfrac{19 - 17.5}{1.65} = 0.91$

d) Look up this value (0.91) in a table of areas under the Normal Curve to find the probability (Table I). In this case the probability of achieving the scheduled date of week 19 is **82%**.

Probability interpretation. If management consider that the probability of 82% is not high enough, efforts must be made to reduce the times or the spread of time of activities on the critical path. It is an inefficient use of resources to try to make the probability of reaching the scheduled date 100% or very close to 100%. In this case management may well accept the 18% chance of not achieving the schedule date as realistic.

Notes:

a) The methods of calculating the Expected Duration and Standard Deviation as shown above cannot be taken as strictly mathematically valid but are probably accurate enough for most purposes. It is considered by some experts that the standard deviation, as calculated above, underestimates the 'true' standard deviation.

b) When activity times have variations the critical path will often change as the variations occur. It is necessary therefore to examine critical and near critical activity paths when tackling an examination question involving variable activity times.

Discrete probabilities

9. Instead of the continuous probabilities, which were derived and used in the example in Para 7 above, on occasions probabilities are sometimes expressed in discrete terms. For example the time estimates for an activity could be given as follows:

	Estimates	
Activity	**Time**	**Probability**
A }	8 weeks	0.6
	11 weeks	0.4

The expected time for activity A would be $(8 \times 0.6) + (11 \times 0.4) = $ **9.2 weeks.**

Discrete probability example. Assume that time estimates have been made for the following network using discrete probabilities thus:

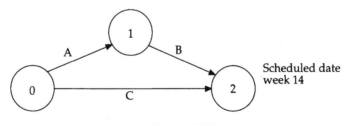

Scheduled date
week 14

Figure 23/8

	Estimates	
Activity	**Time (Weeks)**	**Probability**
A	6	0.5
	10	0.5
B	3	0.4
	5	0.6
C	12	0.6
	14	0.3
	17	0.1

The expected times for the activities are:

$$A = (6 \times 0.5) + (10 \times 0.5) \qquad\qquad = 8$$
$$B = (3 \times 0.4) + (5 \times 0.6) \qquad\qquad = 4.2$$
$$C = (12 \times 0.6) + (14 \times 0.3) + (17 \times 0.1) = 13.1$$

On the basis of the expected times the critical path is C with a duration of **13.1** weeks. However, numerous other possibilities exist and the probabilities of the various completion times and thus of achieving the schedule date of week 14 can be evaluated as follows:

The A, B route can have four durations, each with an associated probability thus:

A, B route

Durations	9	11	13	15	weeks
Probability	0.2	0.3	0.2	0.3	

(These values are found by combining the durations and probabilities of Activities A and B. For example Activity A duration of 6 weeks, probability 0.5, can be combined with Activity B duration of 3 weeks, probability 0.4, to give 9 weeks duration and probability of 0.2 (i.e. 0.5 × 0.4).)

C route

Duration	12	14	17	weeks
Probability	0.6	0.3	0.1	

The A, B route and the C route alternate as the critical path with varying probabilities as shown in the following table.

			A, B route			
			9	11	13	15	Duration
			0.2	0.3	0.2	0.3	Probability
	Duration	**Probability**				
	12	0.6	12 / 0.12	12 / 0.18	13 * / 0.12	15 / 0.18
C route	14	0.3	14 / 0.06	14 / 0.09	14 / 0.06	15 / 0.09
	17	0.1	17 / 0.02	17 / 0.03	17 / 0.02	17 / 0.03

*This means that if the A, B route with 13 weeks duration, probability 0.2, occurs at the same time as the C route duration of 12 weeks, probability 0.6, the critical path would be 13 weeks i.e. the longer duration, with the probability of 0.12 (0.6 × 0.2).

Summary of possible durations

12 weeks probability (0.12 + 0.18)	= 0.30
13 weeks probability (0.12)	= 0.12
14 weeks probability (0.06 + 0.09 + 0.06)	= 0.21
15 weeks probability (0.18 + 0.09)	= 0.27
17 weeks probability (0.02 + 0.03 + 0.02 + 0.03)	= 1.10
	1.00

Thus the probability of achieving 14 weeks or less is **0.63** (0.30 + 0.12 + 0.21) and the probability of exceeding the scheduled date is **0.37**.

Summary

10. a) Basic time analysis of a network involves calculating the critical path, i.e. the shortest time in which the project can be completed.

b) The critical path is established by calculating the EST (Earliest Start Times) and LST (Latest Start Time) for each event and comparing them. The critical path is the chain of activities where the ESTs and LSTs are the same.

c) Float is the spare time available on non-critical activities. There are three types of float; total float, free float and independent float.

d) To calculate the probability of achieving scheduled dates it is necessary to establish what variability is likely to exist for each activity.

e) Given activity time estimates the expected value of the project duration is calculated and an estimate made of the activity standard deviations.

f) The activity standard deviations are combined to form the standard deviation of the overall project so that probability estimates can be made for various project duration possibilities.

Points to note

11. a) The problems of dealing with uncertainty occur again and again in quantitative techniques. Increasingly examination questions are likely to contain variable activity times and associated probabilities.

b) When variable time estimates are used, near critical paths may have more variability than the critical path so the near critical paths may influence probability estimates.

c) The basis of the Standard Deviation estimate given in para 8 is the Normal Distribution. It is known that virtually all of the distribution lies within the mean $\pm 3\sigma$ hence the formula given, i.e.

$$\frac{\text{max} - \text{min}}{6} = \sigma$$

However the tails of the distribution are unlikely to occur very often so the 95% concept may be used in which case the range becomes mean $\pm 2\sigma$. If this is thought to be realistic the revised formula below can be used.

$$\frac{\textbf{Pessimistic time} - \textbf{Optimistic time}}{4} = \sigma$$

Self review questions *Numbers in brackets refer to paragraph numbers*

1. What types of time estimates are made for activity durations? (2)
2. What is the critical path? (3)
3. What are the ESTs and LSTs? (3)
4. How is the critical path determined? (4)
5. What is float? (5)
6. When multiple time estimates of activity durations are available how can an estimate be calculated of the probability of completing the network in a given time? (8)

Exercises with answers

1. Find the critical path of the following network using the EST/LSTs.

Activity	Preceding activity	Duration (days)
1	–	4
2	1	7
3	1	5
4	1	6
5	2	2
6	3	3
7	5	5
8	2, 6	11
9	7, 8	7
10	3	4
11	4	3
12	9, 10, 11	4

2. Calculate the floats of the network in question 1.

3. The standard deviations of the activities on the critical path in question 1 are: 1, 2, 1.5, 3, 2.5 and 3 respectively. Based on these values calculate the probability of achieving a scheduled time of 40 days for the project duration.

Answers to exercises

1.

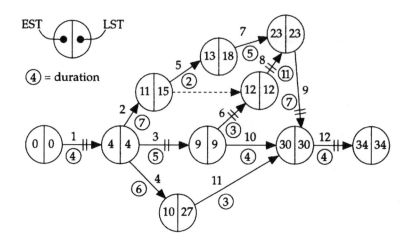

2.

Activity	EST	LST	EFT	LFT	D	Total float LFT-EST-D	Free float EFT-EST-D	Independent float EFT-LST-D
*1	0	0	4	4	4	–	–	–
2	4	4	11	15	7	4	–	–
*3	4	4	9	9	5	–	–	–
4	4	4	10	22	6	12	–	–
5	11	15	13	21	2	8	–	–
*6	9	9	15	15	3	–	–	–
7	13	21	23	23	5	5	5	–
*8	12	12	23	23	11	–	–	–
*9	23	23	30	30	7	–	–	–
10	9	9	30	30	4	17	17	17
11	10	22	30	30	3	17	17	5
*12	30	30	34	34	4	–	–	–

3. s.d. of critical path = $\sqrt{1^2 + 2^2 + 1.5^2 + 3^2 + 2.5^2 + 3^2}$ = **5.61**

$$\therefore Z = \frac{40 - 34}{5.61} = 1.07$$

From Table 1 the probability is nearly **36%**

24 Network analysis – cost scheduling

Objectives

1. After studying this chapter you will
   - understand the principles of least cost scheduling or 'crashing' the network;
   - know the meaning of normal and crash costs, normal and crash times, and cost slopes;
   - be able to use the rules of least cost scheduling;
   - know how to crash a network.

Costs and networks

2. A further important feature of network analysis is concerned with the costs of activities and of the project as a whole. This is sometimes known as PERT/COST.

   Cost analysis objectives. The primary objective of network cost analysis is to be able to calculate the cost of various project durations. The *normal duration* of a project incurs a given cost and by more labour, working overtime, more equipment etc the duration could be reduced but at the expense of higher costs. Some ways of reducing the project duration will be cheaper than others and network cost analysis seeks to find the cheapest way of reducing the overall duration.

   Penalties and Bonuses. A common feature of many projects is a penalty clause for delayed completion and/or a bonus for earlier completion. In examination questions, network costs analysis is often combined with a penalty and/or bonus situation with the general aim of calculating whether it is worthwhile paying extra to reduce the project time so as to save a penalty.

   Cost and networks – basic definitions

   a) Normal cost. The costs associated with a normal time estimate for an activity. Often the 'normal' time estimate is set at the point where resources (men, machines, etc.) are used in the most efficient manner.

   b) Crash cost. The costs associated with the minimum possible time for an activity. Crash costs, because of extra wages, overtime premiums, extra facility costs are always higher than normal costs.

   c) Crash time. The minimum possible time that an activity is planned to take. The minimum time is invariably brought about by the application of extra resources, e.g. more labour or machinery.

   d) Cost slope. This is the average cost of shortening an activity by one time unit (day, week, month as appropriate). The cost slope is generally assumed to be linear and is calculated as follows:

$$\text{Cost slope} = \frac{\text{Crash cost} - \text{Normal cost}}{\text{Normal time} - \text{Crash time}}$$

e.g. Activity A data :

Normal	Crash
Time | Cost	Time | Cost
12 days at £480	8 days at £640

$$\text{Cost slope} = \frac{640 - 480}{12 - 8}$$

$$= \underline{\textbf{£40/day}}$$

e) Least cost scheduling or 'crashing'. The process which finds the least cost method of reducing the overall project duration, time period by time period. The following example shows the process step by step.

Least cost scheduling rules

3. The basic rule of least cost scheduling is simply stated. Reduce the time of the activity on the critical path with the lowest cost slope and progressively repeat this process until the desired reduction in time is achieved. Complications occur when time reductions cause several paths to become critical simultaneously thus necessitating several activities to be reduced at the same time. These complications are explained below as they occur.

Least cost scheduling example

A project has five activities and it is required to prepare the least cost schedules for all possible durations from 'normal time' – 'normal cost' to 'crash time' – 'crash cost'.

Project data activity	Preceding activity	Time (days) Normal	Crash	Costs (£) Normal	Crash	Cost (£) Slope
A	–	4	3	360	420	60
B	–	8	5	300	510	70
C	A	5	3	170	270	50
D	A	9	7	220	300	40
E	B, C	5	3	200	360	80

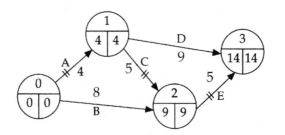

Project network. Figure 24/1

Project durations and costs

a) Normal Duration **14 days**

    Critical path A, C, E

    Project cost (i.e. cost of *all* activities at normal time) = **£1,250**

    (i,e, £360 + 300 + 170 + 220 + 200)

b) Reduce by 1 day the activity on the critical path with the lowest cost slope.

    Reduce activity C at extra cost of £50

    Project Duration **13 days**

    Project cost **£1,300**

Note: *All* activities are now critical.

c) Several alternative ways are possible to reduce the project time by a further 1 day but note 2 or 3 activities need to be shortened because there are several critical paths.

	Possibilities available	
Reduce by 1 day	**Extra Costs**	**Activities critical**
A and B	£60 + 70 = £130	All
D and E	£40 + 80 = £120	All
B, C and D	£70 + 50 + 40 = £160	All
A and E	£60 + 80 = £140	A,D,B,E

An indication of the total extra costs apparently indicates that the second alternative (i.e. D and E reduced) is the cheapest. However, closer examination of the last alternative (i.e. A and E reduced) reveals that activity C is non-critical and with 1 day float. It will be recalled that Activity C was reduced by 1 day previously at an extra cost of £50. If in conjunction with the A and E reduction, Activity C is *increased* by 1 day, the £50 is saved and all activities become critical. The net cost therefore for the 12 day duration is **£1,300 + (140 − 50) = £1,390**. The network is now as follows:

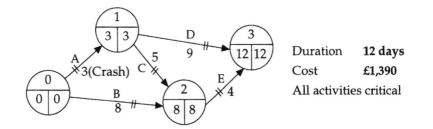

Duration   **12 days**

Cost   **£1,390**

All activities critical

d) the next reduction would be achieved by reducing D and E at an increase of £120 with once again all activities being critical.

    Project duration **11 days**

    Project cost **£1,510**

e) The final reduction possible is made by reducing B, C and D at an increased cost of £160. The final network becomes:

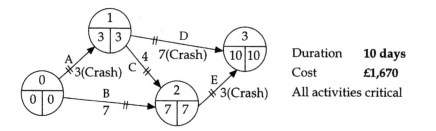

Duration   **10 days**
Cost   **£1,670**
All activities critical

Crashing networks in examinations

4. All the principles necessary to crash networks have already been covered and the following points may save time in an examination.

a) Only critical activities affect the project duration to take care not to crash non-critical activities.

b) The minimum possible project duration is not necessarily the most profitable option. It may be cost effective to pay some penalties to avoid higher crash costs.

c) If there are several independent critical paths then several activities will need to be crashed simultaneously. If there are several critical paths which are not separate i.e. they share an activity or activities, then it may be cost effective to crash the shared activities even though they may not have the lowest cost slopes.

d) Always look for the possibility of *increasing* the duration of a previously crashed activity when subsequent crashing renders it non-critical, i.e. it has float.

Summary

5. a) Cost analysis of networks seeks the cheapest ways of reducing project times.

b) The crash cost is the cost associated with the minimum possible time for an activity, which is known as the **crash time**.

c) The average cost of shortening an activity by one time period (day, weeks etc) is known as the **cost slope**.

d) Least cost scheduling finds the cheapest method of reducing the overall project time by reducing the time of the activity on the critical path with the lowest cost slope.

Points to note

6. a) The total project cost includes **all** activity costs not just those on the critical path.

b) The usual assumption is that the cost slope is linear. This need not be so and care should be taken not to make the linearity assumption when circumstances point to some other conclusion.

c) The example used in this chapter includes increasing the time of a subcritical activity, which has already been crashed, so saving the extra costs incurred. Always look for such possibilities.

d) Dummy activities have zero slopes and cannot be crashed.

Self review questions *Numbers in brackets refer to paragraph numbers*

1   What is the objective of network cost analysis? (2)

2.   What are 'normal' costs and 'crash' costs? (2)

3.   What is the basic rule of least cost scheduling? (3)

Exercises with answers

1.   Calculate the cost slopes and the critical path of the following network:

Activity	Preceding activity	Time Normal	Time Crash	Cost Normal £	Cost Crash £
1	–	5	3	500	620
2	–	4	2	300	390
3	1	7	6	650	680
4	1	3	2	400	450
5	2,3	5	3	850	1,000

2.   Construct a least cost schedule for the network in question 1 showing all durations from normal time – normal cost to crash time – crash cost.

Answers to exercises

1.   Critical path

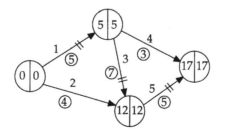

Cost slopes

Activity	Cost slope
1	£60
2	45
3	30
4	50
5	75

2. Total normal cost £2,700 with 17 day duration

	Cost
16 day duration (Activity 3)	£2,730
15 day duration (Activity 1)	£2,790
14 day duration (Activity 1)	£2,850
13 day duration (Activity 5)	£2,925
12 day duration (Activity 5)	£3,000

25 Network analysis – resource scheduling

Objectives

1. After studying this chapter you will
   - understand the principles of Resource Scheduling;
   - know how to draw a Gantt chart;
   - be able to prepare a Resource Aggregation Profile;
   - know how to use Resource Levelling;
   - be able to prepare a Resource Allocation Profile.

Resources and networks

2. The usefulness of networks is not confined only to the time and cost factors which have been discussed so far. Considerable assistance in planning and controlling the use of resources can be given to management by appropriate development of the basic network techniques.

   Project resources. The resources (men of varying skills, machines of all types, the required materials, finance, and space) used in a project are subject to varying demands and loadings as the project proceeds. Management need to know what activities and what resources are critical to the project duration and if resource limitations (e.g. shortage of materials, limited number of skilled craftsmen) might delay the project. They also wish to ensure, as far as possible, constant work rates to avoid paying overtime at one stage of a project and having short time working at another stage.

Resource scheduling requirements

3. To be able to schedule the resource requirements for a project the following details are required.
   a) The customary activity times, descriptions and sequences as previously described.
   b) The resource requirements for each activity showing the classification of the resource and the quantity required.
   c) The resource in each classification that are available to the project. If variations in availability are likely during the project life, these must also be specified.
   d) Any management restrictions that need to be considered, e.g. which activities may or may not be split or any limitations on labour mobility.

Resources scheduling example, using a Gantt chart

4. A simple project has the following time and resource data (for simplicity, only the one resource of labour is considered but similar principles would apply to other types of inter-changeable resources).

Project data

Activity	Preceding activity	Duration (days)	Labour requirements
A	–	1	2 men
B	–	2	1 man
C	A	1	1 man
D	–	5	1 man
E	B	1	1 man
F	C	1	1 man

Resource constraint, 2 men only available

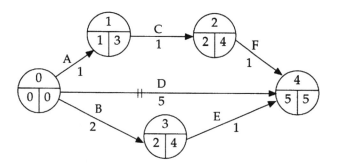

Critical path – Activity D – Duration 5 days
(Without taking account of the resource constraint)

Resource Scheduling Steps

a) Draw the activity times on a Gantt or Bar Chart based on their ESTs.

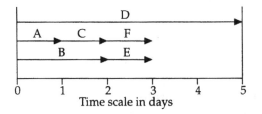

Time scaled network

b) Based on the time bar chart prepare a Resource Aggregation Profile, i.e. total resource requirements in each time period.

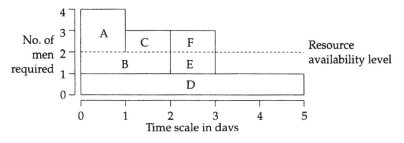

Resource aggregation profile based on ESTs

c) Examination of the above profile shows that at times more resources are required than are available if activities commence at their ESTs. The ESTs/LSTs on the network show that float is available for activities A, C, F, B and E. Having regard to these floats it is necessary to 'smooth out' the resource requirements so that the resources required do not exceed the resource constraint, i.e. delay the commencement of activities (within their float) and if this procedure is still not sufficient then delay the project as a whole. Carrying out this procedure results in the following resource profile.

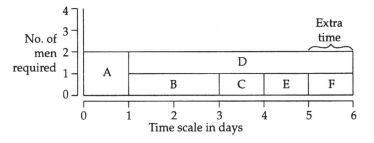

Resource allocation – with 2 man constraints

Note: This procedure is sometimes termed *resource levelling*.

d) Because of the resource constraint of 2 men it has been necessary to extend the project duration by 1 day. Assume that management state that the original project duration (5 days) must not be extended and they require this to be achieved with the minimum extra resources. In such cases a similar process of varying activity start times within their float is carried out, resulting in the following resource profile.

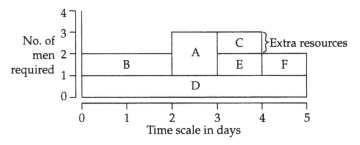

Resource allocation profile – with 5 day constraint

e) The above profile shows that to achieve the 5 day duration it is necessary to have 3 men available from day 2 to day 4.

Summary

5. a) To enable resource scheduling to be carried out the resource requirements for each activity must be specified.

b) In addition the various resources involved (men, machinery, etc.) must be classified and the availability and constraints specified.

c) After calculating the critical path in the usual manner a Resource Aggregation Profile(s) is prepared, i.e. the amount of the resource(s) required in each time period of the project based on the ESTs of each activity.

d) If the resource aggregation indicates that a constraint is being exceeded, and float is available the resource usage is 'smoothed', i.e. the start of activities is delayed.

Points to note

6. The smoothing of resource profiles is largely a matter of experimentation but if the time for the project is fixed concentrate attention on those activities with free float (described in Chapter 23).

Self review questions *Numbers in brackets refer to paragraph numbers*

1. What data are required to be able to carry out resource scheduling on a network? (3)

2. What is a resource aggregation profile? (4b)

3. What is resource levelling? (4c)

Exercises with answers

1. A project has the following activity durations and resource requirements.

Activity	Preceding activity	Duration (days)	Resource requirements (units)
A	–	6	3
B	–	3	2
C	–	2	2
D	C	2	1
E	B	1	2
F	D	1	1

Assuming no restrictions show the network, critical path and resource requirements on a day by day basis, assuming that starts are made on the EST of each activity.

2. What would be the plan if there are only 6 units of resources?

Answers to exercises

1.

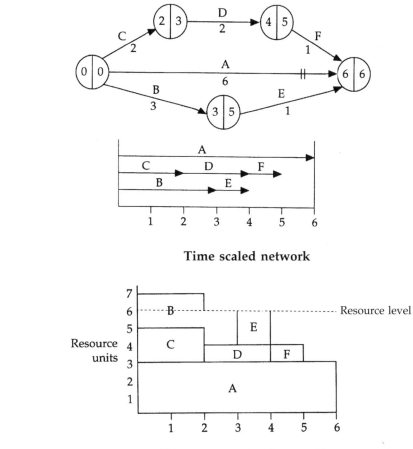

Time scaled network

Resource aggregation profile

2. Delay start of B/E until day 2 resulting in

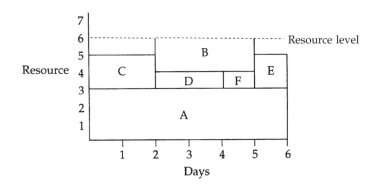

'Smoothed' resource requirements

26 Network analysis – activity on nodes

Objectives

1. After studying this chapter you will
   - understand the characteristics of Activity on Node networks (or Precedence Diagrams);
   - be able to draw Activity on Node networks;
   - know how to identify the Critical Path in such networks;
   - understand how computers can assist in network analysis.

Activity on node networks

2. This type of network, sometimes known as *precedence diagram*, is an alternative form of presentation based on similar estimates to those previously described.

   Activity on node network characteristics.

   a) The activities are shown as boxes, not as arrows.
   b) No events appear.
   c) Boxes are linked by lines to indicate their sequence or precedence.
   d) A box appears for the *start* and another for the *finish* of a project.
   e) No dummies are necessary.

Comparison with conventional networks

3. The Activity on Node approach is best illustrated by comparison with the procedures for drawing conventional networks.

<table>
<tr><th>Example</th><th>Conventional network</th><th>Activity on node network</th></tr>
<tr><td>A Activity X depends upon Activity Y</td><td></td><td></td></tr>
<tr><td>B Activity Z depends upon Activities X and Y</td><td></td><td></td></tr>
<tr><td>C Activity C depends upon Activity A. Activity D depends upon Activities A and B.</td><td></td><td></td></tr>
<tr><td>D Activity B can start immediately Activity A is Complete. Durations are 8 days and 6 days respectively.</td><td></td><td></td></tr>
</table>

Notes:

a) Events as such do not appear in Activity on Node (A/N) networks.

b) The lines shown in A/N networks indicate precedences and are not activities. The activities are represented by the square boxes.

c) From example C it will be seen that dummies are not necessary in A/N networks.

d) The time (6 days) shown on the A/N network in Example D is known as the Dependency time.

Activity on node example

4. Based on the conventions described above a full A/N network is shown below. This uses the same activities, sequences and durations already used in in Para 4, Chapter 25.

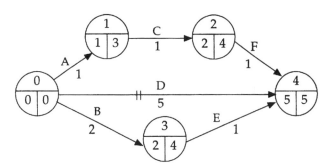

Conventional network (from Chapter 25)

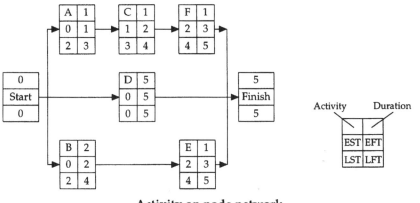

Activity on node network

Notes on Activity on node network.

a) The lines joining the boxes are *not* activities, they are precedences as originally specified.

b) The ESTs/LSTs in the boxes are calculated by the same process as previously described but note that they are not the same values as shown in the events of the conventional network.

c) The EFT (Earliest Finish Time) and LFT (Latest Finish Time) are calculated by adding the activity time on to the EST and LST respectively.

d) The critical path is found by the usual method of comparing the EST and the LST (or EFT and LFT).

Networks and computers

5. So far only the manual handling of networks has been considered but in practice most network analysis is handled by computers. Most installations have standard packages to deal with this technique and there are a number of advantages in utilising these facilities.

a) By virtue of their speed computers can handle large networks with ease. Beyond a certain number of activities (say 50–100) networks become very difficult to handle manually and errors are almost certain to occur.

b) Full activity and event descriptions can be handled and printed out.

c) 'Input' data has thorough validation checks and the network logic ('loops', 'danglers', etc.) is tested.

d) Once the data has been processed output can be presented in various ways, e.g. in activity or event sequence, scheduled data sequence, activities by float sequence, EST, LST, EFT or LFT sequence, etc.

e) When changes in activity times, resources, or scheduled dates occur, the program can be quickly rerun to produce the new results.

f) More comprehensive time analysis and resource allocation can take place.

g) At one time the networks had to be drawn manually and the data fed into the computer. Packages are now available which allow the user to build up a network using the graphics facilities of the computer. The package will then draw out the network diagram and calculate the project duration and isolate the critical path. More comprehensive packages will undertake cost scheduling and resource scheduling.

Summary

6. a) Activity on node networks are an alternative method of presentation to conventional networks.

b) A/N networks do not show events but show the links (precedences) between activities.

c) Forward and backward passes are made as usual. The forward pass gives the same EST as in conventional networks, but the backward pass gives the LST of the activity in the node rather than the LST of the succeeding activity as in the conventional networks.

d) The critical path is established in the normal manner by comparing EST/LST, or EFT/LFT.

e) Computers are widely used for practical problems because of their greater speed, calculating ability and the existence of many standard packages.

Points to note

7. a) It is claimed that A/N networks are easier to draw and so can be used by more people.

    b) At least one of the professional bodies occasionally have questions involving A/N networks.

Self review questions *Numbers in brackets refer to paragraph numbers*

1   What are the features of activities on Node networks? (2)

2   What do the arrows represent in Activity on Node networks? (3)

3   How is the critical path established in A/N networks? (4)

4   Why are computers invariably used for solving practical network analysis problems? (5)

Exercises with answers

1. Draw an activity on node diagram for the following project.

Activity	Preceding activity	Duration (days)
1	–	4
2	1	7
3	1	5
4	1	6
5	2	2
6	3	3
7	5	5
8	2, 6	11
9	7, 8	7
10	3	4
11	4	3
12	9, 10, 11	4

2. Calculate the EST/LST and LFT values for each box.

Answers to exercises

1. & 2.

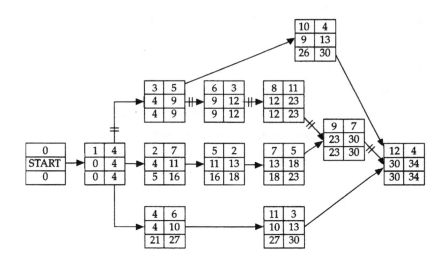

Note: Contrast the above network with Question 1, Chapter 23.

Assessment and revision Chapters 22 to 26

Examination questions with answers

Answers commence on page 547

A1. The network for a small building project is shown below, together with the time, in days, required to complete each task.

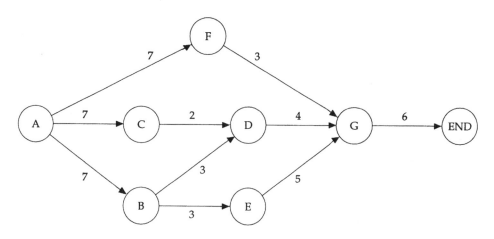

You are required

a) to list the *possible* paths through the network and the length of the path (in days) in each case;

b) to state the critical path and the project duration;

c) calculate the total Float for each activity

d) if D takes 8 days, rather than 4, to explain how much, if at all, the project will be delayed.

CIMA Quantitative Methods

A2. Beeseley Builders have been awarded a contract to build an office block. The project has been broken down into a number of activities.

Activity	Immediately preceding activity	Duration in months	Total cost (£000)
A	–	8	100
B	–	2	75
C	A	3	135
D	A	7	70
E	B	5	160
F	C,D	9	255
G	D	2	30
H	D,E	4	90
I	G,H	3	55

The overheads on this project are £5,000 per month.

Required:

a) Construct a network diagram for this project and hence determine the minimum project duration and its associated cost.

b) Beeseley are subsequently offered a bonus of £25,000 if they can complete the project in 20 months (or less). Their site manager is aware that certain activities can be speeded up, if necessary. These activities, with their associated durations and total costs, are shown in the following table.

Activity	Duration in months	Total cost (£000)
A	6	125
B	1	90
D	5	85
E	3	200
F	7	275
H	2	95

Activities C, G and I cannot be speeded up.

Find the minimum cost project schedule and hence determine whether or not Beeseley should accept the bonus offer.

c) What would your recommendation be if the bonus was £15,000? Give reasons for your recommendation.

ACCA Decision Making Techniques

A3. Preface Retailers is a high-technology retailer and mail-order business. In order to improve its processes the company decides to install a new microcomputer system to manage its entire operation (i.e. payroll, accounts, inventory). Terminals at each of its many stores will be networked for fast, dependable service. The specific activities that Preface will need to accomplish before the system is up and running are listed below. The table also includes the necessary increased staffing to undertake the project.

Activity	Preceding Activities	Duration (days)	Increased staff
A. Build insulated enclosure	–	4	1
B. Decide on computer system	–	1	3
C. Electrical wiring of room	A	3	2
D. Order and collect computer	B	2	1
E. Install air-conditioning	A	4	2
F. Install computer	D, E	2	2
G. Staff testing	B	5	1
H. Install software	C, F	2	1
I. Staff training	G, H	3	1

Required:

a) Draw a network diagram for the project and determine the critical path and its duration.

b) Assuming that all activities start as soon as possible, draw a progress chart for the project, showing the times at which each activity takes place and the manpower requirements.

c) The union has decided that any staff employed on the project must be paid for the duration of the project whether they work or not, at a rate of £500 per day.

Assuming that the same staff are employed on the different activities, determine the work schedule that will minimise labour costs through not necessarily the project time. What is the cost associated with this schedule?

Comment on the validity of this assumption.

ACCA Decision Making Techniques

A4. A particular project comprises 12 activities which have the following durations and precedences:

Activity	Duration (days)	Immediately preceding activities
A	3	–
B	5	A
C	7	L
D	2	L
E	9	–
F	6	B, C
G	5	E
H	6	B, C
I	3	A
J	4	I
K	4	D, F
L	4	–

Required:

a) i)   Represent the project by means of a network diagram.

    ii) Show the earliest and latest times for each activity.

    iii) Determine the critical path and minimum completion time of the project.

b) Calculate the total floats for each activity and explain how they can be of use when allocating resources.

c) Due to a strike the labour force on five activities G, H, I, J, K cannot be utilised. However the non-unionised labour which is used for activities A and B can also be used on these five activities without affecting activity durations.

Activity	G	H	I	J	K
Labour force used	A	B	A	B	B

so that, for example, activities A, G, I cannot overlap in time.

Schedule these five activities outlining the effect, if any, on the project duration time.

ACCA Decision Making Techniques

Examination questions without answers

B1. The following tasks are to be completed on vehicles at a service station. Assume that all the jobs must be done, and that an unlimited number of men is available.

	Tasks (not necessarily in order)	Preceding task
A	Driver arrives and stops	None
B	Driver selects brands of oil and petrol	A
C	Fill petrol tank	B
D	Prepare bill	C and L
E	Receive payment and give stamps	D
F	Wash windscreen	A
G	Polish windscreen	F
H	Check tyre pressure	A
J	Inflate tyres	H
K	Open bonnet	A
L	Check oil equipment	K
M	Add oil	B and L
N	Add distilled water to battery	K
P	Fill radiator	K
Q	Close bonnet	M, N and P
R	Driver departs from forecourt	E, G, J and Q

The radiator, sump and battery are all located under the bonnet for the benefit of this exercise.

Draw a network diagram for the above.

B2. A manufacturing company is planning to introduce a new product for sale in the market. The table below provides a list of the activities required to plan and control this marketing project effectively.

	Activity	Immediate predecessors	Expected duration (days)
A	Initial discussions	–	6
B	Product design	A	22
C	Market survey	A	17
D	Market evaluation	C	3
E	Product costing	B	9
F	Sales plan	C	12
G	Product pricing	D, E	4
H	Prototype construction	F, G	22
I	Market information preparation	B	15
J	Prototype testing	H, I	17

a) Draw a network to represent the various activities of the marketing project.

b) Determine the critical path and minimum project time.

c) It can be assumed that the expected duration of some of the activities are normally distributed, although the remaining activities are know with certainty. The standard deviation of those activities which are uncertain are:

Activity	B	E	H	I	J
Standard deviation (days):	3	1	2.5	2.5	2

Determine a 90% confidence interval for the duration of the critical path activities for the marketing project. State any assumption made.

d) If the project takes longer than 84 days, the company will incur the loss of some potential income. Determine the probability, to two significant figures, that this loss will take place.

ACCA Quantitative Analysis

B3. Your managerial colleagues have recently heard of the techniques of Critical Path Analysis (CPA) and Program Evaluation and Review Technique (PERT).

They are considering a project which can be divided into activities A to H. Estimates have been made of the time and cost each will take under normal conditions and under 'crash conditions'.

Activity	Preceding activities	Normal days	Program cost £	Crash days	Program cost £
A	–	3	160	1	240
B	–	4	230	2	270
C	A, B	4	280	2	300
D	B	5	240	2	290
E	C, D	8	440	4	600
F	E	6	320	4	450
G	A	18	700	10	1,500
H	G,F	6	320	5	400

Site costs are estimated at £300 per day.

Activity G, whose normal time was 18 days costing £700, can be reduced by one day at a time at an extra cost of £100 per day down to 10 days costing £1,500. The other activities can be completed either in normal time and cost or in crash time and cost.

You are required to:

a) identify and state the different paths through the network.

b) calculate and state the normal time for completion and the normal costs;

c) calculate and state the minimum time for completion and the minimum cost of completion in this time;

d) explain clearly the purpose, advantages and disadvantages of both CPA and PERT.

CIMA Quantitative Techniques

B4. Consider the activities required to complete the processing of a customer's order.

Activities	Preceding time activities	Average time in days	Normal variable cost per day £
1. Receipt of order, checking credit rating, etc.	–	2	5
2. Preparation of material specification, availability of material, etc.	1	4	10
3. Inspection, packing etc.	2	1	7
4. Arrangement of transport facilities etc.	1	5	5
5. Delivery	3, 4	3	2

The time for activities 1, 3 and 5 are fixed; for activity 2 there is a 0.5 probability that it will require 2 days and a 0.5 probability that it will require 6 days; for activity 4 a 0.7 probability of taking 4 days, 0.2 of taking 6 days and 0.1 of taking 10 days.

You are required to:

a) draw the network (it is very simple) twice, first using an arrow diagram and secondly an activity-on-node presentation, clearly indicating the meaning of any symbols that you use.

b) indicate the critical path, calculate average duration and variable cost under normal conditions;

c) calculate the minimum and maximum times and the probabilities associated with them.

CIMA Quantitative Techniques

27 Financial mathematics

Objectives

1. After studying this chapter you will
    - have been introduced to key aspects of Financial Mathematics;
    - know how to use Arithmetic Progressions;
    - understand Geometric Progressions;
    - know the principles of Simple and Compound Interest;
    - understand the process of Discounting and how to use Discount tables;
    - know what Annuities are and how to use Annuity Tables;
    - be able to define a Perpetuity;
    - know how to deal with growing or declining Cash Flows.

Series

2. Fundamental to many financial calculations is the process of allocating or paying out or receiving money at regular intervals. Typical examples include: depreciation calculations, investing funds, loan repayment, cash flow analysis. These can be represented by series of which the two most common types are arithmetic and geometric progressions.

Arithmetic progression

3. A series of quantities where each new value is obtained by adding a constant amount to the previous value. The constant amount is sometimes called the *common difference*.

    An arithmetic progression has the form

    $$a, a + d, a + 2d, \ldots, a + (x - 1)d$$

    where a is the first value

    d is the common difference

    x is the number of terms in the series.

Example 1

A firm rents its premises and the rental agreement provides for a regular annual increase of £2,650. If the rent in the first year is £8,500, what is the rent in the tenth year?

$$\text{Rent in 10th year} = a + (x - 1)d$$
$$= 8{,}500 + (10 - 1)\ 2{,}650$$
$$= \underline{\underline{£32{,}350}}$$

Example 2

A firm buys a power press for £32,500 which is expected to last for 20 years and to have a scrap value of £7,500. If depreciation is on the straight line method, how much should be provided for in each year?

In such a problem the number of terms in the series is 1 more than the number of years because the cost is the value at the *beginning* of the first year and the scrap value is at the *end* of the final year.

$$a = 32{,}500 \quad x = 21$$
$$7{,}500 = a + (x - 1)d$$
$$7{,}500 = 32{,}500 + 20d$$
$$d = -1{,}250$$

Straight line depreciation is **£1,250 p.a.**

Sum of arithmetic progressions

4. One way to find the sum of an arithmetic progression is to evaluate each of the successive terms and add them up. This could be lengthy if numerous terms are involved, so the following formula can be used.

$$S = n\ [a + \tfrac{1}{2}\ (x - 1)d]$$

where **S** = sum of the progression.

Example 3

How much rent in total did the firm in Example 1 pay for its premises over the ten years?

$$a = 8{,}500 \quad x = 10 \quad d = 2{,}650$$
$$S = 10\ [8{,}500 + \tfrac{1}{2}\ (10 - 1)\ 2{,}650]$$
$$= \underline{\underline{£204{,}250}}$$

Example 4

An employee, who received fixed annual increments had a final salary of £9,000 p.a. after 10 years. If his total salary was £65,000 over the 10 years, what was his initial salary?

$$65{,}000 = 10 \left[a + \tfrac{1}{2} (10 - 1)d \right]$$

$$6{,}500 = \underline{a + 4.5d} \ \ldots \ldots \text{(i)}$$

and from Para 3 we know that

$$9{,}000 = a + (10 - 1)d$$

$$9{,}000 = \underline{a + 9d} \ \ldots \ldots \text{(ii)}$$

(i) and (ii) are simultaneous equations and (ii) − (i) gives

$$2{,}500 = 4.5d$$

$$\therefore \ d = £555.55$$

substituting we obtain $a = £4{,}000$

Initial salary is £4,000 and

annual increment is £555.55

Geometric progressions

5. A series of quantities where each value is obtained by multiplying the previous value by a constant which is called the *common ratio*.

   A geometric progress has the form

$$a, \ ar, \ ar^2, \ ar^3, \ \ldots, \ ar^{x-1}$$

   where   **a** is the first value

   r is the common ratio

   x is the number of terms in the series.

Example 5

   Given the same details as in Example 2 what would be the depreciation rate as a percentage if the depreciation was to be calculated on the reducing balance method?

$$a = 32{,}500 \quad x = 21 \quad \text{scrap value} = 7{,}500$$

$$7{,}500 = ar^{x-1}$$

$$7{,}500 = 32500r^{20}$$

$$r^{20} = \frac{7{,}500}{32{,}500}$$

which may be solved using logs or the scientific function on a calculator

$$\text{i.e. } \log r = \frac{\log 7{,}500 - \log 32{,}500}{20}$$

$$\therefore r = 0.92927$$

$\therefore$ reducing balance depreciation rate $= 1 - 0.92927$ or $= 7\%$

Example 6

A building cost £500,000 and it is decided to depreciate it at 10% p.a. on the reducing balance method. What will be its written down value be after 25 years?

$$a = 500{,}000 \quad x = 26 \quad r = (1 - d)$$

where d = depreciation rate

$$\text{Value after 25 years} = ar^{x-1}$$
$$= 500{,}000(1 - 0.1)^{26-1}$$
$$\therefore \log \text{25th year value} = \log 500{,}000 + 25 \log (1 - d)$$
$$\log \text{25th year value} = 5.69897 + 25(9.95424 - 10)$$
$$\log \text{25th year value} = 4.55497$$
$$\therefore \text{25th year value} = \underline{£35{,}890}$$

Notes:

a) The common ratio, r, in depreciation problems is always obtained by deducting the depreciation rate from 1, i.e. if the depreciation rate is 20% then $r = (1 - 0.2)$, i.e. 0.8.

b) Make sure you can deal with the multiplication and division of negative logarithms.

Sum of geometric progressions

6. In a similar fashion to an arithmetic progression the individual terms could be evaluated and added together but a simple formula exists:

$$S = \frac{a(r^x - 1)}{r - 1}$$

where S is the sum of the progression

Example 7

A company sets up a sinking fund and invests £10,000 each year for 5 years at 9% compound interest. What will the fund be worth after 5 years?

(*Note*: From such a question it can be inferred that the £10,000 is invested at the end of each year so that the last allocation earns no interest.) To clarify the problem the whole series is set out below.

£(10,000 × 1.09⁴) + £ (10,000 × 1.09³) + £ (10,000 × 1.09²) + £ (10,000 × 1.09) + £ 10,000

From this it will be seen that the series is the reverse of the usual order and that the number of terms in the series (x) is 5, i.e. the same as the number of years.

$$a = 10{,}000 \quad r = 1.09 \quad x = 5$$

$$S = \frac{a(r^x - 1)}{r - 1} = \frac{10{,}000(1.09^5 - 1)}{1.09 - 1}$$

$$S = \underline{£59{,}847}$$

Simple and compound interest

7. Money can often be invested to earn interest and the interest may be either *simple interest* or *compound interest*. The principles and formulae behind simple and compound interest are explained below, using the following symbols:

    P = a sum at the present time

    S = a sum arising in the future

    n = number of interest bearing periods usually, but not exclusively, expressed in years

    r = rate of interest invariably expressed in an interest rate per annum

    I = total amount of interest

Simple interest

8. Where interest only accrues on the principal (the original amount invested), this is known as *simple interest*, i.e. the interest is not re-invested to earn more interest.

$$I = S - P$$

$$\text{and} \quad r = \frac{I}{Pn}$$

$$\text{and} \quad I = Prn$$

$$\text{and} \quad S = P + I = P + Prn = P(1 + rn)$$

Example 8

How much will £10,000 amount to at 8% simple interest over 15 years?

$$S = £10,000 \ (1 + 0.08(15) \ )$$

$$= \underline{\underline{£22,000}}$$

Example 9

How long will it take for a sum of money to double itself at 10% p.a. simple interest?

Let $P = 1$ and $S = 2$ and $I = 1$

$$I = Prn$$

$$1 = (1)(0.10)(n)$$

$$n = \frac{1}{0.10} = \underline{\underline{10 \text{ years}}}$$

Compound interest

9. Of considerably more practical use is the principle of compound interest whereby interest is paid on the principal plus the re-invested interest so that the successive terms of the series form a geometric progression. The principles of compound interest form the basis of annuity calculations, sinking funds and discounting so that a clear understanding is vital.

The basic compounding formula is

$$S = P(1 + r)^n$$

Because of its importance and frequency of occurrence, compound interest tables have been calculated which show the compound interest factor for various rates of interest (r) over various numbers of year (**n**). See Table VI.

Example 10

How much will £10,000 amount to at 8% p.a. compound interest over 15 years?

$$S = 10,000(1 + 0.08)^{15}$$

Note: The expression above could be worked out by calculator giving an answer of £31,722. However, tables are more commonly used which, although giving an approximate answer, are quite accurate enough for all practical purposes. Look up in Table VI under 8% for 15 years and the factor is 3.172.

$$S = £10,000 \ (3.172)$$

$$= \underline{\underline{£31,720}}$$

Note: Contrast this amount with Example 8.

Example 11

What compound rate of interest will be required to produce £5,000 after five years with an initial investment of £4,000?

$$5{,}000 = 4{,}000\,(1 + r)^5$$

$$\frac{5{,}000}{4{,}000} = (1 + r)^5$$

$$(1 + r)^5 = 1.25$$

Look in Table VI along the 5 year line for a compound interest factor of 1.25. It will be seen that the factor for 4% is 1.217 and for 5% 1.276. therefore the solution to this problem is between 4% and 5% and is just over $4\frac{1}{2}$%.

Example 12

How long will it take for a given sum of money to double itself at 10% p.a. compound interest? (Compare this with Example 9.)

$$2 = 1(1 + 0.1)^n$$

$$\therefore (1 + 0.1)^n = 2$$

Once again the tables can be used as a short cut method. Look in Table VI down the 10% column to see the factor closest to 2 which is found to be 7 years having a factor of 1.949. The factor for 8 years is 2.144 so that the period required is approximately **7.3 years**.

Discounting

10. The compounding principle, given in Para 9, essentially looks forward from a known present amount (**P**) together with re-invested interest, equalling some terminal value (**S**).

It will be apparent that there are occasions when the future value (**S**) is known and it is required to calculate the present value (**P**). The compounding formula can be restated in terms of discounting to a present value as follows:

$$P = \frac{S}{(1 + r)^n}$$

Note: This formula is the basis of all discounting methods and is particularly useful as the basis of Discounted Cash Flow techniques which are dealt with in the chapters which follow.

Example 13

How much will have to be invested now to produce £20,000 after 5 years with a 10% compound interest rate?

(Alternatively the question could be framed; what is the present value of £20,000 received in 5 years time at a discount rate of 10%?)

$$P = \frac{S}{(1+r)^n}$$

$$P = \frac{20,000}{(1+0.1)^5}$$

$$= \underline{\underline{£12,418}}$$

Note: The expression above has been evaluated by calculator but because discounting is a common process, discount factor tables have been calculated, see Table VII. The discount factor for 10% for 5 years is 0.621.

$$20,000 \times 0.621 = \underline{\underline{£12,420}}$$

It will be seen that the value obtained using the three figure discount factor from the tables is slightly different from the accurate calculator figure.

For all practical purposes the approximate figure obtained by using discount tables is quite acceptable and, as a consequence, discount tables are widely used.

Discounting a series

11. Many problems are not concerned merely with discounting one value but involve a whole series of cash flows which are required to be discounted to a present value.

In such circumstances the formula given in Para 10 becomes

$$P = \sum_{i=1}^{i=n} \frac{A_i}{(1+r)^i}$$

where A_i represents the cash flows arising at the end of year 1,2,3, . . . n.

Example 14

What is the present value of receiving £1,000 in 1 year's time, £2,000 in 2 years time, and £3,000 in 3 years time when the discount rate is 10%?

$$\therefore P = \frac{1,000}{(1+0.1)^1} + \frac{2,000}{(1+0.1)^2} + \frac{3,000}{(1+0.1)^3}$$

The discount factors are 0.909, 0.826, and 0.751 (Table VII)

$$P = 1{,}000 \ (0.909) + 2{,}000 \ (0.826) + 3{,}000(0.751)$$

$$= \underline{\underline{£4.814}}$$

Where the cash flows vary from year to year there is no single all embracing formula which can be used. A separate calculation has to be made for each year and added together. However, when the cash flows are regular, certain short cuts are possible.

Annuities

12. Where the cash flows received or paid out are constant for all years the series is known as an *annuity*. The present value could be calculated using the formula given in Para 11, i.e. discounting each term and adding them together, but the series of expressions can be brought into one formula, thus

$$P = \frac{A[1 - (1+r)^{-n}]}{r}$$

where **A** is the regular cash receipt.

Example 15

What is the present value of an annuity of £500 p.a. received for 10 years when the discount rate is 10%?

$$P = \frac{500\left[1 - \dfrac{1}{(1+0.1)^{10}}\right]}{0.1}$$

$$\approx \underline{\underline{£3{,}072}}$$

Note: Once again, because annuities are commonly encountered, tables have been calculated for the annuity factor, i.e.

$$\frac{1 - (1+r)^{-n}}{r} \text{ (see Table VIII)}$$

If Table VIII is examined for the 10% rate for 10 years the factor 6.145 is obtained

$\therefore$ present value is £500 $\times$ 6.145 = $\underline{\underline{£3{,}072}}$

Because of the frequency with which annuities are encountered a shorthand way of writing such problems is used, i.e. $a_{n\rceil r}$, which means the annuity factor for n years at r rate of interest. Example 15 could have been written as follows:

$$£500 \ a_{\ 10\rceil\ 10\%} \approx £3{,}072$$

Perpetuities

13. Where a steam of cash flows goes on for ever the expression in square brackets in the annuity formula reduces to 1 and the formula dramatically simplifies to

$$P = \frac{A}{r}$$

Example 16

What is the present value of a perpetual annuity of £500 p.a. at 10% (alternatively £500 $a_{\infty \rceil 10\%}$)?

$$P = \frac{500}{0.1}$$

$$= \underline{\underline{£5,000}}$$

Example 17

What is the rate of interest when a £5,000 investment now provides a perpetual annuity of £300 p.a.?

$$r = \frac{A}{P} = \frac{300}{5,000}$$

$$= \underline{\underline{6\%}}$$

Note: This simple concept is important and is used subsequently when investment appraisal and DCF is considered.

Cash flows growing (or declining) at a compound rate

14. On occasions it is required to find the present value of a series of cash flows which are expected to grow (or decline) at a compound rate e.g. a series of dividends which are expected to grow at a rate of 8% p.a.

$$\text{then } P = \frac{A}{(1+g)} a_{n \rceil r_0\%} \qquad \text{where } r_0 = \frac{r-g}{1+g}$$

($a_{n \rceil r_0\%}$ i.e. as is, of course the annuity factor described in Para 12)

Example 18

What is the present value of a dividend stream which is expected to commence in a year's time with a dividend of £440 and thereafter increase at 10% p.a.? The dividends are expected to last 10 years and the discount rate is 32%.

$$P = \frac{A}{1+g} a_{n\,\rceil\,r_0\%} \qquad \text{where } r_0 = \frac{r-g}{1+g}$$

$$r_0 = \frac{0.32 - 0.10}{1.10} = \frac{0.22}{1.10} = \underline{\underline{0.20}}$$

$$\text{and } \frac{A}{1+g} = \frac{440}{1.10} = £400$$

$$\therefore P = £400\, a_{10\,\rceil\,20\%}$$

$$= 400 \times 4.192$$

$$= \underline{\underline{£1,677}}$$

Note: The factor 4.192 is obtained from Table VIII.

Example 19

What would be the present value of Example 18 if the dividend stream was considered to last indefinitely?

As in Example 18

$$r_0 = 0.20 \text{ and } \frac{A}{1+g} = £400$$

$$\therefore P = £400 a_{\infty\,\rceil\,20\%}$$

$$= \frac{400}{0.2}$$

$$= \underline{\underline{£2,000}}$$

Example 20

Assume the same situation as Example 18 except that the dividend was expected to reduce by 10% p.a.

$$g = -0.10$$

$$r_0 = \frac{0.32 + 0.01}{1 - 0.10} = 0.466$$

$$\text{and } \frac{A}{1+g} = \frac{440}{0.9} = £489$$

$$\therefore \; P = £489a_{10\;\rceil\;46.2/3\%}$$

$$P = £489 \times 2.09650$$

$$P = \underline{\underline{£1,025}}$$

Note: When some large value of r_0 is obtained tables may not be obtainable so that the annuity factor has to be found by using a calculator.

Compounding/discounting at intervals other than annual

15. The assumption so far has been that cash flows have been received or paid out at year ends and that interest is always applied annually. These are the normal assumptions in examination questions but on occasions discounting or compounding may be required at some other interval, e.g. monthly, quarterly, half yearly or even continuously. Continuous compounding or discounting requires special tables which are used in exactly the same way as normal tables based on discrete periods. The normal tables (Table VII and VIII) can be simply used for periods other than a year as shown by the following example.

> **Example 21**
>
> How much will have to be invested now to produce £20,000 after 5 years where a rate of 10% p.a. is compounded half yearly? (Alternatively, what is the present value of £20,000 received in 5 years time discounted half yearly at 10% p.a.?)

Calculate the number of interest/discount periods, i.e. $\mathbf{5 \times 2 = 10}$

Calculate interest/discount rate per period, i.e.

$$\frac{\textbf{Interest/discount rate p.a.}}{\textbf{No. of periods p.a.}} = \frac{\mathbf{10}}{\mathbf{2}} = \underline{\underline{\mathbf{5\%}}}$$

Look up the tables for 10 periods at 5% (i.e. treat the years column as 'periods'). From Table VII the factor obtained is 0.614

$$\textbf{Present value} = £20,000 \times 0.614$$

$$= \underline{\underline{£12,280}}$$

Contrast this answer with Example 13.

Note: Whether compounding or discounting a single value or a series, similar principles apply. Calculate the number of interest/discount periods and the rate of interest per period. The more frequently a value is compounded/discounted, the larger/smaller will be the answer.

Nominal and actual interest rates

16. Most interest rates are expressed as per annum figures even when the interest is

compounded or discounted over periods of less than one year. In such cases the given interest rate is called a nominal rate.

Depending on whether the compounding or discounting is done daily, weekly, monthly, quarterly or six monthly the actual rate or annual percentage rate (APR) will vary by differing amounts from the nominal rate. The discrepancy between the nominal rate and the APR gets larger as the frequency of compounding or discounting and the number of years increases. The APR is an important figure and, by law, must be quoted in finance, loan and hire purchase agreements.

Given a nominal annual rate of interest the APR can be calculated using the following formula:

$$\mathbf{APR} = \left(1 + \frac{i}{n}\right)^n - 1$$

where $\mathbf{i}$ = nominal rate per compounding period

$\mathbf{n}$ = number of compounding periods in one year.

Example 22

A finance company loans money at 20% nominal interest but compounds monthly. What is the APR?

$$\textbf{Monthly rate} = \frac{0.20}{12} = \mathbf{0.0166}$$

$$\therefore \ \mathbf{APR} = (1 + 0.0166))^{12} - 1 = 0.2184 \text{ or } \underline{\mathbf{21.84\%}}$$

Note: An alternative name for APR is AER, i.e. Annual Equivalent Rate.

Summary

17. a) Many financial calculations are based on series. The two most common types are arithmetic and geometric progressions.

b) In an arithmetic progression each new value is obtained by adding a constant amount to the previous value.

c) In a geometric progression each new value is obtained by multiplying the previous value by a constant.

d) Where interest accrues only upon the principal it is known as simple interest

e) Where interest accures on the principal plus the re-invested interest it is known as compound interest Many annuity and discounting problems have the principle of compound interest as their basis.

f) The basic compounding principle looks forward in time to some future date, whereas discounting starts with a known future sum and arrives at its value in present day terms.

g) The basic discounting formula

$$P = \sum \frac{A_i}{(1+r)^i}$$

is important and is the basis of the Discounted Cash Flow (DCF) techniques described in subsequent chapters.

h) Annuities are constant amounts received or paid annually for a given number of years.

i) Perpetuities are constant amounts received annually for ever. Clearly this is not a strictly realistic assumption but it is one occasionally made.

j) Although interest and discount rates are normally applied at yearly intervals this is not always the case. The tables (Table VI, VII, VIII) can still be used by calculating the number of interest bearing periods and adjusting the annual interest/discount rate according to the time period required.

Point to note

18. Financial mathematics are the basis of many accounting and business techniques (including depreciation, discounting) so the contents of this chapter should be thoroughly learned.

Self review questions *Numbers in brackets refer to paragraph numbers*

1 What are the two common series widely used in financial matters? (3 and 5)

2 What is simple interest and what is the formula to calculate a future sum given a present sum and a rate of interest? (8)

3 What is compound interest and what is the equivalent formula to 2. above? (9)

4 What is the relationship between discounting and compounding? (10)

5 What is the formula for calculating the present value of a series? (11)

6 What is an annuity? (12)

7 What is a perpetuity and how is the present value calculated? (13)

8 If interest is compounded or discounted at intervals other than annually how is this dealt with? (15)

Exercises with answers

1. Find the value of the 12th term and the sum of the first 20 terms of the progression:

$$4, 8, 16, 32, 64 \dots$$

2. How long will it take for a given sum to treble itself:

   a) at 15% simple interest?

   b) at 15% compound interest?

3. What is the present value of:

a) £200 p.a. received for 10 years at 10%?

b) £200 p.a. received for 10 years, commencing in 5 years time, at 10%?

c) A dividend of £500 p.a. growing at 5% p.a., with a discount rate of 26%, which is expected to last for 10 years?

Answers to exercises

1. This is a geometric progression with a common ratio of 2.

$$\therefore \text{ Value of 12th term} = 4 \times 2^{12-1} = \underline{\mathbf{8,192}}$$

Sum of first 20 terms

$$\text{Sum} = \frac{\mathbf{a}(\mathbf{r^n} - 1)}{\mathbf{r} - 1}$$

$$= \frac{4(2^{20} - 1)}{2 - 1}$$

$$= \underline{\mathbf{4,194,300}}$$

2. a) Let $\mathbf{P} = 1$, $\mathbf{S} = 3$, $\mathbf{I} = 2$

$$\therefore 2 = \mathbf{Prn}$$

$$2 = 1 \times 0.15 \times \mathbf{n}$$

$$\therefore \mathbf{n} = \frac{2}{0.15} = \underline{\mathbf{13.3 \text{ years}}}$$

b)
$$3 = 1(1 + 0.15)^n$$

$$3 = 1.15^n$$

$$\therefore \mathbf{n} = \underline{\mathbf{7.86 \text{ years}}}$$

3. a) £200 $\mathbf{a}_{10\rceil 10}\% = 200 \times 6.145 = £1,229$

b) £1,229 discounted for 5 years @ 10% = £1,229 × 0.621 = **£763.2**

c)
$$\mathbf{r_0} = \frac{\mathbf{r} - \mathbf{g}}{1 + \mathbf{g}} = \frac{0.26 - 0.05}{1.05} = 20\%$$

$$\mathbf{P} = \frac{\mathbf{A}}{1 + \mathbf{g}} = \frac{500}{1.05} = 476$$

$$\therefore \mathbf{P} = 476 \times \mathbf{a}_{10\rceil 20}\% = 476 \times 4.192 = \underline{\mathbf{£1,995}}$$

28 Investment appraisal – background and techniques

Objectives

1. After studying this chapter you will
   - understand the basis of long-term decision making;
   - be able to use the traditional appraisal techniques of Accounting Rate of Return and Payback;
   - know the reasons why Discounted Cash Flow (DCF) techniques are used;
   - be able to calculate and interpret Net Present Value (NPV);
   - understand Internal Rate of Return (IRR);
   - know what is meant by the Excess Present Value Index;
   - understand the effects of inflation on investment appraisal.

Long run decision making

2. Investment decisions are long run decisions where consumption and investment alternatives are balanced over time in the hope that investment now will generate extra returns in the future. There are many similarities between short-run and long-run decision making, for example, the choice between alternatives, the need to consider future costs and revenues and the importance of incremental changes in costs and revenues but there is the additional requirement for investment decisions that, because of the time scale involved, the time value of the money invested must be considered. The time scale also makes the consideration of uncertainty and inflation of even greater importance than when considering short term decisions. Assuming that finance is available the decision to invest will be based on three major factors:

   a) The investor's beliefs in the future. In business the beliefs would be based on forecasts of internal and external factors including: costs, revenues, inflation and interest rates, taxation and numerous other factors.

   b) The alternatives available in which to invest. This is the stage at which the various techniques used to appraise the competing investments would be used. The various techniques are covered in detail in this chapter.

   c) The investor's attitude to risk. Because investment decisions are often on a large scale, analysis of the investor's attitude to risk and the project uncertainty are critical factors in an investment decision and are dealt with in the following chapter.

   Investment decision making is invariably a top management exercise. This is because of the scale and long term nature of the consequences of such decision. Although the final decision would be made by management the analyst would assess the alternatives available, analyse the data using the most appropriate techniques and present the results of this exercise to management in the hope that better decisions are made.

Traditional investment appraisal techniques

3. Two particular methods of comparing the attractiveness of competing projects have become known as the 'traditional techniques'. These are the Accounting Rate of Return and Payback which are described below.

Accounting rate of return

4. This is the ratio of average annual profits, after depreciation, to the capital invested. This is a basic definition only and variations exist, for example

    • profits may be before or after tax

    • capital may or may not include working capital

    • capital invested may mean the initial capital investment or the average of the capital invested over the life of the project.

    Note: An alternative term is Return on Capital Employed (ROCE).

Example 1

Accounting Rate of Return

A firm is considering three projects each with an initial investment of £1000 and a life of 5 years. The profits generated by the projects are estimated to be as follows:

After tax and depreciation profits

Year	Project I £	Project II £	Project III £
1	200	350	150
2	200	200	150
3	200	150	150
4	200	150	200
5	200	150	350
Total	1,000	1,000	1,000

Calculate the accounting rate of return (ARR) on

a) Initial capital

b) Average capital

Accounting rate of return on Initial Capital

	Project I	Project II	Project III
Average profits =	$\dfrac{1,000}{5}$	$\dfrac{1,000}{5}$	$\dfrac{1,000}{5}$
	= £200 p.a.	= £200 p.a.	= £200 p.a.
∴ ARR is	$\dfrac{200}{1,000}$ = 20%	$\dfrac{200}{1,000}$ = 20%	$\dfrac{200}{1,000}$ = 20%

Accounting rate of return on Average Capital

	Project I	Project II	Project III
Average capital =	$\dfrac{1,000}{2}$	$\dfrac{1,000}{2}$	$\dfrac{1,000}{2}$
	= £500	= £500	= £500
∴ ARR is	$\dfrac{200}{500}$ = 40%	$\dfrac{200}{500}$ = 40%	$\dfrac{200}{500}$ = 40%

Note: Average capital is calculated according to the usual accounting convention that the initial investment is eroded steadily to zero over the life of the project so that the average capital invested is

$$\frac{\textbf{Initial Investment}}{2}$$

Advantages and disadvantages of ARR

5. The only advantage that can be claimed for the ARR is simplicity of calculation, but the disadvantages are more numerous.

   Disadvantages:

   a) Does not allow for the timing of outflows and inflows. The three projects in Example 1 are ranked equally even though there are clear differences in timings.

   b) Uses as a measure of return the concept of accounting profit. Profit has subjective elements, is subject to accounting conventions and is not as appropriate for investment appraisal purposes as the cash flows generated by the project.

   c) There is no universally accepted method of calculating ARR.

Payback

6. Numerous surveys have shown that payback is a popular technique for appraising projects either on its own or in conjunction with other methods. Payback can be defined as the period, usually expressed in years which it takes

for the project's net cash inflows to recoup the original investment. The usual decision rules is to accept the project with the shortest payback period. The following example demonstrates the technique.

Example 2

Calculate the payback periods for the following three projects:

Net cash flows

Year	Project I		Project II		Project III	
	Cash flow	Cumulative cash flow	Cash flow	Cumulative cash flow	Cash flow	Cumulative cash flow
0	−1500	−1500	−1500	−1500	−1500	−1500
1	+600	−900	+400	−1100	+300	−1200
2	+500	−400	+500	−600	+500	−700
3	+400	nil	+600	nil	+400	−300
4	–		–		+300	nil
5	–		–		+300	+300
6	–		–		+300	+600

Note: The usual investment appraisal assumptions are adopted for the above table and all subsequent examples, that Year 0 means now, Year 1 means at the end of 1 year, Year 2 the end of 2 years and so on, and that a negative cash flow represents a cash outflow and a positive sign represents a cash inflow).

Payback Periods　**Project I** = 3 years

Project II = 3 years

Project III = 4 years

Advantages and disadvantages of payback

7.　*Advantages*:

a) Simple to calculate and understand.

b) Uses project cash flows rather than accounting profits and hence is more objectively based.

c) Favours quick return projects which may produce faster growth for the company and enhance liquidity.

d) Choosing projects which payback quickest will tend to minimise those risks facing the company which are related to time. However, not all risks are related merely to time.

Disadvantages:

a) Payback does not measure overall project worth because it does not consider cash flows after the payback period. In Example 2, Project III is ranked after Project I and II, even though it produces cash flows over a 6 year period.

b) Payback provides only a crude measure of the timing of project cash flows. In Example 2, Projects I and II are ranked equally, even though there are clear differences in the timings of the cash flows. In spite of any theoretical disadvantages, payback is undoubtedly the most popular appraisal criterion in practice.

Discounted cash flow (DCF)

8. There is growing use of DCF techniques for appraising projects and for assisting investment decision making. The use of DCF overcomes some of the disadvantages of the traditional techniques but it must be stressed that DCF itself has problems and contains many assumptions so that it should be used with care and with an awareness of its limitations. The main DCF techniques of Net Present Value (NPV) and Internal Rate of Return (IRR) are described in this chapter but it is necessary first to consider two features common to all DCF methods: the use of cash flows and the time value of money.

Use of cash flows

9. All DCF methods use cash flows and not accounting profits. Accounting profits are invariably calculated for stewardship purposes and are period orientated (usually monthly, quarterly or annually) thus necessitating accrual accounting with its attendant conventions and assumptions. For investment appraisal purposes a project orientated approach using cash flows is to be preferred for the following reasons.

a) Cash flows are more objective and in the end are what actually count. Profits cannot be spent.

b) Accounting conventions regarding revenue/capital expenditure classifications, depreciation calculations, stock valuations become largely redundant.

c) The whole life of the project is to be considered, therefore it becomes unnecessary and misleading to consider accounting profits which are related to periods.

d) The timing or expected timing of cash flows is more easily ascertained.

What cash flows should be included?

10. The all embracing answer to this question is the net after tax incremental cash flow effect on the firm by accepting the project, i.e. the comparison of cash flows with and without the project. Many of the cash flow items are readily identifiable, e.g. the initial outlay on a new machine, but others are less easily identified yet are nevertheless just as relevant, e.g. the increase or reduction in sales income of an existing product when a new product is introduced. Typical cash flow items include:

a) Cash Inflows

    i)   The project revenues

    ii) Government grants

    iii) Resale or scrap value of assets

    iv) Tax receipts

    v) Any other cash inflows caused by accepting the project.

b) Cash Outflows

    i)   Initial investment in acquiring the assets

    ii) Project costs (labour, materials, etc.)

    iii) Working capital investment

    iv) Tax payments

    v) Any other cash outflow caused by accepting the project.

Notes:

a) The relevant costs in investment decisions, as with all other decisions, are opportunity costs and not historical accounting costs. For example, if a project occupies storage space rented by the firm at £5/m² which could be sublet by the firm at £7/m², then the relevant cash flow is the benefit foregone of £7/m².

b) It will be noted that depreciation is not included. Depreciation is *not* a cash flow but an accounting convention. The capital outlay is already represented by the cash outflow of the initial investment, so to include depreciation would involve double counting. The only role of depreciation in investment appraisal is in determining the tax payments of the project, which are, of course, real cash flows.

c) Similarly, interest payments are not included because the discounting process itself takes account of the time value of money and to include interest payments and to discount would be to double count.

Time value of money

11. Investment appraisal is concerned with long run decisions where costs and income arise at intervals over a period. Monies spent or received at different times cannot be compared directly, they must be reduced to equivalent values at some common date. This could be at any time during the project life but appraisal methods which take account of the time factor use either now, the present time, or the end of the project as the common date.

Both discounting and compounding methods allow for the time value of money and could thus be used for investment appraisal but on the whole discounting methods are more frequently used. In general it is preferable to receive a given sum earlier rather than later because the sum received earlier can be put to use by earning interest or some productive investment within the business, i.e. money has a time productivity.

It should be noted that the time value of money concept applies even if there is zero inflation. Inflation obviously increases the discrepancy in value between monies received at different times but it is not the basis of the concept.

Assumptions in basic DCF appraisal

12. In describing the two main DCF methods certain assumptions are made initially so that the underlying principles can be more easily understood. These are as follows:

a) Uncertainty does not exist

b) Inflation does not exist

c) The appropriate discount rate to use is known

d) A perfect capital market exists, i.e. unlimited funds can be raised at the market rate of interest.

Subsequently these assumptions will be removed and the problems of dealing with uncertainty, inflation, and capital rationing are dealt with. The choice of an appropriate discount rate involves Financial Management and is outside the scope of this book.

Net present value (NPV)

13. The NPV method utilises discounting principles which should be familiar from the previous chapter.

The NPV method calculates the present values of expected cash inflows and outflows (i.e. the process of discounting) and finds out whether in total the present value of cash inflows is greater than the present value of cash outflows.

The formula is

$$\text{NPV} = \sum_{i=n}^{i=0} \frac{C_i}{(1+r)^i}$$

where C_i is the net cash flow in the period, i is the period number, and r is the discount rate.

Where the discount rate is the cost of capital of the firm the usual decision rule, given the assumptions in para 12, is that a project is acceptable if it has a positive NPV.

Example 3

An investment is being considered for which the net cash flows have been estimated as follows:

Year 0	Year 1	Year 2	Year 3	Year 4
−9,500	+3,000	+4700	+4800	+3200

What is the NPV if the discount rate is 20%? Is the project acceptable?

Note: Conventional year end cash flows have been assumed i.e. Year 0 means now, Year 1 means after 1 year and so on.

Solution

From Table VII the discount factors are **0.833, 0.694, 0.579 and 0.482**

$\therefore$ **NPV = −9,500 + (0.833 × 3,000) + (0.694 × 4,700) + (0.579 × 4,800) + (0.482 × 3,200)**

$$\text{NPV} = + \underline{£582}$$

and, given the assumptions contained in the basic DCF model, the investment would be acceptable, because it has a **positive NPV at the firm's cost of capital**.

It will be seen that the signs of the cash flows, negative or positive, must be respected and that the negative cash flow in Year 0, i.e. now, does not need discounting. This is because the cash flow is already in present day terms so discounting is unnecessary.

Meaning of NPV

14. If the NPV of a project is positive this can be interpreted as the potential increase in consumption made possible by the project valued in present day terms. This is illustrated by Table 1 based on Example 3 which shows the position if £9,500 is borrowed at 20% p.a. to finance the project on overdraft terms where interest is paid on the balance outstanding at the end of each year.

Table 1

	Amount owing b/fwd	+	Year's interest	−	Year's cash flow	=	Balance o/s c/fwd
End Year 1	9,500	+	1,900	−	3,000	=	8,400
End Year 2	8,400	+	1,680	−	4,700	=	5,380
End Year 3	5,380	+	1,076	−	4,800	=	1,656
End Year 4	1,656	+	331	−	3,200		

giving a final surplus of £1,213

This shows that if £9,500 is borrowed at 20% the principal and interest could be repaid from the project cash flows leaving a cash balance at the end of Year 4 of £1,213. This balance is known a Net Terminal Value and has a present value of £585 (£1,213 × 0.482) which, allowing for the approximations contained in three figure tables, is the NPV of Example 3.

Internal rate of return (IRR)

15. Alternative names for the IRR include: DCF yield, marginal efficiency of capital, trial and error method, discounted yield and the actuarial rate of return.

The IRR can be defined as the *discount rate which gives zero NPV*. Except by chance the IRR cannot be found directly; it can be found either by drawing a graph known as a *present value profile* or, more normally, by calculations involving linear interpolation. Both methods are illustrated below using the data from Example 3.

Present value profile

16. A present value profile is a graph of the project NPV's at various discount rates. It is normal to plot at least two points: one at a rate which gives a *positive* NPV and one at a rate which gives a *negative* NPV. We know from Example 3 that 20% gives + £582 NPV so this will do for the positive value. Some higher rate is then tried to see if a negative NPV is obtained. In this example a 25% discount rate gives – £322 NPV so, these points are plotted on Figure 28/1 and a line drawn between them to see where it crosses the horizontal axis, which gives the IRR.

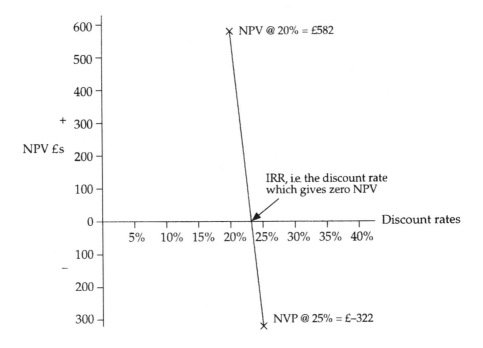

Present value profile. Figure 28/1

Notes on Figure 28/1.

a) At least one discount rate must be chosen which gives a negative NPV so that the present value line crosses the horizontal axis.

b) The present value line crosses the axis at approximately 23% which is a close enough estimate for most practical purposes.

Finding the IRR by linear interpolation

17. Based on the data from Example 3 the IRR can be calculated as follows:

(c)

$$\text{IRR} = 20\% + 5\% \quad \left(\frac{582}{904}\right)$$

$$\quad\text{(a)} \quad \text{(b)} \qquad \text{(d)}$$

$$= \underline{\underline{23.2\%}}$$

Where   **(a)** is a discount rate which gives a positive NPV.   In this example 20% gives £582.

           (b) is the difference between **(a)** and the rate which gives a negative NPV.   In this example 25% − 20% = 5%

           (c) is the positive NPV at the discount rate chosen in **(a)**.   In this example it is £582.

           (d) is the total range of NPV at the rates chosen.   In this example + 582 to − 322 is a range of £904.

Notes:

a) A simple way of obtaining a rough estimate of the IRR is to take the reciprocal of the payback period. This works best with relatively long payback periods. With short payback periods as in Example 3 the method is a poor estimator.

b) As there is a non-linear relationship between discount rates and NPV, linear interpolation does not give a strictly accurate result. However, where the two discount rates that give positive and negative results are fairly close then the result is accurate enough for all practical and examination purposes.

Decision rule using IRR

18. Where the calculated IRR is greater than the company's cost of capital then the project is acceptable, given the assumptions already mentioned. In the majority of cases where there are conventional cash flows, i.e. an initial outflow followed by a series of inflows, then the IRR gives the same accept or reject decision as the NPV which is the case in Example 3 above. This does not follow for all cash flow patterns and the two techniques do not necessarily rank projects in the same order of attractiveness.

NPV and IRR compared

19. In many circumstances either of these decision criteria can be used successfully but there are differences in particular circumstances which are dealt with below.

     a) Accept/reject decisions.

        Where projects can be considered independently of each other and where the cash flows are conventional then NPV and IRR give the same accept/reject decision.

	Accept project	Reject project
NPV	Positive NPV	Negative NPV
IRR	IRR above cost of capital	IRR below cost of capital

b) Absolute and relative measures.

   NPV is an absolute measure of the return on a project whereas IRR is a relative measure relating the size and timing of the cash flows to the initial investment. Thus the NPV reflects the scale of a project whereas the IRR does not.

Example 4

Assume a project has the following cash flows:

		Year 0	Year 5
Project x		−£20,000	+£40,241
NPV @ 10%	=	£4,990	
IRR	=	15%	

∴ Project acceptable by both methods – assuming 10% is the cost of capital

Now assume that the project is scaled up by a factor of 10

		Year 0	Year 5
Project 10x		−£200,000	+£402,410
NPV @ 10%	=	£49,900	
IRR	=	15%	

The NPV method clearly discriminates between Project x and Project 10x whereas the IRR remains unchanged at 15%.

---

c) Mutually exclusive projects.

   An important class of projects is that concerned with mutually exclusive decisions, e.g. where only one of several alternative projects can be chosen. For example, where several alternative uses of the same piece of land are being considered, when one is chosen the others are automatically excluded.

   Mutually exclusive decisions are commonly encountered and make it necessary to rank projects in order of attractiveness and to choose the most profitable. In such circumstances NPV and IRR may give conflicting rankings.

Example 5 (Mutually exclusive projects of differing scale)

A property company wishes to develop a site it owns. Three sizes of property are being considered and the costs and revenues are as follows:

	Year 0 Expenditure £m	Year 1 to perpetuity Rentals p.a. £m
Small development	2	0.6
Medium development	4	1
Large development	6	1.35

The cost of capital is 10% and it is required to rank the projects by NPV and IRR and to select the most profitable.

The projects are mutually exclusive because the building of one size of development excludes the others.

Solution

	Expenditure £m	P.V. of rentals £m	NPV £m	IRR %
Small	2	6	4	30
Medium	4	10	6	25
Large	6	13.5	7.5	22.5

The ranking obtained by NPV and IRR differ and in such circumstances the NPV ranking is preferred (i.e. large development in this example) because it leads to the greatest increase in wealth for the company.

Although safer and simpler to rank by NPV, IRR can be used by adopting an incremental approach and comparing the incremental IRR for each increment with the cost of capital. Example 5 is re-worked using this approach.

	Incremental expenditure £m	Incremental rental £m	Incremental IRR %
Stage 1 (small)	2	0.6	30
Stage 2 (medium – small)	2	0.4	20
Stage 3 (large – medium)	2	0.35	17.5

It will be seen that the IRR of each successive stage, although declining, is greater than the 10% cost of capital so that each successive increment is worthwhile. Although this method leads to the correct conclusion it is cumbersome and the simpler, more direct NPV method is preferable.

Example 6

Mutually exclusive projects, same scale.

Two mutually exclusive investments have cash flows as follows:

	Year 0	Year 1	Year 2	Year 3
Project A	−24,000	+8,000	+12,000	+16,000
Project B	−24,000	+16,000	+10,000	+8,000

The cost of capital is 10%.

The NPV and IRR of these projects are as follows:

	NPV @ 10%	IRR
		%
Project A	+5200	20.65
Project B	+4812	22.8

Thus it will be seen that the ranking differ and, assuming that 10% is the appropriate discount rate, the ranking given by the NPV, i.e. Project A being preferred, gives the maximum wealth to the company.

Note: The conflict in ranking shown above is but a reflection of the differing time profile of the project cash flows. Such time profiles produce different NPV rankings using different discount rates; for example, the NPVs of the above projects at 20% discount rate are £256 and £900 respectively, giving a B – A ranking instead of the A – B ranking at 10%.

Non-conventional cash flows (The multiple rate problem)

20. The projects considered so far have had conventional cash flows, i.e. an initial outflow followed by a series of inflows. Where the cash flows vary from this they are termed ***non-conventional***. The following are examples.

Example 7

Examples of Non-Conventional Cash-flow Patterns.

	Year 0	Year 1	Year 2
Project X	−2,000	+4,700	−2,750
Project Y	+2,000	−4,000	+4,000

Project X has 2 outflows and is thus non-conventional

Project Y has an outflow in a year's time instead of initially and is thus non-conventional.

When a project has non-conventional cash flows it may have

i) one IRR

ii) multiple IRRs

iii) no IRR

Multiple Rates: If the present value profile for Project X in Example 7 is drawn it can be seen that it is a multiple IRR project having two IRRs, at 10% and 25%.

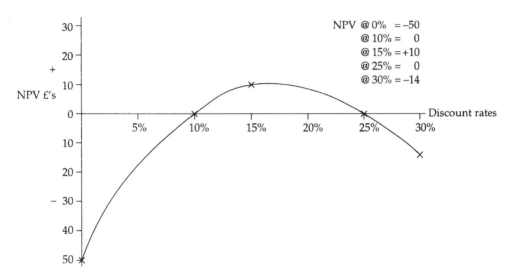

NPV @ 0% = –50
@ 10% = 0
@ 15% = +10
@ 25% = 0
@ 30% = –14

Present Value Profile of Project X. Figure 28/2

No IRR: Project Y in Example 7 above is an example of a project where it is not possible to calculate a real rate of return.

To be able to calculate a real IRR it is necessary to solve for i in the following expression:

$$+2,000 - \frac{4,000}{(1+i)^1} + \frac{4,000}{(1+i)^2} = 0$$

Solving for i produces $i = \sqrt{-1}$, which is not a real number.

In circumstances with non-conventional cash flow patterns which produce multiple IRRs or no real IRR the use of IRR is not recommended.

The NPV method gives clear, unambiguous results whatever the cash flow pattern.

Project X above has positive NPVs at discount rates between 10% and 25% and negative NPV's at lower and higher rates. Project Y has a positive NPV at any discount rate.

Summary of NPV and IRR comparison

21. a) NPV is technically superior to IRR and is simpler to calculate.

b)   Where cash-flow patterns are non-conventional there may be nil or several internal rates of return making the IRR impossible to apply.

c)   NPV is superior for ranking investments in order of attractiveness.

d)   With conventional cash flow patterns both methods give the same accept or reject decision. With conventional independent projects the project with the largest NPV or the greatest IRR would normally be chosen.

e)   Where discount rates are expected to differ over the life of the project such variations can be readily incorporated into NPV calculations, but not in those for the IRR.

f)   Notwithstanding the technical advantages of NPV over IRR, IRR is widely used in practice so that it is essential that students are aware of its inherent limitations.

Excess present value index (EVPI) or profitability index

22.   The EVPI is merely a variant of the basic NPV method and is the ratio of the NPV of a project to the initial investment, i.e.

$$\text{EVPI} = \frac{\text{NPV}}{\text{Initial Investment}}$$

Thus the index is a measure of relative and not absolute profitability. Because of this it suffers from the same general criticisms when used for ranking purposes as the IRR. The EVPI is not suitable for ranking mutually exclusive projects, but because it is a measure of relative profitability it can be used where there are a number of divisible projects (i.e. fractional parts of projects may be undertaken) which cannot all be implemented because of a shortage of capital in the current period. In such circumstances the projects can be ranked in order of their EVPI's and implemented in order of attractiveness until the capital available is exhausted. If, however, the projects being considered covered two or more periods where funds were limited then the EVPI could not be used. In general the EVPI is of limited usefulness and the use of NPV is considered safer.

Inflation and investment appraisal

23.   Inflation can be simply defined as an increase in the average price of goods and services. The accepted measure of general inflation in the UK is the Retail Price Index (RPI) which is based on the assumed expenditure patterns of an average family. General inflation is a factor in investment appraisal but of more direct concern is what may be termed *specific inflation* i.e. the changes in prices of the various factors which make up the project being investigated, e.g. wage rates, sales prices, material costs, energy costs, transportation charges and so on.

Every attempt should be made to estimate specific inflation for each element of the project in as detailed a manner as feasible. General, overall estimates based on the RPI are likely to be inaccurate and misleading.

Synchronised and differential inflation

24. Differential inflation is where costs and revenues change at differing rates of inflation or where the various items of cost and revenue move at different rates. This is normal but the concept of synchronised inflation – where costs and revenues rise at the same rate – although unlikely to be encountered in practice, is useful for illustrating various facets of projects appraisals involving inflation.

Money cash flows and real cash flows

25. Money cash flows are the actual amounts of money changing hands whereas 'real' cash flows are the purchasing power equivalents of the actual cash flows. In a world of zero inflation there would be no need to distinguish between money and real cash flows as they would be identical. Where inflation does exist then a difference arises between money cash flows and their real value and this difference is the basis of the treatment of inflation in project appraisal.

Dealing with inflation

26. The following example will be used to illustrate the way that inflation is dealt with in investment appraisal.

Example 8

A labour saving machine costs £60,000 and will save £24,000 p.a. at current wage rates. The machine is expected to have a 3 year life and nil scrap value. The firm's cost of capital is 10%.

Calculate the project's NPV

a) With no inflation

b) With general inflation of 15% which wage rates are expected to follow (i.e. synchronised inflation).

c) With general inflation of 15% and wages rising at 20% p.a. (i.e. differential inflation).

Solution

a) NPV – No inflation

$$-60,000 + 24,000 \; A_{3 \rceil 10\%}$$
$$= -60,000 + 24,000 \times 2.487 = -£312$$

∴ Project unacceptable as it has a negative NPV at company's cost of capital

b) General inflation 15%, wages increasing at 15%

Wage savings p.a. with no inflation	Wage savings p.a. with 15% inflation
24,000	27,600
24,000	31,740
24,000	36,501

With no inflation the appropriate discounting rate was 10%. With inflation at 15%, the 10% discounting rate is insufficient to bring cash sums arising at different periods into equivalent purchasing power terms. Without inflation £1 now was deemed equivalent to £1.10 a year hence. With a 15% inflation rate the sum required would be £1.10 (1.15) = £1.265, thus the discount rate to be used is 26½%.

Project NPV with 15% synchronised inflation

Year	Cash flow	26½% Discount factors	Present value
0	−60,000	1.000	−60,000
1	+27,600	0.792	21,859
2	+31,740	0.624	19,806
3	+36,501	0.494	18,031
		NPV=	−£304

∴ Project unacceptable

It will be seen that the answers with no inflation and with 15% synchronised inflation are virtually the same, (the difference being due to roundings in three figure tables). This equivalence is to be expected, as with synchronised inflation the firm, in real terms, is no better or no worse off.

c) Project with 15% general inflation and wages rising at 20% p.a. (differential inflation)

Wages per annum

Year 1	24,000 (1.20)	=	£28,800
2	$24,000 (1.20)^2$	=	£34,560
3	$24,000 (1.20)^3$	=	£41,472

Project NPV with differential inflation

Year	Cash flow	26½% Discount factors	Present value
0	−60,000	1.000	−60,000
1	+28,800	0.792	22,810
2	+34,560	0.624	21,565
3	+41,472	0.494	20,487
		NPV=	£4,862

∴ Project acceptable

Thus it will be seen that with differential inflation the project is acceptable. In this case this is to be expected because it was a labour saving project so that, in real terms, the firm is better off if the rate of wage inflation is greater than the general rate of inflation.

Frequently, differential inflation works to the disadvantage of the firm, for example, when costs are rising faster than prices. Each case is different and detailed, individual analysis is required – not generalised assumptions.

Money and real discount rates

27. The 26½% discount rate used in Example 1 was a money discount factor and was used to discount the money cash flows of the project. The relationship between real and money discount factors is as follows:

$$\text{Real discount factor} = \frac{1 + \text{Money discount factor}}{1 + \text{Inflation Rate}} - 1$$

Using the data from Example 1 the real discount factor can be calculated.

$$\text{Real discount factor} = \frac{1 + 0.265}{1 + 0.15} - 1 = 0.1 \text{ i.e.} \quad \underline{10\%}$$

In this case, of course, the real discount factor was already known and the above calculation was for illustrative purposes only.

The real discount factor can be used for project appraisal providing that the money cash flows are first converted into real cash flows by discounting at the general inflation rate as follows.

Example 9

Re-work part (c) of Example 8 using real cash flows and the real discount factor.

Real cash flow evaluation

Year	Money cash flow	General inflation 15% discount factors	Real cash flows	Real discount factors 10%	Present values
0	−60,000	1.000	−60,000	1.000	−60,000
1	+28,800	0.870	25,056	0.909	22,776
2	+34,560	0.756	26,127	0.826	21,581
3	+41,472	0.658	27,289	0.751	20,494
				NPV =	£4,851

From which it will be seen that (table rounding differences apart) the two methods give identical results.

Thus it will be seen that there are two approaches to investment appraisal where inflation is present.

Single discounting   Money cash flows discounted by money discount factor.

Two stages discounting   Money cash flows discounted by general inflation rate and then the real cash flows produced discounted by real discount factor.

The two approaches produce the same answer because the money discount factor includes the inflation allowance. Because of this and because money cash flows are the most natural medium in which estimates will be made, it is recommended that money cash flows should be discounted at an appropriate money discount factor. Take *great care* never to discount money cash flows by a real discount factor or real cash flows by a money discount factor. If real cash flows are directly provided in a question take care to discount once only using a real discount factor.

Taxation and investment appraisal

28.   Because taxation causes a change in cash flows it is a factor to be considered in project appraisal. Indeed in some practical situations the taxation implications are dominant influences on the final investment decision. Any consideration of tax intricacies is clearly outside the scope of this book and the analyst is strongly recommended to seek advice from a tax specialist when dealing with project appraisal. The payment of tax reduces the cash flows of a project so that in outline, the analyst will need to know the amount of the tax liability and when the tax must be paid.

Summary

29.   a)   Investment decisions are long run decisions where consumption and investment opportunities are balanced over time.

b)   The decision to invest is based on many factors including; the investor's beliefs in the future, the alternatives available and his attitude to risk.

c)   The 'traditional' investment appraisal techniques are the accounting rate of return and payback. Payback is shown by recent surveys to be the most widely used technique.

d)   Payback is the number of period's cash flows to recoup the original investment. The project chosen is the one with the shortest payback period.

e)   Discounted Cash Flow (DCF) techniques use cash flows rather than profits and take account of the time value of money.

f)   The formula for Net Present Value is

$$NPV = \sum \frac{C_i}{(1+r)^i}$$

and given the assumption of the basic model, a project is acceptable if it has a positive NPV at the firm's cost of capital.

g)   NPV can be interpreted as the potential increase in consumption made possible by the project valued in present day terms.

h) Internal Rate of Return (IRR) is the discount rate which gives zero NPV and can be found graphically or by linear interpolation.

i) With conventional projects IRR and NPV give the same accept or reject decision. NPV is an absolute measure whereas IRR is a relative one.

j) NPV is a more appropriate measure for choosing between mutually exclusive projects and in general is technically superior to IRR.

k) The discount rate used in DCF calculations is known as the cost of capital.

l) Specific inflation is of more direct concern in investment appraisal and differential inflation is commonly encountered.

m) The general treatment of inflation in investment appraisal means distinguishing between money and real cash flows.

n) The amount and timing of tax payments and other tax effects must be considered.

Points to note

30. a) It is arguable that the existence of a stream of potentially worthwhile investment opportunities is of far greater importance to the firm than the particular appraisal method used.

b) A recent survey of the capital budgeting practices of large companies by Dr R. Pike showed that over 75% of companies used payback as an appraisal method, often in conjunction with other techniques. The same survey showed that only 17% on companies used NPV as their primary evaluation technique in spite of the generally acknowledged technical superiority of NPV over payback. This would seem to suggest that much of the academic preoccupation with refining measurement techniques may be misplaced.

c) Successful investment appraisal is entirely dependent on the accuracy of cost and revenue estimates. No appraisal technique can overcome significant inaccuracies in the estimates.

Self review questions *Numbers in brackets refer to paragraph numbers*

1 Why are investment decisions important? (2)

2 Define the accounting rate of return and give its advantages and disadvantages? (4)

3 What is payback and what is the normal decision rule using payback? (6)

4 Why are cash flows used in DCF calculations and not accounting profits? (9)

5 What cash flows should be included in the appraisal? (10)

6 Why has money a time value? (11)

7 Define NPV and state the basic formula (13)

8 What is Net Terminal Value? (14)

9 What is the Internal Rate of Return and how is it calculated? (15 and 16)

10   What are the major differences between NPV and IRR? (18)

11   What is the multiple rate problem? (19)

12   What is the EVPI or Profitability Index and when can it be used effectively? (22)

13   What type of inflation is most relevant to investment appraisal? (23)

14   What is the general method of dealing with inflation in investment appraisal? (26)

15   Distinguish between money and real discount rates? (27)

16   What is the importance of taxation in investment appraisal? (28)

Exercises with answers

1.   What is the present value of the following series at 10% and its IRR?

0	1	2	3	4
−2,000	+800	+600	+700	+500

2.   Draw a present value profile of the following project and determine the IRR(s). Is the project acceptable at 10% cost of capital?

	Year 0	**Year 1**	**Year 2**
Cash Flows	−4,000	+10,200	−6,300

3.   A firm is considering a project with a cash outlay of £1,000 now and 5 yearly cash inflows of £500.

a)   What is the NPV at 10%?

b)   What is the NPV assuming a general inflation rate of 8% and an increase in cash flows to £510 per annum?

Answers to exercises

1.   NPV @ 10% = − 2,000 + (800 × 0.909) + (600 + 0.826) + (700 × 0.751) + (500 × 0.683)

$$= \underline{\textbf{£90 NPV}}$$

NPV @ 16% = − £140

$$\therefore \textbf{IRR} = 10 + 6\left[\frac{90}{230}\right] = \underline{\underline{\textbf{12.34\%}}}$$

2.

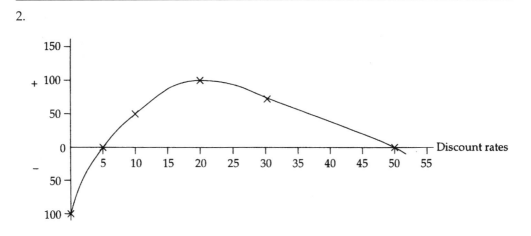

IRR 5% and 50%

Project acceptable at 10% cost of capital.

3.   a)   NPV @ 10% = £895.5

   b)   Discount rate with 8% inflation = 1.10 (1.08) = 1.188 say **19%**

   ∴ NPV of 5 year series of £510 per annum and outlay of £1,000 is

   − 1,000 + 1,559 = **+£559.4**

   This should be contrasted with the value in part (a).

29 Investment appraisal – uncertainty and capital rationing

Objectives

1. After you have studied this chapter you will
    - understand why uncertainty must be considered in investment appraisal;
    - know the time based methods of incorporating inflation into the appraisal;
    - be able to use Probability to assess uncertainty;
    - know how to calculate and use Expected Value;
    - be able to use Discrete and Continuous Probabilistic Analysis;
    - know how to conduct a Sensitivity Analysis of a project;
    - be able to evaluate a Portfolio's Risk;
    - understand the way projects are selected when Capital Rationing exists for Single and Multi-Periods.

Uncertainty in investment appraisal

2. Uncertainty is a major factor to be considered in all types of decision making. It is of particular importance in investment appraisal because of the long time scale and amounts of resources involved in a typical investment decision.

    In general, risky or uncertain projects are those whose future cash flows, and hence the returns on the project, are likely to be variable – the greater the variability, the greater the risk. Unfortunately, elements of uncertainty can exist even if future cash flows are known with certainty. For example, if a lease is being appraised the future cash flows are known and fixed but their value may vary because of changes in the rate of inflation.

    There are three stages of the overall appraisal and decision process in which risk and uncertainty merit special attention:

    a) The risk and uncertainty associated with the *individual project*.
    b) The effect on the overall risk and uncertainty of the firm when the project being considered is combined with the rest of the firm's operations – the *portfolio effect*.
    c) The *decision maker's attitude to risk* and its effect on the final decision.

    These three elements are dealt with below:

Uncertainty and the individual project

3. Various methods of considering the uncertainty associated with projects are described below. They have the general objective of attempting to assess or quantify the uncertainty surrounding a project by some form of analysis that goes

beyond merely calculating the overall return expected from the project. In this way further information is provided for the ultimate decision maker so that a better decision will be made.

It must be emphasised, however that the methods do not of themselves reduce the uncertainties surrounding a proposed investment. If this is feasible, it can only be done by management action.

The methods to be described can be separated into three groups:

a) Time based.

b) Probability based

c) Sensitivity analysis and simulation.

Time based

4. The three methods of incorporating uncertainty which are based on time are Payback, Risk Premium and Finite Horizon. These methods rest on the assumption that project risks and uncertainty are related to time, i.e. the longer the project the more uncertain it is. Whilst it is reasonable to assume that uncertainty does often increase with time it is by no means universally true and there are many projects which are shortlived and risky whilst others are long term and relatively safe. The three methods are described below:

Payback

5. This is the number of periods cash flows required to recoup the original investment. Apart from its use as an accept/reject criterion, payback can also be used as a measure of risk, often in conjunction with a DCF measure such as NPV or IRR. If two projects A and B had approximately the same NPV and A had the shorter payback period then A would be preferred.

    Advantages:

    a) Simplicity of calculation.

    b) General acceptability and ease of understanding

    Disadvantages:

    a) Assumes that uncertainty relates only to the time elapsed.

    b) Assumes that cash flows within the calculated payback period are certain.

    c) Makes a single blanket assumption – that uncertainty is a function of time – and does not attempt to consider the variabilities of the cash flows estimated for the particular project being appraised.

Risk premium

6. On occasions the discount rate is raised above the cost of capital in an attempt to allow for the riskiness of projects. The extra percentage being known as the *risk premium*. Such an inflated discount rate raises the acceptance hurdle for projects

and can be shown to treat risk as a function of time by more heavily discounting later cash flows as demonstrated in the following example.

Example 1

The cash flows for a project are shown below. The cost of capital is 10% and as the project is considered to be risky, a risk premium of 5% is to be added to the basic rate. The effects of the two discount rates are shown.

Table 1

Year	0	1	2	3	4	5	NPV
Estimated cash flows	−10,000	+2,000	+3,000	+2,500	+3,000	+3,500	
10% Discount factors	1,000	0.909	0.826	0.751	0.683	0.621	
PV @ 10%	−10,000	1,818	2,478	1,877	2,049	2,173	+395
15% Discount factors	1.000	0.870	0.756	0.658	0.572	0.495	
PV @ 15%	−10,000	1,740	2,268	1,645	1,716	1,732	−899
Percentage reduction of PVs caused by risk premium		4%	8%	12%	16%	20%	

The bottom line shows the progressively increased discounting which takes place on later cash flows demonstrating that using a risk premium treats uncertainty as being related to time elapsed.

Advantages of a risk premium:

a) Simple to use.

Disadvantages of a risk premium:

a) Makes the implicit assumption that uncertainty is a function of time.

b) By making the same overall, blanket assumption for all projects it does not consider the individual project characteristics nor does it explicitly consider the variability of the project cash flows.

c) It creates the problem of deciding upon a suitable risk premium. Should it be the same for all projects or should it be adjusted for different projects?

Note: One of the theories relating to business profits is that it is the reward for taking uninsurable risks. If this is correct, and it has some intuitive appeal, then a more subtle and long term effect of using a risk premium is that the firm will tend to move towards a portfolio of projects which, although potentially high yielding, have high risks – which is the opposite effect to that intended.

Finite horizon

7. In this method, which is the simplest of all to apply, project results beyond a certain period (e.g. 10 years) are ignored. All projects are thus appraised over the same time period.

Advantage:

a) Simplicity.

Disadvantages:

a) The establishment of any fixed time horizon is arbitrary. Projects do vary in length and this should be reflected in the appraisal.

b) Project cash flows within the time horizon are considered certain.

c) Does not explicitly consider the variabilities of cash flows.

Summary of time-based methods of considering uncertainty

8. These methods are simple to apply and require little, if any, extra calculation. They are for the most part arbitrary and unreliable. Although obviously uncertainty tends to increase with time there is not a straightforward relationship and the time based methods fail to examine the characteristics of individual projects and merely make one blanket assumption covering all projects. The whole purpose of investment appraisal is to distinguish between projects and accepting one overall assumption is likely to mask rather than highlight the differences between the investment opportunities being considered.

Probability based methods of assessing uncertainty

9. The methods to be described rest on the assumption that realistic estimates of the subjective probabilities associated with the various cash flows can be established. For example, the project analyst might ask a manager to make three estimates of the cash flow of a period (optimistic, most likely, pessimistic) instead of just a single estimate, and in addition ask the manager to assess the likelihood of each of three estimates. Following such a request the manager might make the following estimate:

	Cash Flow in Period x	
Optimistic	£8,000	with a probability of 10% (0.1)
Most likely	£4,500	with a probability of 65% (0.65)
Pessimistic	£3,000	with a probability of 25% (0.25)

The estimates thus obtained form a probability distribution of the cash flows. It follows that if the individual cash flows are expected to vary then the overall return for the project will also vary. The main objectives of probability based methods is to demonstrate the likely variations in the result, whether NPV or IRR, due to the estimated variations in the cash flows. In this way the effects of uncertainty are more clearly shown and it is hoped that a more informed decision may be taken. The three methods to be described which use subjective probabilities are Expected Value, Discrete Probabilistic Analysis and Continuous Probabilistic Analysis.

Expected value (EV)

10. This has been covered in detail earlier. It will be recalled that EV is Probability ×

Value and EV can be used for individual cash flows or project NPVs. The following simple example illustrates the technique.

Example 2

The cash flow and probability estimates for a project are shown below.

Calculate:

a) expected value of the cash flows in each period; and

b) expected value of the NPV when the initial project outlay is £11,000 and the cost of capital is 15%.

Cash flow and probability estimates

	Probability		Cash flows Period		
		1	2	3	4
	£	£	£	£	
Optimistic	0.3	5,000	6,000	4,500	5,000
Most likely	0.5	3,500	4,000	3,800	4,500
Pessimistic	0.2	3,200	3,600	3,100	4,000
Expected value of cash flows		3,890	4,520	3,870	4,550

The Expected NPV is found by discounting the expected value of cash flows in the normal manner.

$$\text{ENPV} = -11,000 + (3,890 \times 0.870) + (4,520 \times 0.756) + (3,870 \times 0.658) +$$
$$(4,550 \times 0.572)$$
$$= \underline{£950}$$

The advantages and disadvantages of expected value as a decision criterion have already been covered and these apply equally to the use of Expected Value in investment appraisal. It is worth repeating that expected value, in spite of its limitations, is the decision rule which should normally be employed unless the problem clearly indicates something to the contrary.

Discrete probabilistic analysis (DPA)

11. DPA can be considered as an extension of the expected value procedure described above. As its basis it requires similar estimates of cash flows and associated probabilities, but instead of merely averaging these estimates it uses the component parts of the estimates to show the various outcomes and probabilities possible. The following example illustrates the technique.

Example 3

The NPV of Example 2 was £950 and management consider this somewhat marginal and wish to explore the range of outcomes possible.

Further investigation reveals that two capital costs are possible: the £11,000 as stated with a probability of 0.8 and £15,000 with a probability of 0.2. This results in a new expected NPV of + £150 using the new expected capital cost of £11,800. The full range of outcomes and probabilities is shown in Table 2.

Table 2

	Most likely capital cost P = 0.8 £11,000		Pessimistic capital cost P = 0.2 £15,000	
Optimistic cash flows P = 0.3	3,707	(0.24)	−293	(0.06)
Most likely cash flows P = 0.5	143*	(0.4)	−3,857	(0.1)
Pessimistic cash flows P = 0.2	−1,167	(0.16)	−5,167	(0.04)

The table shows the NPV resulting from each possible combination of the original estimates of cash flows and the capital costs and gives the probability of the combination occurring. For example, the cell marked* is calculated thus:

Present value of most likely cash flows $=$ £11,143
less most likely capital cost   11,000
NPV $=$ 143

The combination has a probability of **0.5 × 0.8 = 0.4**

From Table 2 it will be seen that the outcomes range from +3707 to −£5167 and that the probability of making a loss is **0.16 + 0.06 + 0.1 + 0.04 = 0.36** or alternatively, the probability of at least breaking even is **0.64**.

Management now has more information on which to base a decision.

Advantages of DPA:

a) Simple to apply and understand.

b) Gives some indication of the range of possible outcomes and their probabilities.

c) Considers the detailed variations in the cash flows and investment required for a project rather than merely making one overall assumption such as that uncertainty is directly related to time elapsed.

Disadvantages of DPA:

a) Uses discrete estimates whereas a continuous distribution may be a better representation of a particular project.

b) Increases the amount of subjective estimation necessary.

Continuous probabilistic analysis (CPA)

12. CPA has the same overall objective as DPA which has been described above. That is, to show the variability of the project outcome which results from the variability of the individual cash flows thus enabling the analyst to make probability assessments of the likelihood of various outcomes. It differs from DPA in that continuous distributions and aspects of statistical theory are used

instead of the discrete estimates which are a feature of DPA.

CPA can be shown diagrammatically as follows:

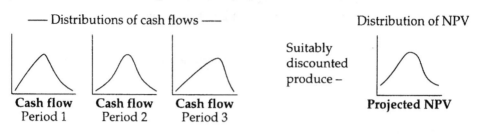

Figure 29/1

To be able to combine the distributions as shown and to be able to make probability statements about the project outcome, it is first necessary to establish the mean (or most likely value) and a measure of the dispersion of each of the individual period's cash flows.

Establishing the means and dispersion of cash flows

13. In general, there is little problem in estimating the mean of the period's cash flows, this being equivalent to the most likely value. A more significant problem is to establish a suitable measure of the variability or dispersion of the period's cash flows. The most useful measure for statistical purposes is the standard deviation but invariably there is insufficient data to calculate the standard deviation in the conventional statistical manner so that some form of subjective estimation becomes necessary.

This could be done as follows:

Assume that the most likely value of the cash flow in a given period was estimated to be £30,000 and it was considered that there was likely to be some variability.

The managers responsible for the estimate could be asked a question similar to the following.

'Given that the most likely value of the cash flow is £30,000 within what limits would you expect the cash flow to be 50% of the time?'

Assume that the answer to the above question was £25,000 to £35,000.

It is known from Normal Area Tables (Table I) that 50% of a distribution lies between the mean $\pm \frac{2}{3} \sigma$ (approximately)

$$\therefore \text{£10,000 (i.e. } 35,000 - 25,000) = \frac{2}{3}\sigma$$

$$\therefore \hat{s} \approx \underline{\text{£7,500}}$$

An alternative to the question asked above would be to ask the manager,

'what is the total range of cash flow that might be expected?'

If the manager was consistent he would answer, '£7,500 to £52,500'.

It is known that the whole of a normal distribution is within the range of the mean $\pm 3 \sigma$ (approximately). Accordingly, the estimate of the standard deviation would be

$$\sigma = \frac{\pounds 52,500 - 7,500}{6} = \pounds 7,500$$

It is clear that the estimation process outlined above is crude and lacks statistical rigour. However, subjective estimation is an unavoidable aspect of all investment appraisals and the procedure does enable some sort of assessment to be made of the probability of achieving various outcomes.

Having obtained the estimates of the means and standard deviations these must be combined to give the mean and standard deviation of the overall project NPV.

Combining the means and standard deviations of the cash flows

14. There is little problem in obtaining the mean of the project NPV. This is simply the means, or most likely values, of the cash flows discounted in the usual manner. The project standard deviation is obtained by combining the discounted standard deviations of the individual cash flows using what is known as the *statistical sum*.

Standard deviations cannot be combined directly but it is possible to add variances, when this is done the square root of the result can be taken thus establishing the standard deviation of the project's NPV, i.e. NPV

$$\therefore \sigma_{NPV}^2 = \sum \left[\frac{\sigma_i}{1 + r^i} \right]^2$$

where σ_i is the standard deviation of the individual cash flows.

$$\sigma_{NPV}^2 = \sum \left[\frac{\sigma_i^2}{(1 + r)^{2i}} \right]^2$$

$$\sigma_{NPV} = \sqrt{\sum \left[\frac{\sigma_i^2}{(1 + r)^{2i}} \right]}$$

This formula is used in the following example.

Example 4

The means and standard deviations of the cash flows of a project are shown below and it is required to calculate:

a) the project NPV (i.e. the mean)

b) the variability of the project NPV (i.e. σ_{NPV})

c) the probability of obtaining

- a negative NPV
- an NPV of at least £20,000

It can be assumed that the cash flows in each period are independent, i.e. variations in one period are independent of variations in other periods and that the cost of capital is 10%.

Project cash flows

Period	0	1	2	3	4	5
Net cash flow (most likely value)	−200,000	+55,000	+48,000	+65,000	+70,000	+40,000
Variability expected (i.e. standard deviation of cash flow)	0	4,000	4,500	3,500	4,500	3,000
Discount Factors @ Cost of Capital of 10%	1.00	0.909	0.826	0.751	0.683	0.621

Solution

a)   The project NPV is found in the usual way, i.e.

$$-200{,}000 + (55{,}000 \times 0.909) + (48{,}000 \times 0.826) + (65{,}000 \times 0.751)$$

$$+ (70{,}000 \times 0.683) + (40{,}000 \times 0.621)$$

$$\therefore \text{ Project NPV (i.e. the mean)} = \underline{\textbf{£11,108}}$$

b)   The standard deviation of the NPV is found by inserting the various estimated cash flow standard deviations into the formula:

$$\sigma_{\text{NPV}} = \sqrt{\sum \left[\frac{4000^2}{(1+0.1)^2} + \frac{4{,}500^2}{(1+0.1)^4} + \frac{3{,}500^2}{(1+0.1)^6} + \frac{4{,}500^2}{(1+0.1)^8} + \frac{3{,}000^2}{(1+0.1)^{10}} \right]}$$

$$\therefore \sigma_{\textbf{NPV}} = \underline{\textbf{£6,847}}$$

It will be seen that the squaring of the denominator has the effect of requiring discount factors at 2, 4, 6, 8 and 10 years instead of the usual 1, 2, 3, 4 and 5 years.

c)   The probability of obtaining a negative NPV (or the probability of any value of NPV) is found by using standard statistical test of normal area i.e. find the 'Z' score or standardised variate and obtain the resulting probability from Normal Area Table I as follows:

$$Z = \left| \frac{£11,108 - 0}{6,847} \right|$$

$$= \underline{\textbf{1.622}}$$

and from the Tables we find that the probability of the NPV being above zero is **0.947 (i.e. 0.5 + 0.4474)** thus there is approximately a **5.3%** chance (1−0.9474) of there being a negative NPV.

The probability of there being at least £20,000 NPV is found by a similar process:

$$Z = \left| \frac{20,000 - 11,108}{6,847} \right|$$

$$= \underline{\mathbf{1.299}}$$

and using the Tables we find that the probability of obtaining at least £20,000 NPV is approximately **9.7%**

The distribution of the project NPV can also be shown diagrammatically as in Figure 29/2.

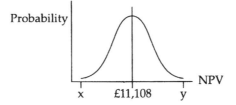

Probability distribution of project. Figure 29/2

Point **x** on the diagram is approximately **£11,108 − (3 × 6,847) = −£9,433** and

point **y** is approximately **£11,108 + (3 × 6,847) = £31,649**

Comparison of projects using CPA

15. Having calculated the means and standard deviations of various project the project distributions can be compared quite simply. The distributions could be drawn on the same graph and visually examined or the relative variability of the distributions could be calculated using their coefficients of variations.

The coefficient of variation is found as follows:

$$\textbf{Coefficient of variation} = \frac{\sigma}{\bar{x}} \ \textbf{100\%}$$

For example, two projects have estimated results as under

Project A	mean =	£80,000	s.d. = £12,500
B	mean =	£130,000	s.d. = £17,500

What are the coefficients of variation and which is the relatively less risky project?

$$\text{Coefficient of variation A} = \frac{12,500}{80,000} \times 100\% = \underline{\mathbf{15.6\%}}$$

$$\text{Coefficient of variation B} = \frac{17,500}{130,000} \times 100\% = \underline{\mathbf{13.5\%}}$$

∴ Project B is relatively less risky assuming that the standard deviation is a reasonable measure of the riskiness of the projects.

Summary of CPA

16. The analysis outlined above can be extended to cover situations where the projects are not independent and/or where the individual distributions are not normal or near normal. All the methods have the same overall objective, which is to find the mean and variability (riskiness) of the project NPV.

    Advantages of CPA:

    a) Produces a distribution of the NPV rather than a single figure.

    b) Enables probability statements to be made about the project's outcome which reflect the variabilities expected in each period's cash flows.

    c) Enables the NPV distributions of competing projects to be compared.

    d) Uses the more realistic assumption of continuous rather than discrete values.

    Disadvantages:

    a) Introduces a further element of subjective estimation.

    b) More complex than DPA, so therefore may not be properly understood or used by decision makers.

Sensitivity analysis

17. This is a practical way of showing the effects of uncertainty by varying the values of the key factors (e.g. sales volume, price, rates of inflation, cost per unit) and showing the resulting effect on the project. The objective is to establish which of the factors affect the project most. When this is done it is management's task to decide whether the project is worthwhile, given the sensitivity of one or more of the key factors. It will be seen that this method does not ask for subjective probability estimates or likely outcomes, but attempts to provide the data upon which judgements may be made. The method is illustrated by the following example.

Example 5

Assume that a project (using single valued estimates) has a positive NPV of £25,000 at a 10% discounting rate. This value would be calculated by the normal methods using particular values for sales volume, sales price, cost per unit, inflation rate, length of life, etc.

Once the basic value (i.e. the NPV of £25,000) has been obtained the sensitivity analysis is carried out by flexing, both upwards and downwards, each of the factors in turn.

An abstract of the results of a sensitivity analysis for the project above might be as follows:

Table 3
Sensitivity analysis abstract

Original NPV = £25,000

A Element to be varied	B Alteration from basic	C Revised NPV £	D Increase + Decrease – £	E Percentage change	F Sensitivity factor, i.e. $\frac{E}{B}$
Sales	+15%	46,000	+21,000	84	5.6
Volume	+10%	33,000	+8,000	32	3.2
(Basic value	−10%	17,000	−8,000	32	3.2
8,000 units in Period 1	−15%	14,000	−11,000	44	2.9
8,500 in Period 2, etc.)	−20%	9,000	−16,000	64	3.2
Sales	+20%	42,000	+17,000	68	3.4
Price	+10%	31,000	+6,000	24	2.4
(Basic value	−10%	17,000	−8,000	32	3.2
£6 unit in Period 1	−15%	11,000	−14,000	56	3.73
£6.25 in Period 2, etc.)	−20%	2,000	−23,000	92	4.6
Cost/Unit	+25%	−12,000	−37,000	148	5.9
(Basic value	+10%	6,000	−19,000	76	7.6
£2.50 in Period 1	−5%	34,000	+9,000	36	7.2
£2.60 in Period 2, etc.)	−10%	47,000	+22,000	88	8.8

From such an analysis the more sensitive elements can be identified. Once identified further analysis and study can take place on these factors to try to establish the likelihood of variability and the range of values that might be expected so as to be able to make a more reasoned decision whether or not to proceed with the project.

Advantages of Sensitivity Analysis:

a) Shows the effect on project outcome of varying the value of the elements which make up the project (e.g. Sales, Costs, etc.).

b) Simple in principle.

c) Enables the identification of the most sensitive variables.

Disadvantages of Sensitivity Analysis:

a) Gives no indication of the likelihood of a variation occurring.

b) Considerable amount of computation involved.

c) Only considers the effect of a single change at a time which may be unrealistic.

Risk and the portfolio effect

18. So far in this chapter we have studied the risks associated with each project considered in isolation. This is an important matter but it will be apparent that of greater significance to the firm is the aggregate risk from all projects accepted which could be termed its port-folio of projects.

The effect on the firm of the risks of individual projects may be neutralised or enhanced when all the individual projects are considered together. A simple example would be where a firm is operating in a cyclical industry with variable (i.e. risky) returns on its existing projects. A new project is being considered which, although variable or risky, is expected to follow a different cyclical pattern to existing operations. When existing operations are experiencing low activity the new project is expected to have substantial activity so that the overall risk to the firm from its portfolio, including the new project, will be minimised. This is, of course, a major reason why firms diversify their operations.

The analysis of this aspect of risk and uncertainty was developed for stock market investment portfolio analysis by Markowitz and others. There are many restrictive assumptions behind the analysis and there are some difficulties in applying it to project investment within the firm but the general reasoning is valid and is of considerable importance.

Assessing the portfolio risk

19. The general procedure for assessing the extent to which the proposed project(s) add to or subtract from the risk of existing operations is to calculate the covariance between the returns of the project(s) and returns of existing operations and to use the covariance(s) to obtain the coefficient of correlation between the project(s) returns and the returns of existing operations.

The interpretation of the correlation coefficients is as follows:

$$\text{Coefficient of correlation} = -1 \text{ risk fully neutralised}$$
$$= 0 \text{ risk unaltered}$$
$$= +1 \text{ risk fully enhanced}$$

The following example illustrates the general procedure.

Example 6

A firm with £100,000 to invest is considering two projects, X and Y each requiring an investment of £100,000. The returns from the proposed projects and from existing operations under three possible views of expected market conditions are shown in Table 4 together with the calculated standard deviations of returns, i.e. the measure of riskiness used by the company.

Market state	I	II	III
Probability of market state	0.3	0.4	0.3
Rate of return Project X	20%	20%	$-1\frac{2}{3}\%$
Standard deviation of returns, Project X = 22%			
Rate of Return Project Y	-2%	15%	27%
Standard deviation of returns, Project Y = 15%			
Rate of Return of existing operations	-9%	$18\frac{1}{4}\%$	28%
Standard deviation of returns on existing operations = 18%			

Table 4

The firm considers that the risk and return of their existing operations are similar to the market as a whole and that a reasonable estimate of a risk free interest rate is 8%. Which, if either, of the two proposed investments should be initiated and why?

Solution

The first stage is to calculate the expected returns for X and Y and existing operations. Expected Returns $\overline{R}$

Project X	$\overline{R}_x = (0.20 \times 0.3) + (0.20 \times 0.4) + (-0.01667 \times 0.3)$	= 0.135
		= 13.5%
Project Y	$\overline{R}_y = (-0.02 \times 0.3) + (0.15 \times 0.4) + (0.27 \times 0.3)$	= 0.135
		= 13.5%
Existing operations	$\overline{R}_0 = (-0.09 \times 0.3) + (0.1825 \times 0.4) + (0.28 \times 0.3)$	= 0.13
		= 13%

It will be seen that the expected returns of Project X and Y are the same and as the standard deviation of Project Y is lower than Project X then Project Y is the preferred project if the projects are considered in isolation from existing operations.

However, this is too superficial a view and further analysis is required on the effects of adding either project to the existing portfolio.

A project's risk can be separated into two elements – systematic and unsystematic risk. The unsystematic risk is the diversifiable risk which can be reduced or eliminated when the project is part of an appropriate portfolio. The systematic risk is the proportion which cannot be eliminated (it applies to the economy or market as a whole) and thus the project's returns must be considered against this residual element.

Portfolio analysis can be used to find the minimum required return for Project X and Y given their risk levels by calculating the covariances between project returns and existing operations and using these values to calculate the correlation coefficients thus;

Covariance between Project X return (R_x) and Company Return (R_0)

	$(R_x - \overline{R}_x)^*$	×	$(R_0 - \overline{R}_0)^*$	×	Market state probability	=	Covariance
State I	0.065	×	−0.22	×	0.3	=	−0.00429
State II	0.065	×	0.0525	×	0.4	=	0.001365
State III	−0.15167	×	0.15	×	0.3	=	−0.00682
					Covariance	=	−0.009745

* These values are found by deducting the calculated expected return $\overline{R}$ from the actual return given in Table 4. For example the value −0.15167 is found as follows: $(-0.01667 - 0.135) = -0.15167$.

The value of the covariance is then used to find the correlation coefficient between Project X returns and returns from existing operations (0).

$$\text{Correlation coefficient between } \mathbf{X} \text{ and } \mathbf{0} = \frac{\text{Covariance}(x, 0)}{\sigma_x.\sigma_0} = \frac{-0.009745}{0.22 \times 0.18}$$

$$\therefore \textbf{ Correlation } (x,0) = \underline{\mathbf{-0.246}}$$

Covariance between Project Y return ($\mathbf{R_y}$) and Company Return ($\mathbf{R_0}$)

	$(\mathbf{R_y} - \mathbf{\bar{R}_y})$	×	$(\mathbf{R_0} - \mathbf{\bar{R}_0})$	×	Market state probability	=	Covariance
State I	−0.155	×	−0.22	×	0.3	=	0.01023
State II	0.015	×	0.0525	×	0.4	=	0.000315
State III	0.135	×	0.15	×	0.3	=	0.006075
							+0.01662

$$\text{Correlation coefficient between } \mathbf{Y} \text{ and } \mathbf{0} = \frac{\text{Covariance}(y, 0)}{\sigma_y.\sigma_0} = \frac{+0.01662}{0.15 \times 0.18}$$

$$\therefore \textbf{ Correlation } (x,0) = \underline{\mathbf{-0.246}}$$

These values can be used to calculate the required return from project X and Y given that the risk free interest rate is 8%, denoted by ($\mathbf{R_F}$)

$$\text{Required return of Project X} = \mathbf{R_F} + \frac{\mathbf{R_0} - \mathbf{R_F}}{\sigma_0} \times \sigma_x \times \textbf{Correlation}(x, o)$$

$$= 0.08 + \frac{0.13 - 0.08}{0.18} \times 0.22 \times -0.246$$

$$= 0.08 - 0.0015 = \underline{\mathbf{6.5\%}}$$

$$\textbf{Required return of Project Y} = \mathbf{R_F} + \frac{\mathbf{R_0} - \mathbf{R_F}}{\sigma_0} \times \sigma_y \times \textbf{Correlation}(y, o)$$

$$= 0.08 + \frac{0.13 - 0.08}{0.18} \times 0.15 \times -0.615$$

$$= 0.08 - 0.0256 = \underline{\mathbf{10.56\%}}$$

Based on the Portfolio analysis, Project X is the preferred project for the following reasons.

a) Project X provides the greatest excess of actual return over minimum return, i.e. 13.5%, c.f. 6.5%, whereas Project Y is 13.5%, c.f. 10.56. Thus Project X maximises the company's wealth.

b) The negative correlation coefficient of Project X, −0.246 means that its pattern of returns to some extent neutralises the overall portfolio risk when Project X is combined with current operations. Although Project Y has a lower individual risk its correlation coefficient of +0.615 means that it enhances risk when combined with current operations.

Note: It will be seen that the decision following analysis of the Portfolio effects is the opposite to that when the project's riskiness is considered in isolation.

It is feasible to work manually through a problem with as few projects as

Example 8 but the number of relationships rises dramatically as the number of projects increases so it is likely that the application of the above principles to any practical sized problem would require computer assistance.

Decision maker's attitude to risk

20. Having dealt with the riskiness of individual projects and of combinations of projects, the third aspect of risk in investment appraisal can now be considered; that of the decision maker's attitude to risk and its influence on the final investment decision.

Surveys and studies have shown that individuals differ in their attitudes to risk and that for serious decisions such as investment appraisals in business, decision makers are *risk averters*. This means that in general, decision makers would prefer a less risky (less variable) investment even though it may have a lower expected value than a higher return yet riskier investment.

This may be demonstrated by the following.

Two investments are being considered.

        Investment A − Return of £100,000 with a probability of 1, i.e. certainty

        Investment B $\begin{cases} \text{Return of £300,000 with a probability of 0.5} \\ \text{Return of zero with a probability of 0.5} \end{cases}$

The expected returns are:

Investment A   **£100,000**
Investment B   (£300,000 × 0.5) + (0 × 0.5) = **£150,000**

It will be apparent that virtually every investor would prefer the certainty of Investment A to the uncertainty or risk involved in Investment B even though it has the higher expected value. Such behaviour is risk aversion.

Implicit in such behaviour is an assumption about the utility or satisfaction derived from money. The utility function of a risk averter declines as the level of income or wealth rises, i.e. a declining marginal utility. A 'risk neutral' investor regards each increment of income or wealth as having the same value whereas a 'risk seeker' is a person whose utility function increases as his level of income or wealth increases. These three possibilities are shown in Figure 29/3.

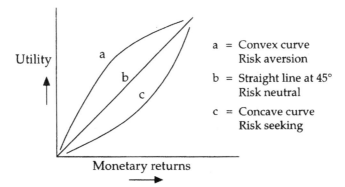

Utility Function of Money. Figure 29/3

471

Utility and certainty equivalents

21. Utility theory applied to decision making in risky conditions states that individuals attempt to optimise something termed utility and assumes that for any individual a formal, quantifiable relationship can be established between utility and money.

    The theoretical way that an individual's utility function is established is by the use of certainty equivalents. These are derived in the following fashion.

    Assume that an individual owns an sweepstake ticket which offers a 50% chance of winning £200,000 and a 50% chance of winning nothing. He would be asked what amount of cash he would accept for the lottery ticket. If he would sell the ticket for £75,000 this is the certainty equivalent, i.e. the value at which he is indifferent between the certain £75,000 and the chance of winning £200,000. This certainty equivalent would be assigned a utile value of 0.5 (a utile is the unit of utility). From such questions the person's utility function can be derived and then asked to make choices in investment decisions.

    However, whilst these processes have theoretical appeal they are virtually impossible to apply in practice because of the extreme difficulty in establishing any form of meaningful utility function. However, risk aversion and varying individual attitudes to risk are very real phenomena although difficult to quantify. Accordingly, it is essential that the project analyst produces some form of risk analysis, both for projects in isolation and in combination, so that the decision maker has more information upon which to make a decision. There are variable or risky elements of every project and to ignore this aspect in the project appraisal can be positively misleading to the decision maker.

Capital rationing – definition

22. This is where the firm is unable to initiate all projects which are apparently profitable because insufficient funds are available. Under the assumptions given for the basic DCF model, a perfect capital market was presumed, i.e. as much finance as required could be raised at the market rate of interest. In imperfect capital market conditions, capital may be raised, but at increasing rates of interest; but there will be some point where there is an absolute limit to the amount that could be raised. This is known as *external capital rationing*. Alternatively, the effects of capital rationing may develop for internal purposes, for example, it may be decided that investment should be limited to the amount that can be financed solely from retained earnings or kept within a given capital expenditure budget. The external and internal factors which impose quantitative limits have led to two opposing view-points developing, known as the 'hard' and 'soft' views of capital rationing.

    The 'hard' view is that there is an absolute limit on the amount of money a firm may borrow or raise externally whereas the 'soft' view is that rationing by a quantitative limit such as an arbitrary capital expenditure budget should only be seen as a temporary, administrative expedient because such a limit is not determined by the market (the assumption being that any amount of funds is available, at a price) and such a limit would not be imposed by a profit maximising firm.

Whatever the causes of the limited capital supply available for investment purposes it means that, not only must each project cover the cost of capital, but that the project or batch of projects selected must maximise the return from the limited funds available, i.e. some form of ranking becomes necessary.

Before considering solution methods some definitions need to be considered.

a) Single period capital rationing – the term used where there is a limit on the funds available now but where it is anticipated that funds will be freely available in subsequent periods.

b) Multi-Period Capital rationing – where the limitation of funds extends over a number of periods or possibly indefinitely.

c) Divisible projects – projects where the whole project or any fraction may be under-taken. If a fractional part is undertaken, then it is assumed that the initial outlay and subsequent cash inflows and outflows are reduced pro rata. Although for most industrial projects this seems somewhat hypothetical, the assumption of divisibility is frequently made in solving capital rationing problems, particularly in examination questions.

d) Indivisible projects – where the whole project must be undertaken or not at all.

Project selection under capital rationing

23. Where capital rationing exists the normal DCF decision rule, i.e. accept all projects which have a positive NPV at the cost of capital is insufficient to make the appropriate project selection.

The objective where capital rationing exists is to maximise the return from the batch of projects selected having regard to the capital limitation. This means that the investment decision changes from simply being 'accept or reject' to what is in effect a ranking problem. Ways of achieving this objective are shown below for the following rationing possibilities.

- single period capital rationing with divisible projects.
- single period capital rationing with divisible projects where some are mutually exclusive
- single period capital rationing with indivisible projects
- multi period capital rationing with divisible projects.

Single period capital rationing – divisible projects

24. This is the simplest situation and the solution method is to rank the projects in order of their EVPI (i.e. NPV per £ of outlay as described earlier) and to choose projects, or fraction of a project, until the supply of capital for investment is exhausted.

Example 7

CR Ltd has a cost of capital of 15% and has a limit of £100,000 available for investment in the current period. It is expected that capital will be freely available in the future. The investment required, the NPV at 15% and the EVPI for each of the 6 projects currently being considered are shown below.

What projects should be initiated?

Project	Outlay £	NPV @ 15% £	EPVI $\left(\dfrac{\text{NPV}}{\text{Outlay}}\right)$
A	20,000	8,000	0.4
B	40,000	28,000	0.7
C	35,000	37,500	1.07
D	50,000	31,500	0.63
E	15,000	3,500	0.23
F	45,000	−5,000	−0.11

Solution

Ranking by EVPI is C, B, D, A and E. Project F cannot be considered because it fails the initial hurdle of achieving a positive NPV.

∴ Optimal Investment Plan

Project	Fraction undertaken	Investment £	NPV £
C	1.00	35,000	37,500
B	1.00	40,000	28,000
D	0.50	25,000	15,750
		£100,000	£81,250

It will be seen that this solution method uses the well known management accounting principle of maximising return per unit of the limiting factor – in this case NPV per £ of capital available for investment. It will be recalled that this principle is appropriate where there is a single constraint only – in this case investment finance for one period.

Single period capital rationing with mutually exclusive divisible projects

25. Where two or more of the projects are mutually exclusive the solution method of ranking of EVPI can still be used but the projects have to be divided into groups each containing one of the mutually exclusive projects. This is shown below.

Example 8

Assume the same data as Example 7 except that projects B and D are mutually exclusive.

What projects should be initiated?

Solution

It is necessary to divide the projects into two groups, rank by EVPI, select projects up to the capital limit and to compare the total NPV obtainable from each group.

	Group I			Group II	
Project	Investment £	EVPI	Project	Investment £	EVPI
A	20,000	0.4	A	20,000	0.4
B	40,000	0.7	C	35,000	1.07
C	35,000	1.07	D	50,000	0.63
E	15,000	0.23	E	15,000	0.23

Ranking the groups and choosing the projects up to the investment limit produces the following:

	Group I				Group II		
Project	Fraction	Investment £	NPV £	Project	Fraction	Investment £	NPV £
C	1.00	35,000	37,500	C	1.00	35,000	37,500
B	1.00	40,000	28,000	D	1.00	50,000	31,500
A	1.00	20,000	8,000	A	$\frac{3}{4}$	15,000	6,000
E	$\frac{1}{3}$	5,000	1,167				
		£100,000	£74,667			£100,000	£75,000

It will be seen that, by a narrow margin, Group II with the proportions indicated, has the greater NPV and would be chosen.

Single period capital rationing – indivisible projects

26. Where projects have to be accepted in their entirety or not at all, then EVPI ranking procedure does not necessarily produce the optimal solution. Providing that relatively few projects are involved a trial and error approach can be used to find a solution. Where projects are indivisible then it is likely that some of the capital available for investment may be unused and in such circumstances a full analysis should include the returns from external investment of under-utilised funds.

Example 9

Lloyds Ltd has a cost of capital of 10% and has a limit of £100,000 available for investment in the current period. Capital is expected to be freely available in future periods. The following *indivisible projects* are being considered.

Initial Project	Investment	NPV @ 10%
	£	£
A	35,000	17,500
B	40,000	22,500
C	65,000	38,000
D	48,000	31,500
E	23,000	9,000

It is required to calculate the optimal investment plan

a) where there are no alternative investments available for any surplus funds.

b) where surplus funds can be invested to produce 12% in perpetuity.

Solution

a) Various combinations are tried to see which combination produces the maximum NPV.

Table 5 shows a few examples.

Table 5

Project combinations	Total outlay for combinations	Surplus funds	Total NPV of combinations
£	£	£	
AC	100,000	–	55,500
ABE	98,000	2,000	49,000
AD	83,000	17,000	49,000
BD	88,000	12,000	54,000
BE	63,000	37,000	31,500
CE	88,000	12,000	47,000
DE	71,000	29,000	40,500

It will be seen from Table 5 that the best investment plan is A and C which utilises all the funds available and produces a combined NPV of £55,500.

b) When surplus funds can be invested externally each of the combinations in Table 5 which have surplus funds must be examined to see if the project NPV plus the return on external investment is greater than £55,500.

Each £1,000 invested at 12% in perpetuity yields £200 NPV, i.e.

$$\left[\frac{1,000 \times .12}{.1}\right] - 1,000 = £200$$

The project combinations and total NPV (Projects + External Investment) are shown in Table 6.

Combi-nation	Total project outlay £	Funds externally invested £	External investment NPV £	+	Project NPV £	=	Total NPV £
ABE	98,000	2,000	400	+	49,000	=	49,400
AD	83,000	17,000	3,400	+	49,000	=	52,400
BD	88,000	12,000	2,400	+	54,000	=	56,400*
BE	63,000	37,000	7,400	+	31,500	=	38,900
CE	88,000	12,000	2,400	+	47,000	=	49,400
DE	71,000	29,000	5,800	+	40,500	=	46,300

Table 6

* When external investment is considered then projects **BD** should be initiated and £12,000 invested externally to produce a total NPV of £54,500. It will be seen that this is slightly better than the **AC** combination shown in Table 5.

Note: Although ranking by EVPI in conditions of single-period capital rationing with indivisible projects does not necessarily produce the correct ranking it usually provides an excellent guide to the best group of projects.

Multi-period capital rationing

27. This has been previously defined to be where investment funds are expected to be limited over several periods. In such circumstances it becomes difficult to choose the batch of projects (some starting immediately, some one period hence, two periods hence, etc.) which yield the maximum return and yet which remain within the capital limits. The problem becomes one of optimising a factor (e.g. NPV) where resources are limited, i.e. the funds available over the periods being considered. This will be recognised as a situation where Linear Programming (LP) can be used (covered previously) and LP has been used successfully in solving Multi-Period Capital Rationing problems.

Multi-period rationing – LP solution

28. To use LP as a solution method means making the assumption that projects are divisible, i.e. fractional parts of a project can be undertaken. This is not necessarily a realistic assumption, but it is one frequently made for examinations purposes. The following example will be used to illustrate the LP formulation solution and interpretation of multi-period capital rationing problems.

Example 10

Trent Ltd has a cost of capital of 10% and is considering which project or projects it should initiate. The following projects are being considered:

Estimated cash flows

Project	Year 0	Year 1	Year 2	Year 3	Year 4
A	−15,000	−25,000	30,000	30,000	20,000
B	−25,000	−15,000	30,000	29,000	30,000
C	−35,000	−15,000	40,000	44,000	30,000

Capital is limited to £40,000 now and £35,000 in Year 1. The projects are divisible.

Solution

Step 1. Calculate the project NPVs in the usual manner.

These are as follows:

$$A = £23,245$$

$$B = £28,414$$

$$C = £37,939$$

Step 2. Formulate the problem in LP terms which means defining the objective function and the constraints. The objective for Trent Ltd is to maximise NPV and this may be expressed as:

$$\text{Maximise } 23,245X_A + 28,414X_B + 37,939X_C$$

where X_A is the proportion of Project A to be initiated
X_B is the proportion of Project B to be initiated
X_C is the proportion of Project C to be initiated

The constraints in this problem are the budgetary limitations in Periods 0 and 1.

Capital at time 0

$$15,000X_A + 25,000X_B + 25,000X_C \leq 40,000$$

Capital at time 1

$$25,000X_A + 15,000X_B + 15,000X_C \leq 35,000$$

In addition it is necessary to specify the formal constraints regarding the proportions of projects accepted to ensure that a project cannot be accepted more than once or that 'negative' projects are accepted.

$$\text{i.e. } X_A, X_B, X_C \leq 1$$

$$X_A, X_B, X_C \geq 0$$

The whole formulation is thus:

Maximise	$23,245X_A$	$+$	$28,414X_B$	$+37,939X_C$	
Subject to	$15,000X_A$	$+$	$25,000X_B$	$+35,000X_C$	$\leq$ 40,000
	$25,000X_A$	$+$	$15,000X_B$	$+15,000X_C$	$\leq$ 35,500
	X_A				$\leq$ 1
			X_B		$\leq$ 1
				X_C	$\leq$ 1
	X_A				$\geq$ 0
			X_B		$\geq$ 0
				X_C	$\geq$ 0

Step 3. Solve the LP Problem.

The above formulation can then be solved by the Simplex method. It will be apparent that even with a highly simplified problem such as this manual solution methods are exceedingly tedious and accordingly it is unlikely that a student would have to work through the method in examinations.

However, formulation of the problem and interpretation of results are possible topics.

The solution of the above problem is as follows:

Project	Fraction accepted
A	0.988
B	0
C	0.719

Value of objective function £50,244

Shadow prices 1st constraint (i.e. the £40,000 Year 0 budget) = 0.922
2nd constraint (i.e. the £35,000 Year 1 budget) = 0.3755

Step 4. Interpretation of Solution.

The solution indicates that 0.988 of Project A and 0.719 of Project C should be initiated. This investment plan uses all the funds available in Year 0 and 1.

The shadow prices indicate the amount by which the NPV of the optimal plan (i.e. £50,244) could be increased if the budgetary constraints could be increased. For every £1 relaxation of the constraint in Period 0, £0.922 extra NPV would be obtained. The shadow prices indicate that extra funds in Period 0 are worth approximately three times those in Period 1. This fact may give a management some guidance in their considerations of various alternative sources of capital.

Note: It will be remembered the assumption made in this example is that these projects are divisible. It is appreciated that this is not necessarily a very practical assumption, but it appears to be one frequently made in examinations. Being divisible the NPV is scaled down by the proportion of the project accepted. This is a way of checking the result obtained. In this example the value of NPV obtained is £50,244, i.e.

$$(0.988 \times £23,245) + (0.719 \times 37,939).$$

Where projects are not divisible the only feasible solution method is Integer Programming which is outside the scope of the syllabuses at which this book is aimed.

Reservations over the LP method of solving capital rationing problems

29. There is no doubt that in the right circumstances LP can be useful method of dealing with multi-period capital rationing problems. There are, however, numerous assumptions and limitations which must be kept in mind if the use of the technique is being considered. The major ones are as follows:

a) Is the assumption of linearity for all functions realistic?

b) Are projects truly divisible and capable of being scaled in a linear fashion?

c) Are all investment opportunities included for each of the periods contained in the model?

d) Are all projects and constraints independent of one another as assumed in the LP model?

e) Are all cash flows, resources, constraints known with the certainty assumed in the model?

   (The way that uncertainty is ignored is possibly the most significant reservation)

f) Is the choice of discount rate a realistic one? Under 'hard' capital rationing (i.e. externally imposed), the opportunity cost of funds cannot be known until the investment plan is formulated which, of course, requires the cost of funds to be known – a classic circular argument!

Summary

30. a) Uncertainty and risk are important factors to be considered in investment appraisal. Three aspects are of special concern: individual project uncertainty, the 'portfolio' effect and the decision makers' attitude to risk.

b) Individual project uncertainty can be analysed by three groups of techniques: time based, probability based, and sensitivity analysis and simulation.

c) The main problem with the time based methods is that they do not explicitly consider the variability of cash flows.

d) The probability based methods use subjective probabilities and range from expected value through to methods employing statistical analysis based on the properties of distributions.

e) Arguably of more importance than individual project risk is the aggregate risk of the firm's portfolio of projects. New projects may, to some extent, neutralise or enhance existing risks.

f) Using the covariance of project returns and returns on existing operations the correlation coefficients of new and existing projects can be calculated.

g) The decision maker's attitude to risk is of critical importance but is extremely difficult to quantify. In general, decision makers are risk averters.

h) Capital rationing is where all apparently profitable projects cannot be initiated because of shortage of capital.

i) The decision rule where capital rationing exists is to maximise the return from the project(s) selected rather than simply accept/reject decisions of projects in isolation.

j) Single period rationing with divisible projects is dealt with by ranking in order of EVPI, having due regard to mutually exclusive projects. Where the projects are indivisible then a trial and error combination approach can be used.

k) Multi-period capital rationing with divisible projects is usually solved by LP which produces the optimal solution quantities (i.e. the projects to be initiated) the value of the objective function (i.e. the total NPV) and the shadow costs (i.e. opportunity costs of the binding constraints).

l) Although useful there are a number of reservations of using LP for solving capital rationing problems.

Points to note

31. a) The treatment of such matters as inflation and uncertainty have been dealt with in the context of investment appraisal. However the principles and techniques described have a much wider application than just investment decisions. For example, the uses of expected values, sensitivity analysis and the concept of the real as opposed to the nominal value of money are applicable in virtually every area of planning and decision making.

b) Because of the amount of data involved and the complexity of the techniques used, computers are widely employed for investment appraisals, particularly for risk evaluation and sensitivity analysis.

Self review questions *Numbers in brackets refer to paragraph numbers*

1 In which stages of the appraisal and decision process should uncertainty and risk be considered? (2)

2 What are the time based methods of considering uncertainty? What is their underlying assumption and their major limitation? (4–8)

3 Describe the method of using Expected Value in project appraisals? (10)

4 How does discrete probabilistic analysis extend the expected value technique? (11)

5 What is continuous probabilistic analysis and, if used, how are the means and dispersions of the cash flows established? (12–13)

6 How is the standard deviation of the NPV established and how is this used? (14)

7   What is the objective of sensitivity analysis and how is it carried out? (17)

8   Why is it important to consider not only the risks of individual projects but the aggregate risk of combinations of projects? (18)

9   What is the general procedure for assessing the portfolio risk? (19)

10   What are risk averters? (20)

11   What is a certainty equivalent? (21)

12   What is the difference between 'hard' and 'soft' capital rationing? (22)

13   How would the investment decision be made if single period rationing existed with divisible projects? (24)

14   How would the answer to 13 change if some of the projects were mutually exclusive? (25)

15   What is multi-period capital rationing and what is a possible solution method? (27)

16   What reservations exist regarding the use of LP to solve capital rationing problems? (29)

Exercises with answers

1.   Calculate the Expected NPV of the following project which has an outlay of £10,000 and a 10% cost of capital.

		Cash flows		
	P	Year 1	2	3
Optimistic	0.2	£5,500	£7,000	£5,500
Most Likely	0.5	4,000	5,500	5,000
Pessimistic	0.3	3,000	4,500	3,000

2.   The standard deviations of the cash flows in question 1 have been estimated as follows

	Year 1	Year 2	Year 3
Estimated s.d.	£400	£500	£350

What is the probability of obtaining an expected NPV of zero or less?

3.   A firm has the following 5 divisible projects. There is a limit of capital for this period only of £150,000. Which projects should be selected?

Project	Outlay	NPV
A	£80,000	£15,000
B	£50,000	£15,000
C	£60,000	£25,000
D	£50,000	£23,000
E	£30,000	£8,000

Answers to exercises

1. Expected cash flows

	Year	
1	**2**	**3**
4,000	5,500	4,500

$$\therefore \text{NPV} = -10,000 + 11,558 = 558$$

2.
$$\sigma_{\text{NPV}} = \sqrt{\sum \left[\frac{400^2}{(1+.1)^2} + \frac{500^2}{(1+.1)^4} + \frac{350^2}{(1+.1)^6} \right]} = £881$$

$$\therefore Z = \frac{558 - 0}{881} = 0.633$$

$$\text{Probability} = 0.5 - 0.2357$$

$$\text{P(zero or less)} = \mathbf{26.43\%}$$

3.

Project	EVP1
A	0.1875
B	0.300
C	0.417
D	0.46
E	0.26

∴ Projects:

		Cumulative	
		Outlay	**NPV**
	All **D**	£50,000	£23,000
	All **C**	110,000	48,000
	$\frac{4}{5}$ **B**	150,000	60,000

30 Matrix algebra

Objectives

1.   After studying this chapter you will
     - know how to define a Matrix;
     - be able to Add, Subtract and Multiply Matrices and Vectors;
     - understand Zero and Unity Matrices;
     - know how to Invert a Matrix;
     - understand Probability Transition Matrices.

Presentation of information

2.   Accountants and businessmen are well used to presenting information in tabular form or writing information in rows and columns. Often one set of information in tabular form, e.g. depreciation by category of fixed asset, is worked upon to produce a subsequent set of information based upon the first table, e.g. depreciation by category of fixed asset spread over cost centres.

   Although not generally referred to as such the above are examples of matrices. An understanding of matrixces and matrix algebra may provide short cut methods of calculation and will provide further insights into improved methods of presenting and manipulating data.

Matrix definition

3.   A matrix is a rectangular array of numbers whose value and position in the matrix is significant. A matrix is usually, but not always, shown in brackets thus,

$$\begin{bmatrix} 1 & 8 \\ 3 & 7 \end{bmatrix}$$

Example 1

The size of a matrix is given by the number of rows and the number of columns, i.e. rows × columns. The symbols most commonly used being **m** (columns) × **n** (rows). The matrix in Example 1 is a 2 × 2 matrix, and as the number of columns equals the number or rows, it is known as *Square matrix*.

The following are further examples of matrices:

$$\begin{bmatrix} x_1 & x_2 \\ x_3 & x_4 \\ x_5 & x_6 \end{bmatrix} \qquad \begin{bmatrix} 5 & 9 & 16 \\ 2 & 8 & 4 \end{bmatrix} \qquad \begin{bmatrix} 8 & 12 & 6 \\ 9 & 5 & 1 \\ 7 & 0 & 2 \end{bmatrix}$$

Example 2	Example 3	Example 4
(a 3 × 2 matrix)	(a 2 × 3 matrix)	(a 3 × 3 matrix, i.e. square)

The standard notation for an element in a matrix is as follows

$$x_{i,j} = \text{the element in Row } i \text{ Column } j$$

thus

$$x_{2,3} = \text{the element in the second row, column three.}$$

Vectors

4. A single row matrix is called a row vector and a single column matrix is called a column vector.

$$\begin{bmatrix} 16 \\ 5 \\ 8 \\ 2 \end{bmatrix}$$

$$[9 \quad 5 \quad 8 \quad 2]$$

Example 5	Example 6
(row vector)	(column vector)

Matrix algebra

5. The particular rules applying the manipulation of data in matrix form are given in the following sections on *matrix addition, matrix subtraction, matrix multiplication* and *matrix inversion*. As a form of shorthand matrices are often referred to by capital letters, for example.

$$\mathbf{A} = \begin{bmatrix} 1 & 6 \\ 5 & 2 \\ 8 & 3 \end{bmatrix}$$

$$\mathbf{C} = \begin{bmatrix} 3 & 2 \\ 1 & 6 \end{bmatrix}$$

Example 7

Matrix addition

6. The only rule is that matrices to be added (or subtracted) must be the same size as one another, i.e. they must have the same number of columns and the same number of rows, for example

$$\mathbf{A} = \begin{bmatrix} 1 & 11 & 2 \\ 6 & 2 & 9 \end{bmatrix}$$

$$\mathbf{B} = \begin{bmatrix} 2 & 0 & 7 \\ 5 & 9 & 6 \end{bmatrix}$$

$$\mathbf{A} + \mathbf{B} = \begin{bmatrix} 1 & 11 & 2 \\ 6 & 2 & 9 \end{bmatrix} + \begin{bmatrix} 2 & 0 & 7 \\ 5 & 9 & 6 \end{bmatrix}$$

$$= \begin{bmatrix} 1+2 & 11+0 & 2+7 \\ 6+5 & 2+9 & 9+6 \end{bmatrix}$$

$$A + B = \begin{bmatrix} 3 & 11 & 9 \\ 11 & 11 & 15 \end{bmatrix}$$ **Example 8**

It will be noted that numbers in the same locations have been added giving a matrix with the same dimensions as those added, i.e. a 2 × 3 matrix. *Note* that **A** + **B** = **B** + **A**, i.e. it does not matter in which sequence the matrices are added. It follows that any number of matrices can be added together providing that they are the same size. If the matrices are not the same size they cannot be added, for example.

$$\text{if } X = \begin{bmatrix} 1 & 4 \\ 8 & 2 \end{bmatrix} \text{ and } Y = \begin{bmatrix} 6 \\ 5 \end{bmatrix}$$ **Example 9**

X + **Y** has no meaning because the matrices involved are not the same size.

Matrix subtraction

7. Matrix subtraction uses the same general rules as matrix addition.

For example, using the same matrices as in Example 8 calculate **A** − **B**

$$\begin{bmatrix} 1 & 11 & 2 \\ 6 & 2 & 9 \end{bmatrix} - \begin{bmatrix} 2 & 0 & 7 \\ 5 & 9 & 6 \end{bmatrix}$$

$$A - B = \begin{bmatrix} 1-2 & 11-0 & 2-7 \\ 6-5 & 2-9 & 9-6 \end{bmatrix}$$ **Example 10**

$$= \begin{bmatrix} -1 & 11 & -5 \\ 1 & -7 & 3 \end{bmatrix}$$

It will be seen that minus numbers appear in the final matrix. If minus numbers appear in the original matrix then the usual rules of arithmetic apply, for example minus a minus equal a plus. It is important to note that whereas **A** + **B** = **B** + **A**, **A** − **B** ≠ **B** − **A**, for example reversing Example 10 above to calculate **B** − **A** gives

$$\begin{bmatrix} 2 & 0 & 7 \\ 5 & 9 & 6 \end{bmatrix} - \begin{bmatrix} 1 & 11 & 2 \\ 6 & 2 & 9 \end{bmatrix}$$

$$B - A = \begin{bmatrix} 2-1 & 0-11 & 7-2 \\ 5-6 & 9-2 & 6-9 \end{bmatrix}$$ **Example 11**

$$= \begin{bmatrix} 1 & -11 & 5 \\ -1 & 7 & -3 \end{bmatrix}$$

It will be seen that this is a different result to that in Example 10.

Matrix multiplication

8. There are two aspects of matrix multiplication, the multiplication of a matrix by a single number, called a *scalar*, and the multiplication of a matrix by another matrix.

Scalar multiplication

9. A scalar is an ordinary number such as 3, 6, 8.2 etc. The rule for this is simple – multiply each element in the matrix by the scalar, for example.

$$\text{Let } \mathbf{A} = \begin{bmatrix} 5 & 2 \\ 8 & 3 \end{bmatrix}$$

and it is required to find $4 \times \mathbf{A}$ **Example 12**

$$4 \times \mathbf{A} = 4 \times \begin{bmatrix} 5 & 2 \\ 8 & 3 \end{bmatrix} = \begin{bmatrix} 20 & 8 \\ 32 & 12 \end{bmatrix}$$

Matrix multiplication

10. The main rule to be remembered when it is required to multiply a matrix by another matrix is that the number of columns in the 1st matrix *must equal* the number of rows in the 2nd matrix, i.e. a 2×3 matrix can be multiplied with a 3×2 matrix (i.e. the number of columns, 3 = number of rows, 2) but a 2×3 matrix cannot be multiplied with another 2×3 matrix. The method of matrix multiplication will be shown using the following matrices as a basis.

$$\mathbf{A} = \begin{bmatrix} 3 & 1 \\ 2 & 4 \\ 7 & 4 \end{bmatrix} \qquad \text{i.e. a } 3 \times 2 \text{ matrix}$$

$$\mathbf{B} = \begin{bmatrix} 8 & 0 & 5 & 4 \\ 3 & 2 & 11 & 1 \end{bmatrix} \qquad \text{i.e. a } 2 \times 4 \text{ matrix}$$

Calculate $\mathbf{AB}$, i.e. $\mathbf{A} \times \mathbf{B}$

		Rows		Columns
First check if feasible	A =	3	×	2
	B =	2	×	4

$\therefore$ No. of columns of $\mathbf{A}$ = No. of rows of $\mathbf{B}$. $\therefore$ Multiplication is feasible.

The new matrix $\mathbf{AB}$ is produced by the following steps

$$\mathbf{AB} = \begin{bmatrix} 3 \times 8 + 1 \times 3 & 3 \times 0 + 1 \times 2 & 3 \times 5 + 1 \times 11 & 3 \times 4 + 1 \times 1 \\ 2 \times 8 + 4 \times 3 & 2 \times 0 + 4 \times 2 & 2 \times 5 + 4 \times 11 & 2 \times 4 + 4 \times 1 \\ 7 \times 8 + 4 \times 3 & 7 \times 0 + 4 \times 2 & 7 \times 5 + 4 \times 11 & 7 \times 4 + 4 \times 1 \end{bmatrix}$$

$$\mathbf{AB} = \begin{bmatrix} 27 & 2 & 26 & 13 \\ 28 & 8 & 54 & 12 \\ 68 & 8 & 79 & 32 \end{bmatrix}$$

which, of course would normally be shown in the usual more compact form

$$\mathbf{AB} = \begin{bmatrix} 27 & 2 & 26 & 13 \\ 28 & 8 & 54 & 12 \\ 68 & 8 & 79 & 32 \end{bmatrix}$$

The steps in obtaining **AB** were:

Multiply 1st element in 1st row in **A** by 1st element in 1st column in **B** (i.e. 3×8).

Multiply 2nd element in 1st row in **A** by 2nd element in 1st column in **B** (i.e. 1×3).

(This multiplication process would be continued until the **nth** element in 1st row of the first matrix had been multiplied by the **n** element in the first column of the second matrix).

All these products are added to give the 1st element in 1st row and 1st column of the new matrix **AB**, i.e.

$$(3 \times 8) + (1 \times 3) = 27$$

Then every number in the 2nd row of matrix **A** is multiplied with every number in the 1st column of matrix **B**, i.e. (2×8) and (4×3), and these are added to give the second element in the first column of matrix **AB**, i.e. 28.

This process is continued until every row of matrix **A** has been multiplied by the 1st column of matrix **B** When this is done, the new matrix, **AB**, has its first column. The process of multiplying each row of matrix **A** with each column of matrix **B** continues until all the elements of the new matrix, **AB**, are calculated.

Note: The size of **AB** is 3×4, i.e. it has the number of rows of **A** and the number of columns of **B**

i.e **A × B = AB**

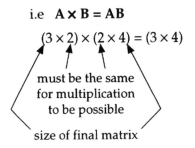

$(3 \times 2) \times (2 \times 4) = (3 \times 4)$

must be the same
for multiplication
to be possible

size of final matrix

Notes:

a) **A** × **B** does *not* equal **B** × **A**. This is unlike ordinary multiplication of numbers where $3 \times 6 = 18$ and so does $6 \times 3 = 18$.

b) If **A** × **B** possible it does not follow that **A** × **B** is possible.

Vector multiplication

11. The process to be followed when multiplying a matrix by a vector is the same as detailed in Para 10. This will be shown by the following example:

$$A = [5 \quad 6 \quad 7] \qquad B = \begin{bmatrix} 4 & 3 \\ 1 & 2 \\ 8 & 9 \end{bmatrix}$$

A is a 1 × 3 matrix (or a row vector)

B is a 3 × 2 matrix

$$\therefore AB = [5 \quad 6 \quad 7] \times \begin{bmatrix} 4 & 3 \\ 1 & 2 \\ 8 & 9 \end{bmatrix}$$

$$= [5 \times 4 + 6 \times 1 + 7 \times 8 \qquad 5 \times 3 + 6 \times 2 + 7 \times 9]$$

$$AB = [82 \quad 90] \text{ i.e. a } 1 \times 2 \text{ matrix or row vector}$$

An example of matrix multiplication

12. A group operates a chain of filling stations in each of which are employed cashiers, attendants and mechanics as shown.

	Types of filling station		
	Large	**Medium**	**Small**
Cashier	4	2	1
Attendants	12	6	3
Mechanics	6	4	2

(Matrix **A**, i.e. 3 × 3)

The number of filling stations are

	Southern England	Northern England
Large stations	3	7
Medium stations	5	8
Small stations	12	4

(Matrix **B**, i.e. 3 × 2)

How many of the various types of staff are employed in Southern England and in Northern England?

Solution

A is a 3 × 3 matrix, B is a 3 × 2 matrix ∴ **AB** is feasible and will be a 3 × 2 matrix.

$$\mathbf{A} \qquad \times \qquad \mathbf{B} \qquad = \qquad \mathbf{AB}$$

$$\begin{bmatrix} 4 & 2 & 1 \\ 12 & 6 & 3 \\ 6 & 4 & 2 \end{bmatrix} \qquad \begin{bmatrix} 3 & 7 \\ 5 & 8 \\ 12 & 4 \end{bmatrix} \qquad \begin{bmatrix} x_{11} & x_{12} \\ x_{21} & x_{22} \\ x_{31} & x_{32} \end{bmatrix}$$

$$\begin{aligned}
x_{11} &= (4 \times 3) + (2 \times 5) + (1 \times 12) &= 34 \\
x_{12} &= (4 \times 7) + (2 \times 8) + (1 \times 4) &= 48 \\
x_{21} &= (12 \times 3) + (6 \times 5) + (3 \times 12) &= 102 \\
x_{22} &= (12 \times 7) + (6 \times 8) + (3 \times 4) &= 144 \\
x_{31} &= (6 \times 3) + (4 \times 5) + (2 \times 12) &= 62 \\
x_{32} &= (6 \times 7) + (4 \times 8) + (2 \times 4) &= 82
\end{aligned}$$

∴ AB is

	South	North
Cashiers	34	48
Attendants	102	144
Mechanics	62	82

Zero matrix

13. In matrix algebra a zero is represented by the zero matrix which is any square matrix in which every element is zero. As with normal numbers if a matrix is multiplied by a zero matrix we obtain a zero matrix, i.e.

$$\begin{bmatrix} 3 & 4 \\ 5 & 6 \end{bmatrix} \times \begin{bmatrix} 0 & 0 \\ 0 & 0 \end{bmatrix} = \begin{bmatrix} (3 \times 0) + (4 \times 0) & (3 \times 0) + (4 \times 0) \\ (5 \times 0) + (6 \times 0) & (5 \times 0) + (6 \times 0) \end{bmatrix}$$

$$= \begin{bmatrix} 0 & 0 \\ 0 & 0 \end{bmatrix}$$

Unity matrix

14. In matrix algebra unity is any square matrix whose top left to bottom right diagonal consists of 1s where all the rest of the matrix consists of zeros. This matrix is important and is always given the symbol I thus

$$\mathbf{I} = \begin{bmatrix} 1 & 0 \\ 0 & 1 \end{bmatrix} \quad \text{or } \mathbf{I} = \begin{bmatrix} 1 & 0 & 0 \\ 0 & 1 & 0 \\ 0 & 0 & 1 \end{bmatrix} \quad \mathbf{I} = \begin{bmatrix} 1 & 0 & 0 & 0 \\ 0 & 1 & 0 & 0 \\ 0 & 0 & 1 & 0 \\ 0 & 0 & 0 & 1 \end{bmatrix}$$

Matrices are only equal where they are the same size and have the same elements in the same place, i.e.

$$\begin{bmatrix} 1 & 0 \\ 0 & 1 \end{bmatrix} \neq \begin{bmatrix} 1 & 0 & 0 \\ 0 & 1 & 0 \\ 0 & 0 & 1 \end{bmatrix}$$

As with normal numbers where a number multiplied by one equals itself

$(3 \times 1 = 3)$ so with matrices. A matrix multiplied by the unity matrix equals itself, i.e.

$$AI = A \text{ and } IA = A$$

$$A = \begin{bmatrix} 1 & 6 \\ 2 & 3 \end{bmatrix} \text{ for example}$$

$$AI = \begin{bmatrix} 1 & 6 \\ 2 & 3 \end{bmatrix} \times \begin{bmatrix} 1 & 0 \\ 0 & 1 \end{bmatrix} = \begin{bmatrix} 1 \times 1 + 6 \times 0 & 1 \times 0 + 6 \times 1 \\ 2 \times 1 + 3 \times 0 & 2 \times 0 + 3 \times 1 \end{bmatrix}$$

$$= \begin{bmatrix} 1 & 6 \\ 2 & 3 \end{bmatrix}$$

Similarly $IA = \begin{bmatrix} 1 & 0 \\ 0 & 1 \end{bmatrix} \times \begin{bmatrix} 1 & 6 \\ 2 & 3 \end{bmatrix} = \begin{bmatrix} 1 \times 1 + 0 \times 2 & 1 \times 6 + 0 \times 3 \\ 0 \times 1 + 1 \times 2 & 0 \times 6 + 1 \times 3 \end{bmatrix}$

$$= \begin{bmatrix} 1 & 6 \\ 2 & 3 \end{bmatrix} \text{ thus proving that } AI = IA = A$$

Note: The unit matrix, **I**, must always be square.

Matrix inversion

15. In matrix algebra the function of division is changed to that of *inversion*. The inverse (or reciprocal) of a matrix has the same property as that of the inverse of an ordinary number. The inverse of 8 is $\frac{1}{8}$ so that

$$8 \times \tfrac{1}{8} = 1 = \tfrac{1}{8} \times 8$$

In matrix algebra the inverse of a matrix is denoted A^{-1} and

$$A \times A^{-1} = I = A^{-1} \times A$$

Only *square* matrices can have inverses which follows from

$$A \times A^{-1} = A^{-1} \times A$$

which expression implies that rows and columns are equal.

Finding the inverse

16. Several methods exist for finding A^{-1} given **A**, but the following step by step method is simplest and is quite manageable for the size of matrices likely to be encountered in examinations.

Assume that it is required to find the inverse of matrix **A**.

$$A = \begin{bmatrix} a_{11} & a_{12} \\ a_{21} & a_{22} \end{bmatrix} = \begin{bmatrix} 1 & 2 \\ 3 & 4 \end{bmatrix}$$

This is done by carrying out row by row operations on A with the objective of transforming it into a unity matrix, **I**. At the same time the same row by row operations are carried out on a unity matrix which at the end of the operations becomes $\mathbf{A}^{-1}$.

Step 1. Place a unity matrix alongside $\mathbf{A}$

$$\begin{bmatrix} 1 & 2 \\ 3 & 4 \end{bmatrix} \quad \begin{bmatrix} 1 & 0 \\ 0 & 1 \end{bmatrix}$$

Step 2. As a_{11} is already 1 we wish to make a_{21} into zero, i.e. Row 2 $-$ (3 $\times$ Row 1)

$$\text{Row 2} - (3 \times \text{Row 1}) \text{ gives } \begin{bmatrix} 1 & 2 \\ 0 & -2 \end{bmatrix} \begin{bmatrix} 1 & 0 \\ -3 & 1 \end{bmatrix}$$

Step 3. We now require a_{22} to be 1, $\therefore$ we multiply Row 2 by $-\frac{1}{2}$

$$\text{Row 2} \times -\frac{1}{2} \text{ gives } \begin{bmatrix} 1 & 2 \\ 0 & 1 \end{bmatrix} \begin{bmatrix} 1 & 0 \\ 1\frac{1}{2} & -\frac{1}{2} \end{bmatrix}$$

Step 4. Finally we wish to make a_{12} into zero, i.e. Row 1 $-$ (2 $\times$ Row 2)

$$\text{Row 1} - (2 \times \text{Row 2}) \text{ gives } \begin{bmatrix} 1 & 0 \\ 0 & 1 \end{bmatrix} \begin{bmatrix} -2 & 1 \\ 1\frac{1}{2} & -\frac{1}{2} \end{bmatrix}$$

$\mathbf{A}$ becomes $\mathbf{I}$ and the original $\mathbf{I}$ becomes $\mathbf{A}^{-1}$

Note: The row by row operations shown are similar to those used in the Simplex method of solving LP problems given earlier.

To prove that the matrix obtained in Step 4 is $\mathbf{A}^{-1}$ we can multiply it by $\mathbf{A}$ and we should obtain a unity matrix, i.e.

$$\mathbf{A} \times \mathbf{A}^{-1} = \mathbf{I}$$

i.e. $\begin{bmatrix} 1 & 2 \\ 3 & 4 \end{bmatrix} \times \begin{bmatrix} -2 & 1 \\ 1\frac{1}{2} & -\frac{1}{2} \end{bmatrix} = \begin{bmatrix} 1 \times -2 + 2 \times 1\frac{1}{2} & 1 \times 1 + 2 \times -\frac{1}{2} \\ 3 \times -2 + 4 \times 1\frac{1}{2} & 3 \times 1 + 4 \times -\frac{1}{2} \end{bmatrix}$

$$= \begin{bmatrix} 1 & 0 \\ 0 & 1 \end{bmatrix} = \mathbf{I}$$

Notes.

a) Not every square matrix has an inverse, for example $\begin{bmatrix} 1 & 1 \\ 1 & 1 \end{bmatrix}$ has no inverse.

b) The product of two matrices, neither of which is a zero matrix, may give a zero matrix as an answer, for example

$$\mathbf{A} = \begin{bmatrix} 1 & 1 \\ 1 & 1 \end{bmatrix} \text{ and } \mathbf{B} = \begin{bmatrix} 1 & -1 \\ -1 & 1 \end{bmatrix}$$

$$\mathbf{AB} = \begin{bmatrix} 1 & 1 \\ 1 & 1 \end{bmatrix} \times \begin{bmatrix} 1 & -1 \\ -1 & 1 \end{bmatrix}$$

$$= \begin{bmatrix} 0 & 0 \\ 0 & 0 \end{bmatrix}$$

Finding the inverse using determinants

17. An alternative method of finding an inverse uses determinants. A determinant of matrix **A** is denoted by | **A** | or **Det A** and is defined as follows (for a 2 × 2 matrix):

$$\text{If } \mathbf{A} = \begin{bmatrix} a & b \\ c & d \end{bmatrix} \text{ then } | \mathbf{A} | = \mathbf{ad} - \mathbf{bc}$$

For example, matrix **A** from para 16 is

$$\begin{bmatrix} 1 & 2 \\ 3 & 4 \end{bmatrix} \quad \therefore | \mathbf{A} | = (1 \times 4) - (3 \times 2) = \underline{\underline{-2}}$$

A determinant is used as follows:

$$\text{If } \mathbf{A} = \begin{bmatrix} 1 & 2 \\ 3 & 4 \end{bmatrix} \quad \text{then } \mathbf{A}^{-1} = \frac{1}{| \mathbf{A} |} \begin{bmatrix} d & -b \\ -c & a \end{bmatrix}$$

This procedure will be used to invert matrix **A** from Para 16 for which the determinant is −2, as calculated above.

$$\mathbf{A} = \begin{bmatrix} 1 & 2 \\ 3 & 4 \end{bmatrix} \quad \text{and } | \mathbf{A} | = \underline{\underline{-2}}$$

$$\therefore \mathbf{A}^{-1} = -\tfrac{1}{2} \begin{bmatrix} 4 & -2 \\ -3 & 1 \end{bmatrix}$$

which, multiplied in the usual manner, gives

$$\mathbf{A}^{-1} = \begin{bmatrix} -2 & 1 \\ 1\tfrac{1}{2} & -\tfrac{1}{2} \end{bmatrix}$$

which is the same result obtained by using the row by row operations in the preceding paragraph.

Solving simultaneous equations by matrix algebra

18. Matrix algebra can be useful for solving simultaneous equations. To be able to find a unique solution there must be an equal number of equations and unknowns (or more equations than unknowns) so that a square matrix can be established. The solution method is similar to the method of finding the inverse of a matrix, i.e. row by row operations. This will be demonstrated using the following example:

Solve, using matrix algebra, the following simultaneous equations.

$$3x + 4y = 10$$
$$2x + 7y = 11$$

Setting out the problem in matrix form gives

$$\begin{bmatrix} 3 & 4 \\ 2 & 7 \end{bmatrix} \begin{bmatrix} 10 \\ 11 \end{bmatrix}$$

from which it is required to produce

$$\begin{bmatrix} 1 & 0 \\ 0 & 1 \end{bmatrix} \begin{bmatrix} x \\ y \end{bmatrix}$$

giving a numeric answer for x and y

Step 1. Set out the equations in matrix form (i.e. as above)

$$\begin{bmatrix} 3 & 4 \\ 2 & 7 \end{bmatrix} \begin{bmatrix} 10 \\ 11 \end{bmatrix}$$

Step 2. Make a_{11} into 1 by subtracting Row 2.

$$\text{Row 1} - \text{Row 2 gives} \begin{bmatrix} 1 & -3 \\ 2 & 7 \end{bmatrix} \begin{bmatrix} -1 \\ 11 \end{bmatrix}$$

Step 3. Make a_{21} into a zero by subtracting 2 × Row 1

$$\text{Row 2} - (2 \times \text{Row 1}) \text{ gives} \begin{bmatrix} 1 & -3 \\ 0 & 13 \end{bmatrix} \begin{bmatrix} -1 \\ 13 \end{bmatrix}$$

Step 4. Make a_{22} into 1 by dividing Row 2 by 13

$$\text{Row 2} \div 13 \text{ gives} \begin{bmatrix} 1 & -3 \\ 0 & 1 \end{bmatrix} \begin{bmatrix} -1 \\ 1 \end{bmatrix}$$

Step 5. Make a_{12} into a zero by adding to Row 1 three times Row 2

$$\text{Row 1} + (3 \times \text{Row 2}) \text{gives} \begin{bmatrix} 1 & 0 \\ 0 & 1 \end{bmatrix} \begin{bmatrix} 2 \\ 1 \end{bmatrix}$$

$$\therefore x = 2 \quad y = 1$$

which can be checked by substituting in the original equations,

$$3x + 4y = 10$$
$$2x + 7y = 11$$

$$\text{i.e. } 3 \times 2 + 4 \times 1 = 10$$
$$2 \times 2 + 7 \times 1 = 11$$

Probability transition matrices

19. These are matrices in which the individual elements are in the form of probabilities, e.g. 0.5, 0.70, 0.1, etc.

The probabilities represent the probability of one event following another event i.e. the probability of the transition from one event to the next. The probabilities of the various changes, applied to the initial state by matrix multiplication, give a forecast of the succeeding state. This process can be repeated indefinitely but usually it will be found that after several cycles the values will stabilise within a restricted range.

Applications of this process, known as the *Markov* process, occur in the marketing area when brand switching is being investigated and in forecasting rates of breakdown in machinery so as to determine appropriate replacement programmes. It can also be applied to other problems where one event succeeds another and the transition probabilities are known.

An application of probability transition matrices.

Assume that two products, Gleem and Sparkle, currently share the market with shares of 60% and 40% each respectively.

Each week some brand switching takes place. Of those who bought Gleem the previous week, 70% buy it again whilst 30% switch to Sparkle. Of those who bought Sparkle the previous week 80% buy it again whilst 20% switch to Gleem.

How can this information be used to find the proportion of the market the brands will eventually hold?

The first step is to construct the probability transition matrix of the switching probabilities.

$$
\begin{array}{cc}
 & \textbf{Gleem} \quad \textbf{Sparkle} \\
\begin{array}{c} \text{Gleem} \\ \text{Sparkle} \end{array} &
\begin{bmatrix} 0.7 & 0.2 \\ 0.3 & 0.8 \end{bmatrix}
\end{array}
$$

The columns show the switching probabilities and total 1.

Next a vector is constructed representing the initial market shares in %.

$$
\begin{array}{c} \text{Gleem} \\ \text{Sparkle} \end{array}
\begin{bmatrix} 60 \\ 40 \end{bmatrix}
$$

The matrix and vector are multiplied to give the new market share for the following week.

$$
\begin{bmatrix} 0.7 & 0.2 \\ 0.3 & 0.8 \end{bmatrix}
\begin{bmatrix} 60 \\ 40 \end{bmatrix}
=
\begin{bmatrix} 50 \\ 50 \end{bmatrix}
$$

i.e. the market shares are now 50 : 50 for Gleem and Sparkle.

To obtain the forecast for the following week the original branch switching probabilities must be multiplied by the new market share.

$$
\begin{bmatrix} 0.7 & 0.2 \\ 0.3 & 0.8 \end{bmatrix}
\begin{bmatrix} 50 \\ 50 \end{bmatrix}
=
\begin{bmatrix} 45 \\ 55 \end{bmatrix}
$$ which is the latest market share position.

This process can be continued indefinitely if required but it will be found that the market shares settle down around particular values. In this example, after five cycles, Gleem's share is 40.625 and Sparkle's share is 59.375 which are close to the eventual share of the market. If it is required to find these equilibrium values this can be done as follows.

At equilibrium

$$\begin{bmatrix} 0.7 & 0.2 \\ 0.3 & 0.8 \end{bmatrix} \begin{bmatrix} G \\ S \end{bmatrix} = \begin{bmatrix} G \\ S \end{bmatrix}$$

where G and S represent the equilibrium shares.

$$\text{As } G + S = 1 \qquad S = 1 - G$$

and substituting

$$0.7G + 0.2(1 - G) = G$$
$$\therefore 0.5G = 0.2$$
$$\therefore G = 0.4$$

$\therefore$ Equilibrium shares are Gleem **40**% and Sparkle **60**%.

Summary

20. a) A matrix is a rectangular array of numbers. Where the number of columns is the same as the number of rows it is a square matrix.

   b) Matrix addition and subtraction. Each matrix must be the same size and corresponding elements are added or subtracted.

   c) Matrix multiplication by scalar. Any size matrix can be multiplied by a number, a scalar, each element being multiplied by the scalar.

   d) Matrix multiplication. A matrix of size $m \times n$ can be multiplied by a matrix of size $n \times p$ producing a matrix of size $m \times p$. Each row of matrix **A** is multiplied by each column in matrix **B**. $AB \neq BA$.

   e) Two special matrices are the zero matrix, i.e. any square matrix where all the elements are zero, and the unity matrix (**I**), i.e. any square matrix with 1s in a diagonal from top left to bottom right with the rest of the elements being zero.

   f) Matrix inversion. The inverse of a matrix, denoted A^{-1} or C^{-1}, etc., is found by carrying out row by row operations on the original matrix and a unity matrix with the objective of changing the original matrix into **I**. The inverse is the matrix which has replaced the original unity matrix.

Point to note

21. There is a strong relationship between the usual algebraic method of solving LP problems, the Simplex method, and matrix inversion.

Self review questions *Numbers in brackets refer to paragraph numbers*

1   How is the size of a matrix described? (3)

2   What is a vector (4)

3   What is the rule regarding the size of matrices that are to be added? (6)

4   Does $\mathbf{A} - \mathbf{B} = \mathbf{B} - \mathbf{A}$? (7)

5   What is scalar multiplication? (8)

6   What is the size rule for matrix multiplication? (10)

7   How is matrix multiplication carried out? (10)

8   What is the unity matrix? (14)

9   Does $\mathbf{A} \times \mathbf{A}^{-1} = \mathbf{A}^{-1} \times \mathbf{A}$? (15)

10 What are the steps in matrix inversion? (16)

11 What are the essential conditions for simultaneous equations to be solved by matrix algebra. (18)

12 What are probability transition matrices and what are they used for?

Exercises with answers

1.

$$\mathbf{X} = \begin{bmatrix} 3 & 11 & 6 \\ 9 & -3 & 8 \\ 5 & 0 & 9 \end{bmatrix} \qquad \mathbf{Y} = \begin{bmatrix} 1 & 2 & 0 \\ 0 & -4 & 5 \\ 5 & -8 & 7 \end{bmatrix}$$

Calculate

a) $\mathbf{X} + \mathbf{Y}$

b) $\mathbf{X} - \mathbf{Y}$

2. Multiply the matrices $\mathbf{X}$ and $\mathbf{Y}$ given in question 1.

3. Let

$$\mathbf{M} = \begin{bmatrix} 6 & 4 \\ 3 & 1 \end{bmatrix}$$

Find $\mathbf{M}^{-1}$ using determinants and check your answer using row by row operations.

Answers to exercises

1.

$$X + Y = \begin{bmatrix} 3+1 & 11+2 & 6+0 \\ 9+0 & -3+(-4) & 8+5 \\ 5+5 & 0+(-8) & 9+7 \end{bmatrix} = \begin{bmatrix} 4 & 13 & 6 \\ 9 & -7 & 13 \\ 10 & -8 & 16 \end{bmatrix}$$

$$X - Y = \begin{bmatrix} 2 & 9 & 6 \\ 9 & 1 & 3 \\ 0 & 8 & 2 \end{bmatrix}$$

2.

$$X \times Y = \begin{bmatrix} 33 & -86 & 97 \\ 49 & -34 & 41 \\ 50 & -62 & 63 \end{bmatrix}$$

3.

$$M = \begin{bmatrix} 6 & 4 \\ 3 & 1 \end{bmatrix} \quad |M| = -6$$

$$\therefore \ M^{-1} = -\tfrac{1}{6} \begin{bmatrix} 1 & -4 \\ -3 & 6 \end{bmatrix}$$

$$= \begin{bmatrix} -\tfrac{1}{6} & \tfrac{2}{3} \\ \tfrac{1}{2} & -1 \end{bmatrix}$$

Using row by row operations

$$\text{Row } 1 - 6 = \begin{bmatrix} 1 & \tfrac{2}{3} \\ 0 & -1 \end{bmatrix} \begin{bmatrix} \tfrac{1}{6} & 0 \\ -\tfrac{1}{2} & 1 \end{bmatrix}$$

$$\text{New Row } 2x - 1 = \begin{bmatrix} 1 & \tfrac{2}{3} \\ 0 & 1 \end{bmatrix} \begin{bmatrix} \tfrac{1}{6} & 0 \\ \tfrac{1}{2} & -1 \end{bmatrix}$$

$$\text{New Row } 1 - \tfrac{2}{3} \ (\text{New Row 2}) = \begin{bmatrix} 1 & 0 \\ 0 & 1 \end{bmatrix} \begin{bmatrix} -\tfrac{1}{6} & \tfrac{2}{3} \\ \tfrac{1}{2} & -1 \end{bmatrix}$$

Thus proving inversion.

31 Replacement analysis

Objectives

1. After studying this chapter you will

   - know what is meant by a Replacement Decision;
   - be able to calculate the minimum cost point for sudden failure items;
   - understand the principles for the least cost replacement for items that deteriorate.

Replacement decisions

2. Most items of equipment (i.e. components, parts, vehicles, machinery) need replacement at some time or other. Replacement analysis is the process by which the various cost consequences involved are studied so that the optimum replacement decision can be taken. The two most common replacement problems relate to.

   a) Parts or components that work adequately up to a point and then fail, e.g. fan belts, light bulbs, some types of tools.

   b) Items that deteriorate. These are usually relatively expensive items which could be kept functioning with increasing amounts of maintenance, e.g. vehicles, machine tools, boilers.

   These two categories are dealt with below.

Sudden failure items

3. Often these items are inexpensive in themselves but the cost consequence of their failure and/or the installation costs involved in replacing them can be considerable. It is therefore necessary to estimate the various costs involved and choose the least cost position. The three categories of cost are:

   a) The replacement cost of the item; usually the purchase price at the time of replacement.

   b) The consequential costs of failure which might be trivial say, if an electric bulb failed, but could be substantial if a small component failure caused an assembly line stoppage.

   c) The costs involved in the actual replacement of the item. Because of location and/or accessibility problems consideration is often given to group replacement at intervals or on the failure of one item. For example, if a single electric bulb failed in an overhead lamp cluster in a factory then all the bulbs might be replaced at the same time even though many are still functioning.

   As a result of the need to minimise the various costs involved several decision alternatives are usually explored and the least cost alternative chosen. For example, assume that a machine contains 50 components whose maximum life is 4 weeks. The various alternatives are:

a) replace on failure only;

b) replace on failure and all components at end of Week 1;

c) replace on failure and all components at end of Week 2;

d) replace on failure and all components at end of Week 3;

e) replace on failure and all components at end of Week 4.

Using the probabilities of a component failing at the end of each week the expected cost of each alternative can be found and compared.

Illustration of replacement for sudden failure items

4. The following example will be used to illustrate the approach to this type of problem.

Example 1

A special component is used throughout a large machine tool. The component has a limited life and the following data have been collected on failures.

	Week after replacement			
	1	2	3	4
Percentage of components which have failed by the end of that week (cumulative)	20	50	80	100

500 components are in use at any one time and they can be replaced on a mass replacement basis for £2 per component. If they are replaced individually as they fail the cost is £10 representing £1 for the component and £9 labour charges.

It is required to establish the least cost replacement policy comparing individual replacement on failure with mass replacement at the end of the various weeks together with individual replacement of components which have failed in the preceding interval.

Solution

Proportions failing each week

$$
\begin{array}{cc}
\text{Week 1} & 20\% \\
2 & 30\% \\
3 & 30\% \\
4 & \underline{20\%} \\
& 100\%
\end{array}
$$

∴ Average life of a component $= (1 \times 0.2) + (2 \times 0.3) + (3 \times 0.3) + (4 \times 0.2)$

$$= \underline{\underline{\textbf{2.5 weeks}}}$$

$\therefore$ Average number of replacement per week $= \dfrac{500}{2.5} = \underline{\underline{\textbf{200}}}$

$\therefore$ Cost of individual replacement on failure $= 200 \times £10 = \underline{\textbf{£2,000 per week}}$

Mass Replacement

To calculate the costs involved it is necessary to find the number of failures each week which include the proportion of original components which fail plus the replaced components which fail.

The number of individual failures is shown in Table 1 which applies the proportions failing each week to the original batch of 500 and then to the individual failures which are replaced.

Failure table. Table 1

	Week 1	Week 2	Week 3	Week 4
500 batch (original batch)	100	150	150	100
100 batch (week 1 replacement)		20	30	30
150 batch (week 2 replacement)			30	45
150 batch (week 3 replacements)				30
20 batch (week 2 replacements)			4	6
60 batch (30 + 30) (week 3 replacements)				12
Total failures	100	170	214	223

The total cost is then the cost of mass replacement plus the cost of individual failures.

		Total cost	Average weekly cost
Week 1	100 × £10 + £1,000	2,000	2,000
2	270 × £10 + £1,000	3,700	1,350*
3	484 × £10 + £1,000	5,840	1,947
4	707 × £10 + £1,000	8,070	2,018

* Thus the best strategy is to replace en masse in Week 2.

Replacement of items that deteriorate

5. A lorry could be kept operating with satisfactory performance for say, 15 years, but to do so would incur increasing amounts of maintenance so that careful cost analysis is needed to choose the most economical replacement time.

    There are two major cost consequences involved: the annual capital loss each year (the difference between the market value at the beginning and end of the year) and the maintenance charges. The two costs are accumulated and averaged over the number of years to find the minimum average annual cost.

This is illustrated by the following example.

Example 2

A lorry cost £50,000 and it is required to find the least cost point to replace it with a new vehicle. The following data have been estimated.

Year	1	2	3	4	5	6	7	8
Resale value at year end	£36,000	28,000	22,500	17,500	13,000	10,000	8,000	6,000
Annual maintenance cost	£2,000	2,600	3,200	4,600	7,000	11,000	13,000	15,000

Ignore the time value of money and inflation.

Year	Annual capital loss £	Cumulative capital loss £	Annual maintenance cost £	Cumulative maintenance cost £	Cumulative total cost £	Average annual cost £
1	14,000	14,000	2,000	2,000	16,000	16,000
2	8,000	22,000	2,600	4,600	26,600	13,300
3	5,500	27,500	3,200	7,800	35,300	11,767
4	5,000	32,500	4,600	12,400	44,900	11,225*
5	4,500	37,000	7,000	19,400	56,400	11,280
6	3,000	40,000	11,000	30,400	70,400	11,733
7	2,000	42,000	13,000	43,400	85,400	12,200
8	2,000	44,000	15,000	58,400	102,400	12,800

* Best replacement point is at the end of Year 4 although it would make little difference, based on the estimates given, if replacement was at the end of Year 5.

Summary

6. a) Replacement analysis seeks the least cost point at which to replace items.

    b) The main costs involved with sudden failure items are: cost of the replacement item, consequential costs, installation costs.

    c) For sudden failure items, typically individual replacement on failure is compared with group replacement and the least cost policy chosen.

    d) The main costs involved with deteriorating items are the capital loss each year and the (increasing) annual maintenance costs.

    e) The average annual cost is calculated and the least cost position chosen.

Points to note

7. a) Simple replacement analysis assumes that there are no changes in performance or technology. In practice there are frequently advantages to be gained in the potential improvement in performance from a new item. In such cases the extra output or contribution becomes a relevant factor in the analysis.

b) Particularly for expensive items, liquidity and taxation implications frequently are dominant factors in replacement decisions.

Self review questions *Numbers in brackets refer to paragraph numbers*

1   What is the objective of replacement cost analysis? (2)
2   What are the three categories of cost associated with sudden failure items? (3)
3   What is the procedure for finding the minimum cost point for items that deteriorate? (5)

Exercises with answers

1.   A part is used in an equipment handling system. The component has a limited life and data have been collected on failures.

	Week after replacement		
	1	**2**	**3**
Cumulative percentage of failures	30	75	100

1,000 components are in use and they can be replaced en masse for £5 per component. If replaced individually they cost £30 per component.

Calculate:

a)   average component life
b)   average number of replacement per week.
c)   cost of individual replacement per week.

2.   Calculate the cost of mass replacement at the end of each week for the components in question 1 and determine the best replacement strategy.

Answers to exercises

1.   a)   Average component life $= (1 \times 0.3) + (2 \times 0.45) + (3 \times 0.25) = \underline{\textbf{1.95}}$

b)   Average number of replacements per week $= \dfrac{1,000}{1.95} = \underline{\textbf{513}}$

c)   Cost of individual replacement $= 513 \times £30 = \underline{\textbf{£15,390}}$

Failure table

	Week 1	**Week 2**	**Week 3**
Original batch (1,000)	300	450	250
300 Week 1 batch		90	135
450 Week 2 batch			135
90 Week 2 batch			27
Total failures	300	540	547

Total cost of mass replacement

	Total cost	Average cost/week
Week 1 £5,000 + 300 × 30	£14,000	£14,000
Week 2 £5,000 + 840 × 30	£30,200	£15,100
Week 3 £5,000 + 1,387 × 30	£46,610	£15,537

Therefore best replacement strategy is replace en masse each week.

Assessment and revision
Chapters 27 to 31

Examination questions with answers

Answers commence on page 551

A1. a) A new machine is expected to last for six years and to produce annual (year-end) savings of £10,000. What is the maximum sum worth paying for the machine now, assuming compound interest at 10% per annum?

   b) A new machine costs £25,000 and will replace existing equipment whose scrap value is £5,000. The new machine will produce (year-end) savings in running costs of £3,000 a year. Assuming compound interest at 10% per annum, analyse how long the machine must last in order for it to be worth buying. [The old machine could carry on working but would produce no savings in running costs.]

CIMA Quantitative Methods

A2. The managers of a company are considering the installation of double-glazing to cut down heating costs. This year heating costs were £200,000. The company has received a quotation of £50,000 for double-glazing which it is claimed will save 5% of heating costs per annum. Heating costs are expected to rise by 3% per annum on a compound basis. The company has a cost of capital of 12%.

Requirements:

   a) Estimate what the heating costs will be for the next six years.

   b) Estimate the annual savings in heating costs that might be expected from the installation of double-glazing over the next six years.

   c) Calculate the maximum amount that the company will be prepared to pay for the installation of double-glazing, using a discounted cash flow approach.

   d) State, with reasons, whether or not the company should invest in the installation of double-glazing.

CIMA Cost Accounting and Quantitative Methods

A3. A company is considering the launch of a new product, for which an investment in equipment of £150,000 would be required. The project life would be limited to five years by the expected life cycle of the product. It is expected that the equipment could be sold for £10,000 in Year 6.

Market research has indicated a 70% chance of demand for the new product being high and a 30% chance of demand being low. Cash inflows are forecast as follows:

Year	High demand £000	Low demand £000
1	60	50
2	62	50
3	65	50
4	70	50
5	70	50

If the new product is launched now, an existing product, which could otherwise be retained for a further five years, would be discontinued immediately. If retained, cash inflows of £12,000 per annum would be expected for the existing product.

The company's cost of capital (discount rate) is 15% per annum.

Assume that all cash flows occur at year ends.

You are required to:

a) Calculate the expected net present value of the new product.

b) Advise the company whether to launch the new product, or to retain the existing product.

ACCA Management Information

A4. The following data relate to (i) the contractual payment terms (average credit days given) and (ii) how long companies have to wait for their bills to be paid (average credit days taken) in some countries.

Country	Average credit days contracted	Average credit days taken
Denmark	30	50
Finland	30	56
France	60	110
Germany	30	48
Holland	30	52
Ireland	30	60
Italy	60	90
Norway	30	45
Sweden	30	48
Switzerland	30	60
United Kingdom	30	75

(*Source:* Financial Times, *11 February 1992*)

You are required

a) to state and to calculate a simple measure of 'late payment' for each country;

b) to draw a suitable diagram to illustrate this 'late payment' for the eleven countries, clearly showing their rank order;

c) to find the annual cost of 'late payment' for a new company in its first year of trading, which should have received a payment of £10,000 on the first day of each month but is paid exactly two months later. The monthly rate of interest of 1%.

The sum, S, of a geometric progression of N terms, with first term A and common ratio R, is given by

$$S = \frac{A(R^N - 1)}{(R - 1)}$$

CIMA Quantitative Methods

A5. a) St. Swithin's hospital has three departments which service the main operating departments of the hospital. The three are maintenance, laundry and catering. In each case, part of the services provided is within the department itself and/or to the other service departments. Past records show the proportions of such services to be as follows:

Maintenance: 5% to itself, 25% and 10% to laundry and catering respectively.

Laundry: 5% to maintenance and 10% to catering.

Catering: 5% to itself, 10% and 15% to maintenance and laundry respectively.

In a given period, direct costs, allocated to the operating of each of these departments are shown as:

Maintenance	£56,000
Laundry	£250,000
Catering	£35,000

You are required to:

i) draw up the cost equations to derive the gross operating cost of each of the service departments;

ii) demonstrate, both algebraically and numerically, the matrix algebra steps which would lead to obtaining the inverse matrix to solve the unknown values in the equation of (i) above. (Do not attempt to invert the matrix.)

b) You are given the following as the inverse matrix for the problem of (a) above:

1.083	0.067	0.125
0.293	1.034	0.194
0.145	0.116	1.086

(Figures rounded to three decimal places)

You are required, using the inverse matrix, to obtain the gross cost of operation of each service department. Workings must be shown.

CIMA Quantitative Techniques

A6. A retailer is facing increasing competition from new shops that are opening in his area. He thinks that if he does not modernise his premises, he will lose sales. A local builder has estimated that the cost of modernising the shop will be £40,000 if the work is started now. The retailer is not sure whether to borrow the money and modernise the premises now, or to save up and have the work carried out when he has sufficient funds himself. Current forecasts show that if he delays the work for three years, the cost of the modernisation is likely to rise by 4% per annum.

Investigations have revealed that, if he borrows, he will have to pay interest at the rate of 3% per quarter, but if he saves the money himself he will only earn 2% per quarter.

Required:

a) Calculate the equal amounts that would need to be paid at the cost of each quarter if the retailer decides to borrow the money at 3% per quarter for three years.

b) Calculate the equal quarterly amounts that need to be invested into the savings fund (first instalment paid immediately), so that there is sufficient in the fund to carry out the work in three years' time.

c) Discuss the advantages of each of the two ways of financing the modernisation and suggest any other factors that should be taken into consideration before deciding how and when to finance the modernisation.

CIMA Business Mathematics

A7. Hulk Petroleum Company Ltd (HPC) has constructed a collection platform at an oil and gas field known as Gibson 6. This field holds reserves of oil and gas, both of which may be extracted and carried to an onshore terminal along the same pipeline. Gas collected at the platform is converted into liquid petroleum gas (LPG) for transport.

Oil and LPG received at the terminal are processed and transported 300 miles by rail to refineries. Details of annual capacities are:

Pipeline capacity

If the pipeline is used exclusively to transport oil, then a maximum of 100,000 barrels of oil can be transported. 1.4 barrels of LPG can be transported in place of each barrel of oil.

Processing capacity

If processing facilities are used exclusively to process oil, then a maximum of 150,000 barrels of oil can be processed. 0.5333 barrels of LPG can be processed instead of each 1 barrel of oil up to a maximum of 70,000 barrels of LPG. No more than 70,000 barrels of LPG can be processed.

Rail transport capacity

If rail transport capacity is used exclusively to transport oil, then a maximum of 120,000 barrels of oil can be transported. One barrel of LPG can be transported, using specialised wagons, each instead of 1 barrel of oil, up to a maximum of 20,000 barrels of LPG.

Transporting amounts of LPG in excess of 20,000 barrels involves transporting 0.65 barrels of LPG instead of each 1 barrel of oil, using general-purpose wagons.

The initial capital cost of developing extraction capabilities is £40,000 per 1,000 barrels of oil per year and £60,000 per 1,000 barrels of LPG per year. Once developed, an extraction capability can be operated for at least 10 years. Contribution generated by the two products is £8 per barrel for oil and £11.50 per barrel for LPG. HPC evaluates projects using a 12% per annum discount rate and a 10-year time horizon.

Required:

a) Identify the annual output combination of oil and LPG at which net present value is maximised.

b) State whether or not HPC should invest in a new pump to increase the capacity of the Gibson 6 pipeline by 10%. The pump costs £50,000. Support your answer with relevant calculations.

Note: This is a testing question which brings together LP and DCF

CIMA Management Accounting Applications

Examination questions without answers

B1. The Management Services Division of a company has been asked to evaluate the following proposals for the maintenance of a new boiler with a life of seven years.

Proposal 1

The boiler supplier will make a charge of £13,000 per year on a seven-year contract.

Proposal 2

The company will carry out its own maintenance estimated at £10,000 per annum now, rising at 5% per annum with a major overhaul at the end of year 4 costing an additional £25,000.

The discount rate is 10% and all payments are assumed to be made at year ends.

Requirements:

a) Calculate the maintenance cost for each year if the company provides its own maintenance.

b) Calculate the present value of the cost of maintenance, if the company carries out its own maintenance.

c) Calculate the present value of the supplier's maintenance contract.

d) Recommend, with reasons, which proposal should be adopted.

CIMA Cost Accounting and Quantitative Methods

B2. An oil well is currently producing annual (year-end) cash flows of £0.5 million. The best geological evidence suggests that the well has reserves that will last for another 10 years, at the present rate of extraction. A special pump could be installed, at a cost of £0.75 million, that could double the rate of extraction but halve the life of the well. After the well had been exhausted, this special pump could be sold for £100,000. The immediate introduction of the special pump is now being considered. Last year's earning have just been distributed and the existing equipment has no resale value.

You are required

a) to tabulate the annual effect on earnings over the next ten years of introducing the special pump;

b) to compare the net present value of the options if the cost of capital to the company is 8% per annum and to explain the answer;

c) to find whether the pump should be installed if the cost of capital were to be 12%, giving reasons.

CIMA Quantitative Methods

B3. a) In exactly three years from now, a company will have to replace capital equipment which will then cost £0.5 million. The managers have decided to set up a Reserve Fund into which twelve equal sums will be put at quarterly intervals, with the first one being made now. The rate of compound interest is 2% a quarter.

You are required to calculate the quarterly sums required for the Reserve Fund.

b) A fixed-interest ten-year £100,000 mortgage is to be repaid by 40 equal quarterly payments in arrears. Interest is charged at 3% a quarter on the outstanding part of the debt.

You are required

i) to find the sum to which an investment of £100,000 would grow after 10 years, at a quarterly compound interest rate of 3%.

ii) using you answer to (i), or otherwise, to calculate the quarterly repayments on the mortgage;

iii) to find the effective annual rate of interest on the mortgage.

The sum, S, of a geometric series of N terms, with first term A and common ratio R, is given by

$$S = \frac{A(R^N - 1)}{(R - 1)}$$

CIMA Quantitative Methods

B4. A company consists of three service departments (S_1, S_2, S_3) and two production departments (P_1, P_2)

Analysis of the service department activities shows:

Services supplied by	Services supplied to (%)				
	S_1	S_2	S_3	P_1	P_2
S_1		50		30	20
S_2	20		20	20	40
S_3	20			30	50

Cost directly allocated to the service departments are (per annum):

Dept. (£)	Variable (£)	Fixed (£)	Total
S_1	40,000	30,000	70,000
S_2	120,000	80,000	200,000
S_3	60,000	40,000	100,000

You are required:

a) to derive the cost equations to show the gross total operating costs of each of the service departments;

b) to calculate the gross total operating costs of each service department after re-apportionment of the intra-service department costs;

c) if P_1 and P_2 respectively produce 400,000 and 200,000 units of finished product per annum, to determine the service department charge in each of the product costs;

d) given that S_3, produces 90,000 units of service per annum, to comment on whether it would be advisable to accept an offer from an outside supplier to supply this quantity at £0.90 per unit.

CIMA Quantitative Techniques

Statistical and financial tables

Table I[†]

Areas under the Standard Normal curve from 0 to Z

z	0	1	2	3	4	5	6	7	8	9
0.0	0.0000	0.0040	0.0080	0.0120	0.0160	0.0199	0.0239	0.0279	0.0319	0.0359
0.1	0.0398	0.0438	0.0478	0.0517	0.0557	0.0596	0.0636	0.0675	0.0714	0.0754
0.2	0.0793	0.0832	0.0871	0.0910	0.0948	0.0987	0.1026	0.1064	0.1103	0.1141
0.3	0.1179	0.1217	0.1255	0.1293	0.1331	0.1368	0.1406	0.1443	0.1480	0.1517
0.4	0.1554	0.1591	0.1628	0.1664	0.1700	0.1736	0.1772	0.1808	0.1844	0.1879
0.5	0.1915	0.1950	0.1985	0.2019	0.2054	0.2088	0.2123	0.2157	0.2190	0.2224
0.6	0.2258	0.2291	0.2324	0.2357	0.2389	0.2422	0.2454	0.2486	0.2518	0.2549
0.7	0.2580	0.2612	0.2642	0.2673	0.2704	0.2734	0.2764	0.2794	0.2823	0.2852
0.8	0.2881	0.2910	0.2939	0.2967	0.2996	0.3023	0.3051	0.3078	0.3106	0.3133
0.9	0.3159	0.3186	0.3212	0.3238	0.3264	0.3289	0.3315	0.3340	0.3365	0.3389
1.0	0.3413	0.3438	0.3461	0.3485	0.3508	0.3531	0.3554	0.3577	0.3599	0.3621
1.1	0.3643	0.3665	0.3686	0.3708	0.3729	0.3749	0.3770	0.3790	0.3810	0.3830
1.2	0.3849	0.3869	0.3888	0.3907	0.3925	0.3944	0.3962	0.3980	0.3997	0.4015
1.3	0.4032	0.4049	0.4066	0.4082	0.4099	0.4115	0.4131	0.4147	0.4162	0.4177
1.4	0.4192	0.4207	0.4222	0.4236	0.4251	0.4265	0.4279	0.4292	0.4306	0.4319
1.5	0.4332	0.4345	0.4357	0.4370	0.4382	0.4394	0.4406	0.4418	0.4429	0.4441
1.6	0.4452	0.4463	0.4474	0.4484	0.4495	0.4505	0.4515	0.4525	0.4535	0.4545
1.7	0.4554	0.4564	0.4573	0.4582	0.4591	0.4599	0.4608	0.4616	0.4625	0.4633
1.8	0.4641	0.4649	0.4656	0.4664	0.4671	0.4678	0.4686	0.4693	0.4699	0.4706
1.9	0.4713	0.4719	0.4726	0.4732	0.4738	0.4744	0.4750	0.4756	0.4761	0.4767
2.0	0.4772	0.4778	0.4783	0.4788	0.4793	0.4798	0.4803	0.4808	0.4812	0.4817
2.1	0.4821	0.4826	0.4830	0.4834	0.4838	0.4842	0.4846	0.4850	0.4854	0.4857
2.2	0.4861	0.4864	0.4868	0.4871	0.4875	0.4878	0.4881	0.4884	0.4887	0.4890
2.3	0.4893	0.4896	0.4898	0.4901	0.4904	0.4906	0.4909	0.4911	0.4913	0.4916
2.4	0.4918	0.4920	0.4922	0.4925	0.4927	0.4929	0.4931	0.4932	0.4934	0.4936
2.5	0.4938	0.4940	0.4941	0.4943	0.4945	0.4946	0.4948	0.4949	0.4951	0.4952
2.6	0.4953	0.4955	0.4956	0.4957	0.4959	0.4960	0.4961	0.4962	0.4963	0.4964
2.7	0.4965	0.4966	0.4967	0.4968	0.4969	0.4970	0.4971	0.4972	0.4973	0.4974
2.8	0.4974	0.4975	0.4976	0.4977	0.4977	0.4978	0.4979	0.4979	0.4980	0.4981
2.9	0.4981	0.4982	0.4982	0.4983	0.4984	0.4984	0.4985	0.4985	0.4986	0.4986
3.0	0.4987	0.4987	0.4987	0.4988	0.4988	0.4989	0.4989	0.4989	0.4990	0.4990
3.1	0.4990	0.4991	0.4991	0.4991	0.4992	0.4992	0.4992	0.4992	0.4993	0.4993
3.2	0.4993	0.4993	0.4994	0.4994	0.4994	0.4994	0.4994	0.4995	0.4995	0.4995
3.3	0.4995	0.4995	0.4995	0.4996	0.4996	0.4996	0.4996	0.4996	0.4996	0.4997
3.4	0.4997	0.4997	0.4997	0.4997	0.4997	0.4997	0.4997	0.4997	0.4997	0.4998
3.5	0.4998	0.4998	0.4998	0.4998	0.4998	0.4998	0.4998	0.4998	0.4998	0.4998
3.6	0.4998	0.4998	0.4999	0.4999	0.4999	0.4999	0.4999	0.4999	0.4999	0.4999
3.7	0.4999	0.4999	0.4999	0.4999	0.4999	0.4999	0.4999	0.4999	0.4999	0.4999
3.8	0.4999	0.4999	0.4999	0.4999	0.4999	0.4999	0.4999	0.4999	0.4999	0.4999
3.9	0.5000	0.5000	0.5000	0.5000	0.5000	0.5000	0.5000	0.5000	0.5000	0.5000

† From Statistics by SPEIGEL, Copyright 1972 McGraw-Hill Publishing Co Ltd.
Used with permission of McGraw-Hill Book Company.

Table II

Normal distribution

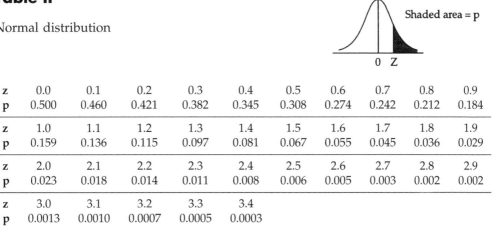

Shaded area = p

0 Z

z	0.0	0.1	0.2	0.3	0.4	0.5	0.6	0.7	0.8	0.9
p	0.500	0.460	0.421	0.382	0.345	0.308	0.274	0.242	0.212	0.184
z	1.0	1.1	1.2	1.3	1.4	1.5	1.6	1.7	1.8	1.9
p	0.159	0.136	0.115	0.097	0.081	0.067	0.055	0.045	0.036	0.029
z	2.0	2.1	2.2	2.3	2.4	2.5	2.6	2.7	2.8	2.9
p	0.023	0.018	0.014	0.011	0.008	0.006	0.005	0.003	0.002	0.002
z	3.0	3.1	3.2	3.3	3.4					
p	0.0013	0.0010	0.0007	0.0005	0.0003					

Table III

Percentage points of the **t** distribution.

The table gives the values for the area in both tails.

The table shows the total of the shaded areas

0 t

Degrees of freedom	.10	.05	.02	.01	Degrees of freedom	.10	.05	.02	.01
v = 1	6.314	12.706	31.821	63.657	v = 21	1.721	2.080	2.518	2.831
2	2.920	4.303	6.965	9.925	22	1.717	2.074	2.508	2.819
3	2.353	3.182	4.541	5.841	23	1.714	2.069	2.500	2.807
4	2.132	2.776	3.747	4.604	24	1.711	2.064	2.492	2.797
5	2.015	2.571	3.365	4.032	25	1.708	2.060	2.485	2.787
6	1.493	2.447	3.143	3.707	26	1.706	2.056	2.479	2.779
7	1.895	2.365	2.998	3.499	27	1.703	2.052	2.473	2.771
8	1.860	2.306	2.896	3.355	28	1.701	2.048	2.467	2.763
9	1.833	2.262	2.821	3.250	29	1.699	2.045	2.462	2.756
10	1.812	2.228	2.764	3.169	30	1.697	2.042	2.457	2.750
11	1.796	2.201	2.718	3.106	40	1.684	2.021	2.423	2.704
12	1.782	2.179	2.681	3.055	60	1.671	2.000	2.390	2.660
13	1.771	2.160	2.650	3.012	120	1.658	1.980	2.358	2.6170
14	1.761	2.145	2.624	2.977	∞	1.645	1.960	2.326	2.576
15	1.753	2.131	2.602	2.947					
16	1.746	2.120	2.583	2.921					
17	1.740	2.110	2.567	2.898					
18	1.734	2.101	2.552	2.878					
19	1.729	2.093	2.539	2.861					
20	1.725	2.086	2.528	2.845					

Both sub-tables are headed **Area in both tables combined**.

Table IV

The χ^2 distribution.

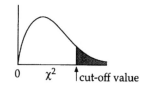

Degrees of freedom	Level of significance	
	5%	1%
v = 1	3.841	6.635
2	5.991	9.210
3	7.815	11.345
4	9.488	13.277
5	11.070	15.086
6	12.592	16.812
7	14.067	18.475
8	15.507	20.090
9	16.919	21.666
10	18.307	23.209
11	19.675	24.725
12	21.026	26.217
13	22.362	27.688
14	23.685	29.141
15	24.996	30.578
16	26.296	31.999
17	27.587	33.409
18	28.869	34.805
19	30.144	36.191
20	31.410	37.566
21	32.671	38.932
22	33.924	40.289
23	35.172	41.638
24	36.415	42.979
25	37.652	44.314

Table V

a)   Table of individual poisson probabilities

	Number of occurrences (x)						
Mean (m)	0	1	2	3	4	5	6
0.1	0.9048	0.0905	0.0045	0.0002	0.000	0.000	0.000
0.2	0.8187	0.1637	0.0164	0.0011	0.0001	0.000	0.000
0.3	0.7408	0.2222	0.0333	0.0033	0.0003	0.000	0.000
0.4	0.6703	0.2681	0.0536	0.0072	0.0007	0.0001	0.000
0.5	0.6065	0.3033	0.0758	0.0126	0.0016	0.0002	0.000
0.6	0.5488	0.3293	0.0988	0.0198	0.0030	0.0004	0.000
0.7	0.4966	0.3476	0.1217	0.0284	0.0050	0.0007	0.0001
0.8	0.4493	0.3595	0.1438	0.0383	0.0077	0.0012	0.0002
0.9	0.4066	0.3659	0.1647	0.0494	0.0111	0.0020	0.0003
1.0	0.3679	0.3679	0.1839	0.0613	0.0153	0.0031	0.0005

Table shows probability of a given number of occurences for a given mean (m).

b)   Table of cumulative poisson probabilities

	Number of occurrences (x)						
Mean (m)	0	1	2	3	4	5	6
0.1	0.9048	0.9953	0.9998	1.00	1.00	1.00	1.00
0.2	0.8187	0.9824	0.9988	0.9999	1.00	1.00	1.00
0.3	0.7408	0.9630	0.9963	0.9996	0.9999	1.00	1.00
0.4	0.6703	0.9384	0.9920	0.9992	0.9999	1.00	1.00
0.5	0.6065	0.9098	0.9856	0.9982	0.9998	1.00	1.00
0.6	0.5488	0.8781	0.9769	0.9967	0.9997	1.00	1.00
0.7	0.4966	0.8442	0.9659	0.9943	0.993	1.00	1.00
0.8	0.4493	0.8088	0.9526	0.9909	0.9986	0.9998	1.00
0.9	0.4066	0.7725	0.9372	0.9866	0.9977	0.9997	1.00
1.0	0.3679	0.7358	0.9197	0.9810	0.9963	0.9994	0.9999

Table shows probability of finding x or fewer occurences for a given mean (m).

Table VI

Compound interest.

Table shows value of £1 at compound interest $(1 + r)^n$.

Interest rates (r) %

Years (n)	1	2	3	4	5	6	7	8	9	10	11
1	1.010	1.020	1.030	1.040	1.050	1.060	1.070	1.080	1.090	1.100	1.110
2	1.020	1.040	1.061	1.082	1.102	1.124	1.145	1.166	1.188	1.210	1.232
3	1.030	1.061	1.093	1.125	1.158	1.191	1.225	1.260	1.295	1.331	1.368
4	1.041	1.082	1.126	1.167	1.216	1.262	1.311	1.360	1.412	1.464	1.518
5	1.051	1.104	1.159	1.217	1.276	1.338	1.403	1.469	1.539	1.610	1.685
6	1.061	1.126	1.194	1.265	1.340	1.419	1.501	1.587	1.677	1.772	1.870
7	1.072	1.149	1.230	1.316	1.407	1.504	1.606	1.714	1.828	1.949	2.076
8	1.083	1.172	1.267	1.369	1.477	1.594	1.718	1.851	1.993	2.144	2.304
9	1.094	1.195	1.305	1.423	1.551	1.689	1.838	1.999	2.172	2.358	2.558
10	1.105	1.219	1.344	1.480	1.629	1.791	1.967	2.159	2.367	2.594	2.839
11	1.116	1.243	1.384	1.539	1.710	1.898	2.105	2.332	2.580	2.853	3.152
12	1.127	1.268	1.426	1.601	1.796	2.012	2.252	2.519	2.813	3.138	3.498
13	1.138	1.294	1.468	1.665	1.886	2.133	2.410	2.720	3.066	3.452	3.883
14	1.149	1.319	1.513	1.732	1.980	2.261	2.578	2.937	3.342	3.797	4.310
15	1.161	1.346	1.558	1.801	2.079	2.397	2.759	3.172	3.642	4.177	4.785
20	1.220	1.486	1.806	2.191	2.653	3.207	3.870	4.661	5.604	6.727	8.062
25	1.282	1.641	2.094	2.666	3.386	4.292	5.427	6.848	8.623	10.835	13.585

Interest rates (r) %

Years (n)	12	13	14	15	16	17	18	19	20	25	30
1	1.120	1.130	1.140	1.150	1.160	1.170	1.180	1.190	1.200	1.250	1.300
2	1.254	1.277	1.297	1.322	1.346	1.367	1.392	1.416	1.440	1.562	1.690
3	1.405	1.443	1.481	1.521	1.561	1.602	1.643	1.685	1.728	1.953	2.197
4	1.573	1.630	1.689	1.749	1.811	1.874	1.939	2.005	2.074	2.441	2.856
5	1.762	1.842	1.925	2.011	2.100	2.192	2.288	2.386	2.488	3.052	3.713
6	1.974	2.082	2.195	2.313	2.436	2.565	2.700	2.840	2.986	3.815	4.827
7	2.211	2.353	2.502	2.660	2.826	3.001	3.186	3.379	3.583	4.768	6.275
8	2.476	2.658	2.853	3.059	3.278	3.511	3.759	4.021	4.300	5.960	8.157
9	2.773	3.004	3.252	3.518	3.803	4.108	4.435	4.785	5.159	7.451	10.604
10	3.106	3.395	3.707	4.046	4.411	4.807	5.234	5.695	6.192	9.313	13.786
11	3.478	3.836	4.226	4.662	5.117	5.624	6.176	6.777	7.430	11.641	17.922
12	3.896	4.334	4.818	5.350	5.936	6.580	7.288	8.064	8.916	14.552	23.298
13	4.363	4.898	5.492	6.153	6.886	7.699	8.599	9.596	10.699	18.190	30.287
14	4.887	5.535	6.261	7.076	7.988	9.007	10.147	11.420	12.839	22.737	39.374
15	5.474	6.254	7.138	8.137	9.265	10.539	11.974	13.589	15.407	28.422	51.186
20	9.646	11.523	13.743	15.366	19.461	23.106	27.393	32.429	38.338	86.736	190.050
25	17.000	21.230	26.462	32.920	40.874	50.658	62.669	77.388	95.396	264.698	705.641

Table VII

Present value factors. Present value of £1 $(1 + r)^{-n}$

Periods (n)	Interest rates (r)%								
	1%	2%	4%	6%	8%	10%	12%	14%	15%
1	0.990	0.980	0.962	0.943	0.926	0.909	0.893	0.877	0.870
2	0.980	0.961	0.925	0.890	0.857	0.826	0.797	0.769	0.756
3	0.971	0.942	0.889	0.840	0.794	0.751	0.712	0.675	0.658
4	0.961	0.924	0.855	0.792	0.735	0.683	0.636	0.592	0.572
5	0.951	0.906	0.822	0.747	0.681	0.621	0.567	0.519	0.497
6	0.942	0.888	0.790	0.705	0.630	0.564	0.507	0.456	0.432
7	0.933	0.871	0.760	0.665	0.583	0.513	0.452	0.400	0.376
8	0.923	0.853	0.731	0.627	0.540	0.467	0.404	0.351	0.327
9	0.914	0.837	0.703	0.592	0.500	0.424	0.361	0.308	0.284
10	0.905	0.820	0.676	0.558	0.463	0.386	0.322	0.270	0.247
11	0.0896	0.804	0.650	0.527	0.429	0.350	0.287	0.237	0.215
12	0.887	0.788	0.625	0.497	0.397	0.319	0.257	0.208	0.187
13	0.879	0.773	0.601	0.469	0.368	0.290	0.229	0.182	0.163
14	0.870	0.758	0.577	0.442	0.340	0.263	0.205	0.160	0.141
15	0.861	0.743	0.555	0.417	0.315	0.239	0.183	0.140	0.123
16	0.853	0.728	0.534	0.394	0.292	0.218	0.163	0.123	0.107
17	0.855	0.714	0.513	0.371	0.270	0.198	0.146	0.108	0.093
18	0.836	0.700	0.494	0.350	0.250	0.180	0.130	0.095	0.081
19	0.828	0.686	0.475	0.331	0.232	0.164	0.116	0.083	0.070
20	0.820	0.675	0.456	0.312	0.215	0.149	0.104	0.073	0.061
21	0.811	0.660	0.439	0.294	0.199	0.135	0.093	0.064	0.053
22	0.803	0.647	0.422	0.278	0.184	0.123	0.083	0.056	0.046
23	0.795	0.634	0.406	0.262	0.170	0.112	0.074	0.049	0.040
24	0.788	0.622	0.390	0.247	0.158	0.102	0.066	0.043	0.035
25	0.780	0.610	0.375	0.233	0.146	0.092	0.059	0.038	0.030

Periods (n)	Interest rates (r)%								
	16%	18%	20%	22%	24%	25%	26%	28%	30%
1	0.862	0.847	0.833	0.820	0.806	0.800	0.794	0.781	0.769
2	0.743	0.718	0.694	0.672	0.650	0.640	0.630	0.610	0.592
3	0.641	0.609	0.579	0.551	0.524	0.512	0.500	0.477	0.455
4	0.552	0.516	0.482	0.451	0.423	0.410	0.397	0.373	0.350
5	0.476	0.437	0.402	0.370	0.341	0.328	0.315	0.291	0.269
6	0.410	0.370	0.335	0.303	0.275	0.262	0.250	0.227	0.207
7	0.354	0.314	0.279	0.249	0.222	0.210	0.198	0.178	0.159
8	0.305	0.266	0.233	0.204	0.179	0.168	0.157	0.139	0.123
9	0.263	0.225	0.194	0.167	0.144	0.134	0.125	0.108	0.094
10	0.227	0.191	0.162	0.137	0.116	0.107	0.099	0.085	0.075
11	0.195	0.162	0.135	0.112	0.094	0.086	0.079	0.066	0.056
12	0.168	0.137	0.112	0.192	0.076	0.069	0.062	0.052	0.043
13	0.145	0.116	0.093	0.075	0.061	0.055	0.050	0.040	0.033
14	0.125	0.099	0.178	0.062	0.049	0.044	0.039	0.032	0.025
15	0.108	0.084	0.065	0.051	0.040	0.035	0.031	0.025	0.020
16	0.093	0.071	0.054	0.042	0.032	0.028	0.025	0.019	0.015
17	0.080	0.060	0.045	0.034	0.026	0.023	0.020	0.015	0.012
18	0.069	0.051	0.038	0.028	0.021	0.018	0.016	0.012	0.009
19	0.060	0.043	0.031	0.023	0.017	0.014	0.012	0.009	0.007
20	0.051	0.037	0.026	0.019	0.014	0.012	0.010	0.007	0.005
21	0.044	0.031	0.022	0.015	0.011	0.009	0.008	0.006	0.004
22	0.038	0.026	0.018	0.013	0.009	0.007	0.006	0.004	0.003
23	0.033	0.022	0.015	0.010	0.007	0.006	0.005	0.003	0.002
24	0.028	0.019	0.011	0.008	0.006	0.005	0.004	0.003	0.002
25	0.024	0.016	0.010	0.007	0.005	0.004	0.003	0.002	0.001

Table VIII

Present value annuity factors.

Present value of £1 received annually for **n** years $\left(\dfrac{1-(1+r)^{-n}}{r}\right)$

Periods (n)	1%	2%	4%	6%	8%	10%	12%	14%	15%
				Interest rates (r) %					
1	0.990	0.980	0.962	0.943	0.926	0.909	0.893	0.877	0.870
2	1.970	1.942	1.886	1.833	1.783	1.736	1.690	1.647	1.626
3	2.941	2.884	2.775	2.675	2.577	2.487	2.402	2.322	2.283
4	3.902	3.808	3.610	3.465	3.312	3.170	3.037	2.914	2.855
5	4.853	4.713	4.452	4.212	3.996	3.791	3.605	3.433	3.352
6	5.795	5.601	5.242	4.917	4.623	4.355	4.111	3.889	3.784
7	6.728	6.472	6.002	5.582	5.206	4.868	4.564	4.288	4.160
8	7.652	7.325	6.733	6.210	5.747	5.335	4.968	4.639	4.487
9	8.566	8.162	7.435	6.802	6.247	5.759	5.328	4.946	4.772
10	9.471	8.983	8.111	7.360	6.710	6.145	5.650	5.216	5.019
11	10.368	9.787	8.760	7.887	7.139	6.495	5.988	5.453	5.234
12	11.255	10.575	9.385	8.384	7.536	6.814	6.194	5.660	5.421
13	12.114	11.343	9.986	8.853	7.904	7.103	6.424	5.842	5.583
14	13.004	12.106	10.563	9.295	8.244	7.367	6.628	6.002	5.724
15	13.865	12.849	11.118	9.712	8.559	7.606	6.811	6.142	5.847
16	14.718	13.578	11.652	10.106	8.851	7.824	6.974	6.265	5.954
17	15.562	14.292	12.166	10.477	9.122	8.022	7.120	6.373	6.047
18	16.328	14.992	12.659	10.828	9.372	8.201	7.250	6.467	6.128
19	17.226	15.678	13.134	11.158	9.604	8.365	7.366	6.550	6.198
20	18.046	16.351	13.590	11.470	9.818	8.514	7.469	6.623	6.259
21	18.857	17.011	14.029	11.764	10.017	8.649	7.562	6.687	6.312
22	19.660	17.658	14.451	12.042	10.201	8.772	7.645	6.743	6.369
23	20.456	18.292	14.857	12.303	10.371	8.883	7.718	6.792	6.399
24	21.243	18.914	15.247	12.550	10.529	8.985	7.784	6.815	6.434
25	22.023	19.523	15.622	12.783	10.675	9.077	7.843	6.873	6.464

Periods (n)	16%	18%	20%	22%	24%	25%	26%	28%	30%
				Interest rates (r)%					
1	0.862	0.847	0.833	0.820	0.806	0.800	0.794	0.781	0.769
2	1.605	1.566	1.528	1.492	1.457	1.440	1.424	1.392	1.361
3	2.246	2.174	2.106	2.042	1.981	1.952	1.923	1.868	1.816
4	2.798	2.690	2.589	2.494	2.404	2.362	2.320	2.241	2.166
5	3.274	3.127	2.991	2.864	2.745	2.689	2.635	2.532	2.436
6	3.685	3.498	3.326	3.167	3.020	2.951	2.885	2.759	2.643
7	4.039	3.812	3.605	3.416	3.242	3.161	3.083	2.937	2.802
8	4.344	4.078	3.837	3.619	3.421	3.329	3.421	3.076	2.925
9	4.607	4.303	4.031	3.786	3.566	3.463	3.366	3.184	3.019
10	4.833	4.949	4.192	3.923	3.682	3.571	3.465	3.269	3.092
11	5.029	4.636	4.327	4.035	3.766	3.656	3.544	3.335	3.147
12	5.197	4.793	4.439	4.127	3.851	3.725	3.606	3.387	3.190
13	5.342	4.910	4.533	4.203	3.912	3.780	3.656	3.427	3.223
14	5.468	5.008	4.611	4.265	3.961	3.824	3.965	3.459	3.249
15	5.575	5.092	4.675	4.315	4.001	3.859	3.726	3.483	3.268
16	5.669	5.162	4.730	4.357	4.033	3.887	3.751	3.503	3.283
17	5.749	5.222	4.775	4.391	4.059	3.910	3.771	3.518	3.295
18	5.818	5.273	4.812	4.419	4.080	3.928	3.786	3.529	3.304
19	5.877	5.316	4.844	4.442	4.097	3.942	3.799	3.539	3.311
20	5.929	5.353	4.870	4.460	4.110	3.954	3.808	3.546	3.316
21	5.973	5.384	4.891	4.476	4.121	3.963	3.816	3.551	3.320
22	6.011	5.410	4.909	4.488	4.130	3.970	3.822	3.556	3.323
23	6.044	5.432	4.925	4.499	4.137	3.976	3.827	3.559	3.325
24	6.073	5.451	4.937	4.507	4.143	3.981	3.831	3.562	3.327
25	6.097	5.467	4.948	4.514	4.147	3.985	3.834	3.564	3.329

Answers to examination questions

Assessment and revision Chapters 1 to 3

A1. P(A entering market) = P(A) = 0.4 and P(Ā) = 0.6
P(B entering market) = P(B) = 0.7 and P(B̄) = 0.3
P(C entering market) = P(C) = 0.2 and P(C̄) = 0.8

   i) No competitor entering market
       = P(Ā) × P(B̄) × P(C̄) = 0.6 × 0.3 × 0.8
       = **0.144**

   ii) A and B to enter market but not C
       = P(A) × P(B) × P(C̄) = 0.4 × 0.7 × 0.8
       = **0.224**

   iii) Only 1 competitor
       = P(A only competitor) + P(B only competitor) + P(C only competitor)
       = (0.4 × 0.3 × 0.8) + (0.6 × 0.7 × 0.8) + (0.6 × 0.3 × 0.2)
       = **0.468**

A2. Let   I = Improvement   NI = No improvement
          T = Treatment received   P = Placebo effect

a) Prob. (I) = Prob. (T & I) + (Prob. (P & I)

$$= \frac{600}{1,000} + \frac{125}{1,000} = \frac{725}{1,000}$$

$$= \mathbf{0.725}$$

c) $\text{Prob. (T/NI)} = \dfrac{\text{Prob. (T \& NI)}}{\text{Prob. (NI)}}$

$$= \frac{0.15}{0.275}$$

$$= \mathbf{0.546}$$

b) $\text{Prob. (I/T)} = \dfrac{\text{Prob. (T \& I)}}{\text{Prob. (T)}}$

$$= \frac{600}{1,000}$$

$$= 0.6$$

Prob. (T) = Prob. (T & I) + Prob. (T & N)

$$= \frac{600}{1,000} + \frac{150}{1,000} = \frac{750}{1,000}$$

$$= 0.75$$

Therefore, Prob. (I/T)

$$= \frac{0.6}{0.75}$$

$$= \mathbf{0.8}$$

d) $\text{Prob. (P/I)} = \dfrac{\text{Prob. (P \& I)}}{\text{Prob. (I)}}$

$$= \frac{125}{1,000}$$

$$= 0.125$$

Therefore, Prob. (P/I)

$$= \frac{0.125}{0.725}$$

$$= \mathbf{0.172}$$

e) 750 volunteers received the treatment of whom 600 showed improvement i.e. 80%, whilst 250 volunteers received the placebo of whom only 125 showed improvement, i.e. 50%. This suggests that the 'treatment' may be effective, but it must be remembered that people tend to get better anyway.

A3. a) Tree diagram

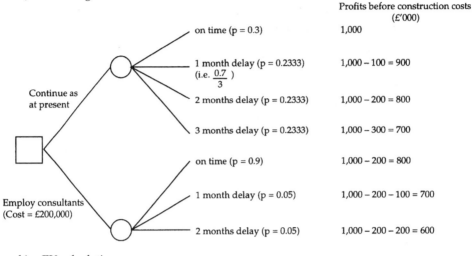

		Profits before construction costs (£'000)
	on time (p = 0.3)	1,000
	1 month delay (p = 0.2333) (i.e. $\frac{0.7}{3}$)	1,000 − 100 = 900
Continue as at present	2 months delay (p = 0.2333)	1,000 − 200 = 800
	3 months delay (p = 0.2333)	1,000 − 300 = 700
	on time (p = 0.9)	1,000 − 200 = 800
Employ consultants (Cost = £200,000)	1 month delay (p = 0.05)	1,000 − 200 − 100 = 700
	2 months delay (p = 0.05)	1,000 − 200 − 200 = 600

b) EV calculations

EV (present arrangement) = £(0.3 × 1,000) + (0.2333 × 900) + (0.2333 × 800) + (0.2333 × 700)

$$= \underline{\mathbf{860}}$$

EV (with consultants) = £(0.9 × 800) + (0.05 × 700) + (0.05 × 600)

$$= \underline{\mathbf{785}}$$

c) On strict EV calculations the firm should not use consultants. However these results are all based on estimates of outcomes and probabilities which may not be fulfilled. Aside from the financial calculations management may consider using the consultants in order to improve the chances of completing on time and thus safeguarding their reputation.

A4. a) The expected number of failures during the machine lifetime

= [(0 × 0.15) + (1 × 0.3) + (2 × 0.25) + (3 × 0.2) + (4 × 0.1)]

$$= \underline{\mathbf{1.8}}$$

b) The following table shows the total cost in £000 for each failure number/order quantity combination.

		No. of failures				
		0	**1**	**2**	**3**	**4**
	0	0	20	40	60	80
	1	5	5	25	45	65
Number ordered	**2**	10	10	10	30	50
	3	15	15	15	15	35
	4	20	20	20	20	20

where, for example, order = 3, failures = 4 results in the cost of 3 spare parts + 1 major breakdown costing (3 × £5,000) + (1 × £20,000) = £35,000

c) Expected values using values from above table and probabilities of failures in question.

Order O = (0 × 0.15) + (20 × 0.3) + (40 × 0.25) + (60 × 0.2) + (80 × 0.1)
= 36

Similar calculations give the following EVs

Order Quantity	0	1	2	3	4
EV ('000)	36	24	18	17*	20

*Order 3 spare parts gives lowest cost.

d) The expected cost using perfect information is:

$(0 \times 0.15) + (5 \times 0.3) + (10 \times 0.25) + (15 \times 0.02) + (20 \times 0.1)$

$= 9$

∴ EV of perfect information = 17,000 − 9,000 = 8,000

e) The table below shows the utility of each failure number/order quantity

		\multicolumn{5}{c}{Number of Failures}	Expected utility
		0	1	2	3	4	
	0	1.00	.86	.64	.34	0	0.636
Number	1	.97	.97	.81	.57	.26	0.779
ordered	2	.94	.94	.94	.76	.50	0.860
	3	.90	.90	.90	.90	.70	0.880*
	4	.86	.86	.86	.86	.86	0.860
	prob	.15	.3	.25	.2	.1	

*It will be seen that ordering 3 spares produces the highest utility. This agrees with the decision based on EVs.

A5. a) i) Expected profits:

Standard: $(60 \times 0.2) + (40 \times 0.5) + (20 \times 0.3) = £38,000$

Deluxe: $(100 \times 0.2) + (60 \times 0.5) + (0 \times 0.3) = 50,000$

Super: $(120 \times 0.2) + (80 \times 0.5) - (40 \times 0.3) = 52,000$

∴ based on expected profits SUPER is best.

ii) Maximin criterion:

Minimum profit for each model

$$\text{Standard} = \quad 20$$
$$\text{Deluxe} = \quad 0$$
$$\text{Super} = \quad -40$$

∴ based on maximin criterion STANDARD is best.

Expected value is a summary measure based on the forecast profits and probabilities whereas maximin only considers the worst outcome for each model and ignores other possibilities.

b) The expected profit from perfect information is $= 20(0.3) + 80(0.5) + 120(0.2) = £70,000$

∴ the expected value of perfect information is

$$£70,000 - £52,000 = \underline{\mathbf{£18,000}}$$

c) Replacing profits by utilities

		Standard	Deluxe	Super
Excellent	0.2	0.78	0.96	1.00
Moderate	0.5	0.65	0.78	0.89
Poor	0.3	0.52	0.37	0

The expected utility for each model type is:

Standard: $0.2(0.78) + 0.5(0.65) + 0.3(0.52) = 0.637$

Deluxe: $0.2(0.96) + 0.5(0.78) + 0.3(0.37) = 0.693*$

Super: $0.2(1.00) + 0.5(0.89) + 0.3(0) = 0.645$

*using utilities DELUXE is best.

The theoretical advantage of using utilities is that non-monetary factors can be incorporated into decision making. However there is the practical problem of assigning utility values to all possible outcomes.

d) Given that p = prob (moderate model acceptance) then, the expected utility for each model is

Standard: $0.2(0.78) + p(0.65) + (0.8 - p)0.52 = 0.572 + 0.13p$
Deluxe: $0.2(0.96) + p(0.78) + (0.8 - p)0.37 = 0.488 + 0.41p$
Super: $0.2(1.00) + p(0.89) + (0.8 - p)0 = 0.200 + 0.89p$

The Standard model is preferable if

$0.572 + 0.13p > 0.488 + 0.41p$ and $0.572 + 0.13p > 0.200 + 0.89p$
$0.084 > 0.28p$ $0.372 > 0.76p$
$p < 0.3$ $p < 0.489$

Hence $p < 3$

The Deluxe model is preferable if

$p > 0.3$ and $0.488 + 0.41p > 0.200 + 0.89p$
$0.288 > 0.48p$
$0.6 > p$

Hence $0.3 < p < 0.6$

The Super model is preferable if $0.6 < p < 0.8$

A6. a) i) Estimated monthly sales for the year to 31 May 1996

$$= 2.643 \times 1.06 \times \frac{113.6}{110.9} = £2.870m$$

ii) Total settlement discount expected in the year to 31 May 1996
$= 2.870 \times 12 \times 0.01 \times 0.85 = £0.293m$

b) i) Probability that invoice unpaid

At due date = 0.15

At one month overdue = 0.09 (0.15 × 0.6)

At two months overdue = 0.027 (0.09 × 0.3)

There is a 2.7% chance that an invoice will remain unpaid two months after the due date for payment.

ii) Probability that invoice paid between one month and two months overdue

$= 0.09 - 0.027$

$= 0.063$

There is a 6.3% chance that an invoice will be paid between one month and two months overdue.

c) Can be taken from the text.

A7. a) The variable costs of production are £0.20 per cake and the retail price is £0.50 per cake. The entries in the table below are based on the following reasoning:

If production = 20: variable costs are £4. If demand ≥20, the revenue is always £10.00. Contribution is £6.

If production = 60: variable costs are £12.

If demand = 20, revenue = £12. This is £10 from customers + 40(£0.05) from the pig farmer. Contribution = £0.

If demand ≥ 60, revenue is £30, and the contribution £18.

Daily production level	Daily level of demand (cakes)				
	20	60	100	140	180
20	6	6	6	6	6
60	0	18	18	18	18
100	−6	12	30	30	30
140	−12	6	24	42	42
180	−18	0	18	36	54
Probability	0.1	0.2	0.3	0.3	0.1

Contribution (£) for possible combinations of demand and production

Production = 20: Expected contribution = **£6.00**
Production = 60: Expected contribution = $\overline{0 + 18}$ (0.9) = **£16.20**
Production = 100: Expected contribution = −6 (0.1) + 12 $\overline{(0.2)}$ + 30 (0.7) = **£22.80**
Production = 140: Expected contribution = −12 (0.1) + 6 (0.2) + 24 (0.3) + 42 $\overline{(0.4)}$ = **£24.00**
Production = 180: Expected contribution = −18 (0.1) + 0 + 18 (0.3) + 54 (0.1) = **£19.80**

Based on the estimates available the best production level is 140 units.

However, there is little to choose between 100 and 140 and slight changes in estimated demand or costs could alter the recommendation. In addition, it is worth mentioning that the decision is based on expected value with its inherent assumptions given in the text.

Assessment and revision Chapters 4 to 9

A1.

Number of calls	Mid-points	Frequency		
	x	f	fx	fx²
0–9	4.5	0	0	0
10–19	14.5	5	72.5	1,051.25
20–29	24.5	10	245	6,002.5
30–39	34.5	20	690	23,805
40–49	44.5	15	667.5	29,703.75
50–59	54.5	10	545	29,702.5
60–69	64.5	4	258	16,641
70 +	74.5	0	0	0
		64	2,478	106,906

The mean number of calls $= \dfrac{\Sigma fx}{\Sigma f}$

$$= \dfrac{2,478}{64}$$

$\therefore \bar{x} = \textbf{38.72 calls}$

The standard deviation, s, is given by

$$s = \sqrt{\dfrac{\Sigma fx^2}{\Sigma f} - \bar{x}}$$

$$= \sqrt{\dfrac{106,906}{64} - 38.72^2}$$

$$= \sqrt{171.17}$$

hence s = **13.08 calls**

b) The 95% confidence limits for the mean number of calls per day are

$$\bar{x} \pm 1.96 \frac{s}{\sqrt{n}}$$

that is

$$38.72 \pm \frac{1.96 \times 13.08}{8}$$

or **35.52** and **41.92 calls**

The average cost of each call is £20 and there are 256 working days per year, and so the confidence limits on total annual cost are

$$35.52 \times 20 \times 256 \text{ and } 41.92 \times 20 \times 256$$

that is **£181,862** and **£214,630**

The best estimate of the **annual total cost is £198,246** (based on the mean number of calls per day). Further, we can be 95% confident that the total will lie between £181,862 and £214,630.

A2. a)

Payment in days	Mid-point	Number of customers		
	x	f	fx	fx^2
5–9	7	4	28	196
10–14	12	10	120	1,440
15–19	17	17	289	4,913
20–24	22	20	440	9,680
25–29	27	22	594	16,038
30–34	32	16	512	16,384
35–39	37	8	296	10,952
40–44	42	3	123	5,292
		100	2,405	64,895

$$\text{Arithmetic mean} = \frac{\sum fx}{\sum f} = \frac{2,405}{100} = 24.05 \text{ days}$$

b)

$$\text{Standard deviation} = \sqrt{\frac{\sum fx^2}{\sum f} - \left(\frac{\sum fx}{\sum f}\right)^2}$$

$$= \sqrt{\frac{64,895}{100} - \left(\frac{2,405}{100}\right)^2}$$

$$= \textbf{8.4}$$

c) Histogram to show payment record of 100 customers

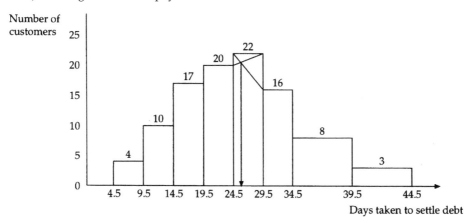

d) Out of 100, 16 lie in the class 30 to 34 days and 8 lie in the class 35 to 39 days. Therefore, the best estimate that an unpaid invoice chosen at random will be between 30 and 39 days old is

$$\frac{2{,}478}{64} = 0.24$$

A3. a) i) The standard error (the standard deviation of the sampling distribution) multiplied by the square root of the sample size is equal to the standard deviation of the population.

ii) As the sample size increases the standard error of the sampling distribution decreases, and vice versa, because the variability of data around the mean decreases the larger the sample size taken.

b) Limits of the mean unit volume with 95% confidence.

$$= 0.25 \pm 1.96 \times \frac{0.004}{\sqrt{50}}$$

$$= 0.25 \pm 0.0011$$

$$= \text{between } 0.2489 \text{ and } 0.2511$$

$\therefore$ 95% confident that the true mean lies between 0.2489 and 0.2511 litres.

A4. a) i) $\Sigma x = 250 \qquad \Sigma x^2 = 13{,}444 \qquad n = 10$

$$\therefore \bar{x} = \frac{350}{10} = 35$$

$$s = \sqrt{\frac{\Sigma x^2}{n} - \bar{x}^2 \left(\frac{n}{n-1}\right)}$$

$$= \sqrt{\frac{13{,}444}{10} - 35^2 \left(\frac{10}{9}\right)}$$

$$= 11.52$$

The 95% confidence interval is found using the t distribution with $n - 1$ degrees of freedom thus

525

$$\bar{x} \pm t_q \frac{s}{\sqrt{n}} = 35 \pm 2.26 \frac{11.52}{\sqrt{10}}$$

$$= 35 \pm 8.23$$

i.e. from 26.77 to 43.23 days sickness.

ii) The number of branches required to give a band width no more than 4 can be found thus

$$1.96 \frac{11.52}{\sqrt{n}} = 2$$

$$\frac{1.96 \times 11.52}{2} = \sqrt{n}$$

i.e. n = 127.4

so **128** branches will be required.

b) This part of the question using the chi-square test.

i) The information given is summarised in the following table:

| | Grouping (in terms of sales) | | | |
	Small	Medium	Large	Total
No sickness	52	15	29	96
Some sickness	38	21	55	114
Total	90	36	84	210

ii) The null hypothesis is that there is no relationship between sales classification and existence of sickness.

Under the null hypothesis, the expected frequencies are:

	Small	Medium	Large
No sickness	41.1	16.5	38.4
Some sickness	48.9	19.5	45.6

Analysing the discrepancy between the observed and expected frequencies

$$\chi^2 = \sum \frac{(O-E)^2}{E} = \frac{10.9^2}{41.1} + \frac{1.5^2}{16.5} + \frac{9.4^2}{38.4} + \frac{10.9^2}{48.9} + \frac{1.5^2}{19.5} + \frac{9.4^2}{45.6}$$

If the null hypothesis is true, this value should have approximately a chi-squared distribution with 2 degrees of freedom. As the 5% chi-squared value is 5.99, the above result is significant and so the sales classification is associated with the existence of days sickness.

Examination of the table in (b)(i) shows that medium and large branches have more sickness.

A5. a) The Poisson distribution is appropriate because there is a small probability of an event occurring there are discrete values and the average of these events (i.e. m = np) is below 10 (50 × 0.1 = 5).

b) The Poisson formula is

$$P(x) = \frac{m^x e^{-m}}{x!}$$

In this example m = np = 50(0.2) = 1

$$\therefore P(x \le 2) = P(x = 0) + P(x = 1) + P(x = 2)$$

$$= \frac{e^{-m}}{0!} + \frac{e^{-m}}{1!} + \frac{m^2 e^{-m}}{2!}$$

$$= \frac{2.718^{-1}}{1} + \frac{2.718^{-1}}{1} + \frac{1^2 \times 2.718^{-1}}{2 \times 1}$$

$$= 0.3679 + 0.3679 + 0.1840 = \underline{\textbf{0.92}}$$

c) Similarly, the following table can be drawn up

p	pa
0	1.000
0.02	0.920
0.05	0.544
0.10	0.125
0.15	0.020

These values are shown on the graph below. The graph is used to estimate two values of interest.

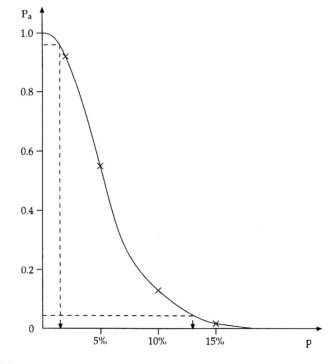

i)   1.5%

ii)   13%

If the percentages of failures is below 1.5% it is likely the batch will be accepted; above 13% it is likely to be rejected. Between these values there is an element of chance about acceptance/rejection.

A6. a)

Table showing observed/expected frequencies

	Manual		Non-manual		Total
	O	E	O	E	
In favour	30	32.08	25	22.92	55
Not in favour	40	37.92	25	27.08	65
Total	70	70	50	50	120

As this is a 2×2 table Yates correction is applied

$$\text{i.e. } \chi^2 = \sum \frac{(|O - E| - \frac{1}{2})^2}{E}$$

$$= \ |30 - 32.08| \ - 0.5 = 1.58$$
$$|40 - 37.82| \ - 0.5 = 1.58$$
$$|25 - 22.92| \ - 0.5 = 1.58$$
$$|25 - 27.08| \ - 0.5 = 1.58$$

$$\therefore \chi^2 = \frac{1.58^2}{32.08} + \frac{1.58^2}{37.92} + \frac{1.58^2}{22.92} + \frac{1.58^2}{27.08}$$

$$= \underline{\mathbf{0.345}}$$

As χ^2 with 1 degree of freedom at 5% = 3.84 the calculated result is not significant and we conclude there is no evidence that manual and non-manual differ in their attitude to the buy-out.

b) For females the chi-square test is invalid because there are two expected frequencies less than 5. For males chi-square is valid and the value is significant at the 5% level (i.e. 3,89 cf 3.84), i.e. for males the job classification does affect their attitude to the buy-out.

c) Key points in report

Percentages in favour:

Male manual	40%	} Manual 43%
Female manual	80%	
Male non-manual	63%	} Non-manual 50%
Female non-manual	20%	
Male	48%	
Female	35%	

Note also that the female results are based on small numbers. On balance the non-manual workers have opposite attitudes; males in favour, females not in favour.

A7. a)

Year	Qtr		Sum in qtrs	Sum of two qtrs	Trend	Actual minus Trend	Actual/ trend
1990/91	1	49					
	2	37	211				
	3	58	212	423	52.875	5.125	1.097
	4	67	213	425	53.125	13.875	1.261
1991/92	1	50	214	427	53.375	−3.375	0.937
	2	38	215	429	53.625	−15.625	0.709
	3	59	216	431	53.875	5.125	1.095
	4	68	218	434	54.25	13.75	1.253
1992/93	1	51	219	437	54.625	−3.625	0.934
	2	40	221	440	55	−15	0.727
	3	60	220	441	55.125	4.875	1.088
	4	70	222	442	55.25	14.75	1.267
1993/94	1	50	223	445	55.625	−5.625	0.899
	2	42					
	3	61					

b) Either

Additive model

		Quarter		
Year	Q1	Q2	Q3	Q4
1990/91			5.125	13.875
1991/92	−3.375	−15.625	5.125	13.75
1992/93	−3.625	−15	4.875	14.75
1993/94	−5.625			
Total	−12.625	−30.625	15.125	42.375
Average	−4.208	−15.313	5.042	14.125 = −0.354
Seasonal variation	−4	−15	5	14 = 0

Or

Multiplicative model

		Quarter		
Year	Q1	Q2	Q3	Q4
1990/91			1.097	1.261
1991/92	0.937	0.709	1.095	1.253
1992/93	0.934	0.727	1.088	1.267
1993/94	0.899			
Total	2.770	0.718	3.280	3.781
Average	0.923	0.718	1.093	1.260 = 3.994
Adjustment factor*	1.0015	1.0015	1.0015	1.0015
Seasonal variation	0.924	0.719	1.095	1.262 = 4.000

* Adjustment factor = 4/(3.994) = 1.0015

c) An explanation of the forecasting method:

i) Plot a graph of the trend.

ii) By eye, establish an appropriate forecast of the trend for the last quarter of 1993/ 94 and the first three quarters of 1994/5. (Note: linear regression or the high/low method may be appropriate methods to establish the forecast.)

iii) Adjust the forecast trend for these quarters for the seasonal variations:

Additive model

Estimated data value = forecast trend value + appropriate seasonal variation value.

Multiplicative model

Multiply each point by the appropriate seasonal factor.

A8. a)
$$a = \bar{y} - b\bar{x} \qquad\qquad b = \frac{\sum(x - \bar{x})(y - \bar{y})}{\sum(x - \bar{x})^2}$$

		A			**B**	**A × B**	
x	$\bar{x}$	$x - \bar{x}$	y	$\bar{y}$	$y - \bar{y}$		$(x - \bar{x})^2$
2	6	−4	60	104.8	−44.8	179.2	16
8	6	2	132	104.8	27.2	54.4	4
6	6	0	100	104.8	−4.8	0	0
8	6	2	120	104.8	15.2	30.4	4
10	6	4	150	104.8	45.2	180.8	16
4	6	−2	84	104.8	−20.8	41.6	4
4	6	−2	90	104.8	−14.8	29.6	4
2	6	−4	68	104.8	−36.8	147.2	16
6	6	0	104	104.8	−0.8	0	0
10	6	4	140	104.8	35.2	140.8	16
60			1,048			804.0	80

$$\frac{60}{10} = 6 \qquad\qquad \frac{1,048}{10} = 104.8$$

Therefore $b = \dfrac{804}{80} = 10.05 \qquad a = 104.8 - (10.05 \times 6)44.5$

Where y = total cost (× 10)
x = age in years

b) Maintenance costs using formula from (a):

Age of vehicle (years)	Cost (£ × 10)
1	54.55
2	64.60
3	74.65
4	84.70
5	94.75
6	104.80
7	114.85
8	124.90
9	134.95
10	145.00

c) A 12-year-old vehicle would have an estimated maintenance cost of:
$$44.5 + (10.05 \times 12) = 165.1 \ (\pounds \times 10), \text{ or } \pounds1,651.00$$
This forecast is an extrapolation beyond the data and consequently is less sound.

A9. a) H_0: No correlation between interest rate and housing sales
The test statistic is

$$T = \frac{-0.81\sqrt{9}}{\sqrt{(1 - (-0.81)^2)}} = -4.14$$

Thus as the critical values for a t distribution with 9 degrees of freedom are 2.26 (and −2.26) we reject the Null Hypothesis and conclude there is a significant correlation between interest rates and sales.

b) i) Scatter diagram.

It will be seen that there is a strong negative (or inverse) correlation between rates and sales.

ii) Where x = mortgage rate in previous year and y = house sales

$\Sigma x = 100$ $\Sigma y = 1{,}055$ $\Sigma x^2 = 1{,}066$ $\Sigma y^2 = 104{,}125$ $\Sigma xy = 9{,}600$ $n = 10$

$$x = \frac{n \sum xy - (\sum x)(\sum y)}{\sqrt{(n \sum x^2 - (\sum x)^2)(n \sum y^2 - (\sum y)^2)}}$$

$$= \frac{10 \times 9{,}600 - 100 \times 1{,}005}{\sqrt{(10 \times 1{,}066 - 100^2)(10 \times 104{,}125 - 1{,}005^2)}}$$

$$= \frac{-4{,}500}{\sqrt{660 \times 31{,}225}} = \underline{-0.991}$$

i.e. there is a very strong negative correlation between the sales index and the previous years mortgage rate.

c) All the values to calculate b have already been found above

$$\therefore b = \frac{-4{,}500}{600} = -6.818 \text{ and } a = 168.68$$

$\therefore$ when x = 11 y = 168.68 $-(6.818 \times 11) = \underline{93.7}$

Thus the prediction for 1993's index = **94**

d) Although the data shows strong relationships any form of forecasting is difficult. Conditions may change making past trends inapplicable.

A10. a) Can be taken from the text.

b)

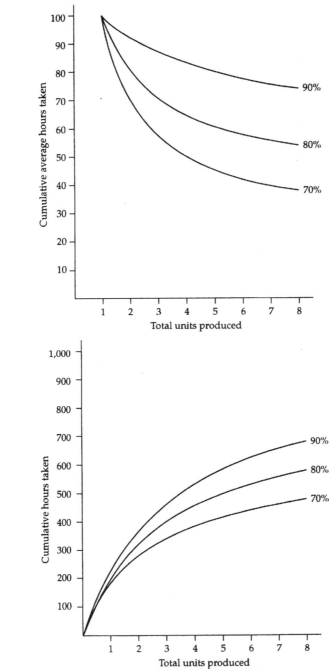

c) Can be taken from the text.

d) The learning curve is traditionally applied to repetitive manual operations where there is learning. The experience curve attempts to show the learning effect on the whole organisations efficiency due to internal learning and covers functions such as marketing distribution administration as well as production. Although more nebulous there clearly are general learning effects throughout the organisation especially in large organisations experiencing rapid growth.

A11. a)   Average weight of a packet = 21 × 50g = 1,050 g

Standard deviation is 5 g for 1 fillet. Therefore, variance of 1 fillet = 25 g

For a packet of 21 fillets, standard deviation = $\sqrt{25 \times 21}$ = **22.91 g**

b)   The probability that a packet will be below the nominal weight of 1 kg is:

$$Z = \frac{1{,}050 - 1{,}000}{22.91} = 2.18$$

From Normal Area Tables, probability = 0.50 − 0.4854 = **0.0146**

c)   Current compensation costs per period with 21 fillets per packet:

Underweight 1.5% of the time = 100,000 packets × 1.5% = 1,500 packets.

Dividing this by 2, we get 750 packets to pay compensation on @ £50 per packet. Therefore, total compensation = £37,500.

By putting one more fillet in a packet, revised average weight of a packet = 22 × 50g = 1,100g.

Revised standard deviation = $\sqrt{25 \times 22}$ = 23.45g

The probability of an underweight packet will be affected as follows:

$$Z = \frac{1{,}100 - 1{,}000}{23.45} = 4.26$$

Therefore, the probability is effectively zero and no compensation will be paid. However, the extra cost per period of putting an additional fillet in each packet is:

$$50p \times 100{,}000 = £50{,}000$$

Therefore, on purely financial grounds, it would not be cost-effective to put an additional fillet in each packet. However, the company may still prefer to do so to preserve goodwill.

A12. a)   Assuming that 1,000 units are manufactured:

	Machine 1 units	Machine 2 units	Total units
Units within target	279	679	958
Units outside target	21	21	42
Total units	300	700	1000

i)   Probability $\dfrac{679}{1000}$ = 0.679 or (0.97 × 0.7)

ii)   Probability $\dfrac{958}{1000}$ = 0.958 or (0.93 × 0.3) + (0.97 × 0.7)

(b)   Confidence interval = $p \pm Z \sqrt{\dfrac{p(1-p)}{n}}$

$$= 0.07 \pm 1.96 \sqrt{\frac{0.07 \times 0.93}{100}}$$

$$= 0.07 \pm 0.05$$

$$= 0.02 \text{ to } 0.12$$

(c) The formula is:

$$n \geq \frac{Z^2 p(1-p)}{1^2}$$

Where n is the required sample size. This is dependent upon:

- Z, which is the standardised normal distribution variable representing the required degree of confidence, measured in number of standard errors.
- p, the initial sample proportion, which is used as the best estimate of the population proportion.
- l, which is the maximum limited (expressed as a decimal) of the population proportion above or below the sample proportion,. The values for Z, p and l are substituted in the above formula to find n.

A13. a) Two possible reasons for the large variation in output each month are:

- seasonal variation;
- production problems in some months.

b) Graph showing relationship between output and costs

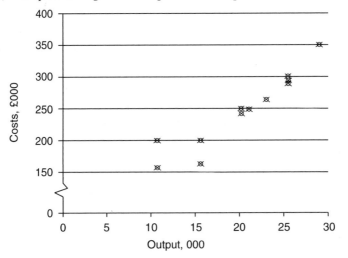

The graph shows there is a strong positive relationship between output and costs. This means that output may be used to predict costs.

c)

x	y	xy	x^2	y^2
16	170	2,720	256	28,900
20	240	4,800	400	57,600
23	260	5,980	529	67,600
25	300	7,500	625	90,000
25	280	7,000	625	78,400
19	230	4,370	361	52,900
16	200	3,200	256	40,000
12	160	1,920	144	25,600
19	240	4,560	361	57,600
25	290	7,250	625	84,100
28	350	9,800	784	122,500
12	200	2,400	144	40,000
240	2,920	61,500	5,110	745,200

$$b = \frac{12 \times 61{,}500 - 240 \times 2{,}920}{12 \times 5{,}110 - 240^2} = \frac{37{,}200}{3{,}720} = 10$$

$$a = \frac{2{,}920}{12} - 10 \times \frac{240}{12} = 43.333$$

$$y = 43.333 + 10x$$

This means that when output is zero, costs are £43,333, and that for every one unit increase in output, costs will rise by £10. This assumes linearity.

d) $$r = \frac{12 \times 61{,}500 - 240 \times 2{,}920}{\sqrt{(12 \times 5{,}110 - 240^2)(12 \times 745{,}200 - 2{,}920^2)}}$$

$$= \frac{37{,}200}{\sqrt{(3{,}720)(416{,}000)}}$$

$$= \frac{37{,}200}{39{,}338.53} = \underline{\mathbf{0.946}}$$

A correlation coefficient of 0.946 indicates that there is a strong positive correlation (linear relationship) between output and costs. Output will be a useful predictor of costs.

e) $x = 20$; $y = 43.333 + 10(20) = 243.33$; or $y = £243{,}333$

$x = 40$; $y = 43.333 + 10(40) = 443.33$; or $y = £443{,}333$

As the correlation coefficient is high, using linear regression should provide accurate forecasts.

The first forecast is an interpolation, which makes it a valid forecast. The second forecast is an extrapolation, which could reduce the accuracy of the forecast. It is not known whether the linear relationship holds for such a high value of output.

A14. a)

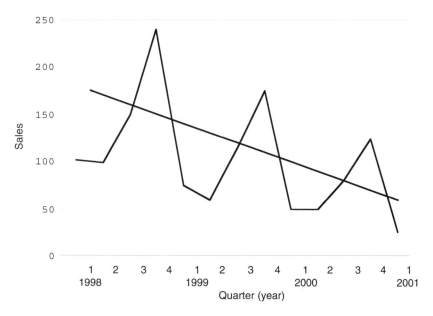

b) Since the seasonal variations are percentages a *Multiplicative model* is required.

Calculations:

2001 Q$_2$: Given that 1998 Q$_1$ = 1, the time period for this quarter equals 14. Using S = 180 − 10T, the trend is 180 − 10(14) = 40 units.

Seasonal factor = 0.6
∴ Forecast = 40 × 0.6 = 24 units

Q$_3$: Given that 1998 Q$_1$ = 1, the time period for this quarter equals 15. Using S = 180 − 10T, the trend is 180 − 10(15) = 30 units.

Seasonal factor = 1
∴ Forecast = 30 × 1.0 = 30 units

Q$_4$: Given that 1998 Q$_1$ = 1, the time period for this quarter equals 16. Using S = 180 − 10T, the trend is 180 − 10(16) = 20 units.

Seasonal factor = 1.8
∴ Forecast = 20 × 1.8 = 36 units

Given the clear-cut long-term trend it is likely that the forecasts are reasonably reliable.

c) Given the reliability and consistency of the trend and seasonal variations the sales position for the brand looks very bad.

Assessment and revision Chapters 10 to 15

A1. a) Total profit = Total revenue − Total cost

$$= -\frac{x^2}{2} + 1,500x - 330x - 415,000$$

$$= -\frac{x^2}{2} + 1,170x - 415,000$$

b) i)
$$MR = \frac{dTR}{dx} = -x + 1,500$$

$$MP = \frac{dTP}{dx} = -x + 1,170$$

ii) Profit maximising output occurs at a turning point of the marginal profit function.

Turning point is where −x + 1,170 = 0

i.e. where x = 1,170

This is a maximum because the second derivative is negative.

i.e. $\frac{d^2TP}{dx^2} = -1$

c) Marginal revenue = −x + 1,500

At 1,000 units = 500
1,100 units = 400
1,200 units = 300
1,300 units = 200

Marginal cost = Marginal revenue − Marginal profit
= 330

$$(\text{or Marginal cost} = \frac{dTC}{dx} = 330)$$

Graph showing profit maximising output.

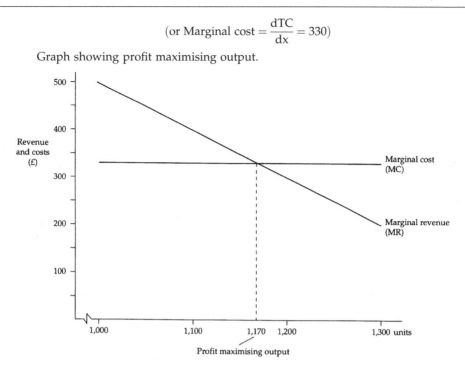

A2   a)   i)   Demand = 25,000

For Apex

Carrying cost = 0.2(8) = £1.6 per unit

Delivery cost = £50

$$\therefore \text{EOQ} = \sqrt{\frac{2 \times 50 \times 25,000}{1.6}} = 1,250 \text{ units}$$

∴ Best policy for Apex is to obtain 1,250 units a time, 20 times a year, i.e. every 15 working days.

Total annual cost = holding cost + delivery cost + cost of units

$$\therefore \text{Total annual cost} = £\frac{1,250}{2}(1.6) + £50\ (20) + £8\ (25,000)$$

$$= \underline{£202,000}$$

For Bemax

With no delivery charge best to order smallest delivery, i.e. 5,000 units.

$$\therefore \text{Total annual cost} = £\frac{5,000}{2}(1.59) + 0 + £7.95(25,000)$$

$$= \underline{£202,725}$$

ii)   Apex has lower cost (just!).

iii)   Can be taken from text.

b) With only 15,000 from Apex, 10,000 must be obtained from Bemax, i.e.

$$\frac{15,000}{25,000}(£202,000) + \frac{10,000}{25,000}(£202,725) = £202,290$$

$\therefore$ Additional cost of £290.

A3   a)   An order cost comprises

Purchasing labour + stationery + delivery

$$£\left(15(10) + \frac{1,425}{10} + 80\right) = \mathbf{£265.625} \text{ as stated.}$$

b)
Annual demand	$= 4,800 \times 50$	$= \mathbf{240,000}$
Unit storage cost	$= 0.15(8.5)$	$= \mathbf{£1.275}$
Buffer stock	$= 20,000 - 3\,(4,800)$	$= \mathbf{5,600}$

Total annual stock = storage + replenishment + cost of goods

$$= £\left(\frac{24,000}{2} + 5,600\right)1.275 + 10(265.625) + 240,000(8.5)$$

$$= \mathbf{£2,065,096}$$

c)

$$\therefore \text{EOQ} = \sqrt{\frac{2 \times 265.625 \times 240,000}{1.275}} = \underline{\mathbf{10,000}}$$

Annual cost using 10,000 order quantity

$$= £\left(\frac{10,000}{2} + 5,000\right)1.275 + 24\,(265.625) + 240,000\,(8.5)$$

$$= \underline{\mathbf{£2,059,890}}$$

i.e. savings of **£5,206**

d)   The discount point is 100,000 units so there is no point in ordering in larger quantities.

$\therefore$ Annual cost = storage + replenishment + cost of goods + extra storage

$$= £\left(\frac{10,000}{2} + 5,000\right)1.21125 + 24\,(265.625) + 240,000\,(8.5) + 25,000$$

$$= \underline{\mathbf{£2,059,890}}$$

$\therefore$ Worth taking discount.

A4.   a)   Can be taken from the text.

b)   First assign random numbers in proportion to probabilities.

Time to failure (weeks)	Random numbers (RN)
7	0
8	1
9	2, 3
10	4, 5, 6
11	7, 8
12	9

The simulation takes place using the RN in the question for each of the tubes and assuming that there are new tubes at time O.

Replacement costs

$$1 \text{ tube} = £400 + 0.5(900) + 100 = 950$$

$$2 \text{ tubes (simultanously)} = 800 + {}^5/_6 \,(900) + 160 = 1,710$$

$$3 \text{ tubes (simultanously)} = 1,200 + 900 + 200 = 2,300$$

i) Replacing each tube as it fails

	Remaining lifetime (weeks)			
Week	Tube 1	Tube 2	Tube 3	Cost
0	⑪	⑪	⑩	2,300
10	1	1	⑪	950
11	⑦	⑪	10	1,900*
18	⑦	4	3	950
21	4	1	⑨	950
22	3	⑪	8	950
25	⑧	8	5	950
30	3	3	⑦	950
33	⑪	⑩	4	1,900*
37	7	6	⑨	950
43	1	⑪	3	950
44	⑨	10	2	950
46	7	8	⑪	950
				15,600

* If it is assumed that in weeks 11 and 33 there was simultaneous replacement the costs would be £1,710 in each week making £15,220 in total.

ii) Replacing each tube as it fails

	Remaining lifetime (weeks)			
Week	Tube 1	Tube 2	Tube 3	Cost
0	⑪	⑪	⑩	2,300
10	⑦	⑪	⑪	2,300
17	⑦	⑪	⑨	2,300
24	⑧	⑩	⑦	2,300
31	⑪	⑪	⑨	2,300
40	⑨	⑩	⑪	2,300
49	⑧	⑨	⑩	2,300
				16,100

c) The estimates above are not very reliable as there are only 3 tubes over a short period.

d) Simulation when replacing units more than eight weeks old.

	Remaining lifetime (weeks)			
Week	Tube 1	Tube 2	Tube 3	Cost
0	⑪	⑪	⑩	2,300
10	⑦	⑪	⑪	2,300
17	⑦	4	4	950
21	3	⑪	⑨	1,710
24	⑧	8	6	950
30	2	⑩	⑦	1,710
32	⑪	8	5	950
37	6	3	⑨	950
40	⑨	⑪	6	1,710
46	3	5	⑪	950
49	⑧	⑩	8	<u>1,710</u>
				<u>16,190</u>

This system is more expensive than the others.

A5. a) Can be taken from the text.

b) i) 99% confidence interval for

$$= x \pm Z\frac{s}{\sqrt{n}}$$

$$= 3.64 \pm 2.58 \times \frac{0.68}{\sqrt{40}}$$

$$= 3.64 \pm 0.28$$

$$= 3.36 \text{ to } 3.92 \text{ weeks}$$

ii) Sample size

$$= \frac{(Zs)^2}{l}$$

$$= \frac{(1.96 \times 0.68)^2}{0.15} = \underline{\mathbf{79}}$$

c) Re-order level = maximum usage × maximum lead time
= 150 × 10 = 1,500

Safety stock = re-order level − (average usage × average lead time)
= 1,500 − (120 × 8) = 540

Average stock = safety stock + (EOQ × 0.5)
= 540 + (2,900 × 0.5) = **1,990**

A6. a) Can be taken from the text.

b) Production requirements:

	Product A		Product B	
Sales	24,600	units	9,720	units
+ stock increase	447	units	178	units
= production (net of rejects)	25,047	units	9,898	units
÷ net of rejection %	0.99		0.98	
= total production	25,300	units	10,100	units

× material usage (net of wastage)	1.8 kg/unit	3.0 kg/unit
= materials (before wastage)	45,540 kg	30,300 kg
÷ net of wastage %	0.95	0.89
= Material Z required	47,937 kg	34,045 kg
	= 81,982 kg total	

$$\text{Economic Order Quantity (EOQ)} = \sqrt{\frac{2 \times 30 \times 81,982}{0.63}}$$

$$= \mathbf{2,794\ kg}$$

c) Average stock investment

$$= \frac{2,794 + 1,000}{2}$$

$$= 2.397\ \text{kg} \times £3.50/\text{kg}$$

$$= \underline{\mathbf{£8,390}}$$

Annual stock holding costs $= £8,390 \times 0.18$

$$= \underline{\underline{\mathbf{£1,510}}}$$

Assessment and revision Chapters 16 to 21

A1. Maximise $z = 1{,}500x + 2{,}000y$
     Subject to (budget) $5x + 10y \le 160$ (in £ '000)
            (maintenance time) $25x + 10y \le 40$ (hrs)
            (space) $50x + 50y \le 1{,}000$ (m²)

                   where x = no. of x machines
                       y = no. of y machines

b)

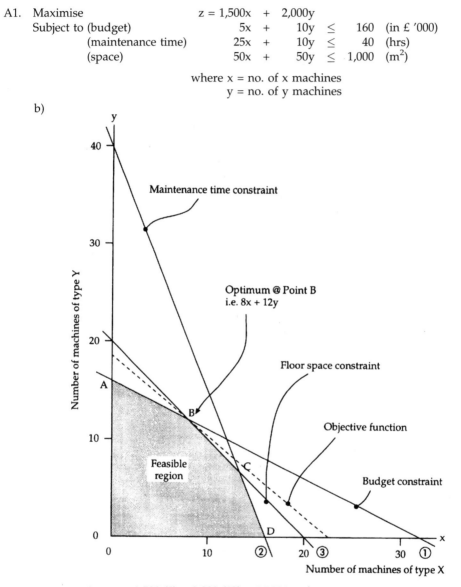

$\therefore$ optimum $= 1{,}500\,(8) + 2{,}000\,(12) = \underline{\mathbf{36{,}000\ units}}$

c) The floor space and budget are fully utilised but there is 80 hours unused for maintenance, i.e. $(400 - (25(8) + 10(12)))$

A2. a) LP model is:

Maximise contribution P = 5.50x + 6.50y

Subject to:

0.4x	+	0.3y	≤	2,200	(Mat A)
0.2x	+	0.5y	≤	2,500	(Mat B)

where x = quantity of Product X
y = quantity of Product Y

b) Three possibilities can be explored and the contributions compared.

No 1 where both materials are used to their limits the simultaneous equations are

0.4x	+	0.3y	=	2,200	①
0.2x	+	0.5y	=	2,500	②

Solving:

0.4x	+	1.0y	=	5,000	② × 2 = ③
		0.7y	=	2,800	③ − ① = ④
		y	=	4,000	

Substituting for y in ①:

$$x = \frac{2,200 - (0.3 \times 4,000)}{0.4}$$

$$= 2,500$$

∴ a feasible solution is:

Product X 2,500 units
Product Y 4,000 units

which gives £39,750 contribution (i.e. 2,500 × 5.50 + 4,000 × 6.50)

No 2 where x = 0

where x = 0, y could be 5,000 units (maximum imposed by limitation on **Material B**)
Relevant equation is 0.2x + 0.5y = 2,500.
Contribution would be:

Product Y 5,000 × £6.50 = £32,500.

No 3 where y = 0, x could be 5,500 units (maximum imposed by limitation on **Material A**)

Relevant equation is 0.4x + 0.3y = 2,200.
Contribution would be:

Product X 5,000 × £5.50 = £30,250.

∴ Option 1, where x = 2,500 and y = 4,000 gives the most contribution.

A3. a) The Simplex formulation is:

Maximise	15.6x	+	9.2y	+	13.4z			
Subject to (Material A)	0.3x	+			0.2z	≤	500	
(Material B)	0.2x	+	0.5y	+	0.4z	≤	1,000	
(Material C)	0.5x	+	0.1y	+	0.1z	≤	800	
(Material D)			0.4y	+	0.3z	≤	600	

x, y, z ≥ 0

where x = tonnes of Xtragrow
y = tonnes of Youngrow
z = tonnes of Zupergrow

b) i) Based on the final tableau:

optimum production = 1,279.1 tonnes Xtragrow
1,023.3 tonnes Youngrow
581.4 tonnes Zupergrow

ii) Contribution = £37,158.10 and 16.3 tonnes of D unused.

iii) The dual values are   Material A = £29.54 per tonne
Material B = £17.07 per tonne
Material C =   £6.65 per tonne
Material D =   £0 per tonne

These are the amounts of contribution which would be gained by having one more tonne of the particular material (or lost by having one tonne less).

iv) Although A could be sold at £15 per tonne and costs only £10 per tonne it earns £29.54 per tonne contribution so it should not be sold unless the availability could be increased.

A4. Sensitivity analysis identifies the range of values for contributions usages and limitations over which the optimal solution will apply.

Comments on dual values can be taken from the text.

a) The formation is

Maximise	$250x$	$+$	$170y$	$+$	$150z$		
Subject to (Desktops)	x	$+$	y			$\leq$	500
(Laptops)					z	$\leq$	250
(80386 chips)	x					$\leq$	120
(80286 chips)			y	$+$	z	$\leq$	400
(Prod. time)	$5x$	$+$	$4y$	$+$	$3z$	$\leq$	2,000

$$x, y, z \geq 0$$

where x = production of Desktops − 386s
y = production of Desktops − 286s
z = production of Laptops − 286s

b) i) Interpretation   Production levels
120 Desktop-386
200 Desktop-286
200 Laptop-286

Profit = 120 (250) + 200 (170) + 200 (150) = **£94,000**

Two of the constraints are not binding and have unused amounts (i.e. because they do not have dual values)

i.e. Desktops   180 spare
Laptops   50 spare

ii) Sensitivity Analysis interpretation

Objective function:

The upper and lower limits indicate the range of profits for the three products for which the original solution still applies.

e.g. a Laptop – 286 (i.e. X_3) can range in profit from £127.50 to £170 without changing solution (assuming X_1 and X_2 remain at their original values)

Constraints:

Similarly the limits indicate the range for the resource limits over which the dual values apply.

e.g. Production time could be increased from 2000 to 2180 giving 180 × £20 extra profit.

A5. a) Contribution = Price − Variable cost − Transportation cost

Contribution Table

	Customer A	Customer B	Customer C
Manchester	£35	£37	£33
London	30	26	36

b) Allocate to factory offering greatest contribution.

		Total contribution £
Customer A from Manchester	= 3,000 × £35 =	105,000
Customer B from Manchester	= 4,200 × £37 =	155,400
Customer C from Manchester	= 5,300 × £36 =	190,800
		451,200

Capacities	Manchester	7,200
	London	5,300

c) When capacity is 10,000 and demand is 12,500 a dummy source with zero contribution is required.

Initial Tableau

Destination

	A	B	C	Total
M	35	37	33	4,000
L	30	26	36	6,000
Dummy	0	0	0	2,500
Total	3,000	4,000	5,300	12,500

As usual make the initial allocations where there are the greatest contributions, i.e. 4,000 M to B then 5,300 L to C then fill in remaining gaps to meet the row and column totals. This results in the following table.

Destination

	A	B	C	Total
M	35	37 / 4,000	33	4,000
L	30 / 700	26	36 / 5,300	6,000
Dummy	0 / 2,300	0 / 200	0	2,500
Total	3,000	4,000	5,300	12,500

Check the shadow costs

$$M + B = 37$$
$$L + A = 30$$
$$L + C = 36$$
$$D + A = 0$$
$$D + B = 0$$

where M = 0
B = 37
D = −37
A = +37
L = −7
C = +43

The shadow costs of unused routes are

$$M + A = \quad 0 + (+37) \quad = \quad +37$$

$$M + C = \quad 0 + (+43) \quad = \quad +43$$

$$L + B = -7 + 37 \quad\quad = \quad +30$$

The shadow costs are checked against the actual contributions to see if there are any positive values

For cell M:A = 35 − 37 = −12
cell M:C = 33 − 43 = −10
cell L:B = 26 − 30 = −4

∴ initial allocation is optimum giving a total contribution of:

$$(700 \times 30) + (4{,}000 \times 37) + (5{,}300 \times 36) = \textbf{£359{,}800}$$

This results in a shortage of 2300 for A and 200 for B.

iii) To satisfy 2 out of 3 customers would mean either A or B being satisfied as well as C thus:

Customer A: if this customer had the shortfall then the London/B cell should increase by 200, resulting in a decrease in contribution of 200 × 4 = £800.

Customer B: if this customer had the shortfall then the Manchester/A cell should increase by 2,300, resulting in a decrease in contribution of 2,300 × 2 = £4,600.

Hence it is best to satisfy the orders of B and C which would give a total contribution of £800 less than the £359,800 achieved without this extra constraint.

A6. a) i) The contributions per project are as follows:

	A £	B £
Fee	1,700	1,500
Researchers:		
Qualified	600	360
Junior	112	210
Expenses	408	310
	1,120	880
Contribution	580	620

The iso-contribution line is 580A + 620B

ii) Constraints used in the linear programming model:

20A + 12B ≤ 1,344 (Qualified researchers)
8A + 15B ≤ 1,120 (Junior researchers)
B ≤   60 (Maximum type-B projects)
A ≥   20 (Minimum type-A projects)

iii) Profit-maximising product mix

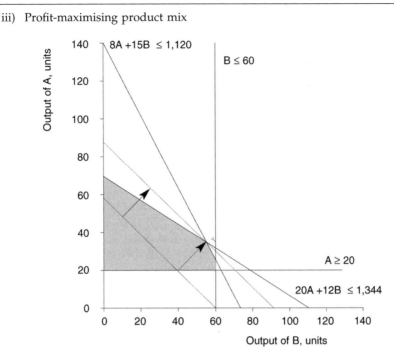

As always the solution is at a vertex of the feasible region indicated by Point A which is 33 Type A projects and 57 Type B projects.

(b) Project contributions

		£
A:	33 @ £580	19,140
B:	57 @ £620	35,340
	Total contribution	54,480
	Less: fixed costs	28,000
	Profit	**26,480**

(c) Can be taken from the text.

Assessment and revision Chapters 22 to 26

A1. a)

Possible paths	Path length (in days)
A – F – G – END	16
A – C – D – G – END	19
A – B – D – G – END	20
A – B – E – G – END	21

b) The critical path is the path with greatest length. The critical path is thus **A – B – E – G – End** and the project duration is **21 days**.

Activity	Activity duration	Earliest		Latest		Total float*
		start	finish	start	finish	
A	7	0	7	0	7	0
B	3	7	10	7	10	0
C	2	7	9	9	11	2
D	4	10	14	11	15	1
E	5	10	15	10	15	0
F	3	7	10	12	15	5
G	6	15	21	15	21	0

* Total Float is calculated thus

Latest Finish Time − Earliest Start Time − Activity Duration
e.g. for Activity F = 15 − 7 − 3 = **5 days**

d) If D takes 4 days longer than the original estimate, the first day will not delay the whole project, as it will use up the one day's float shown above. The next 3 days will, however, extend D's earliest finish time to 18 days, thereby delaying task G and the whole project by 3 days.

A2 a)

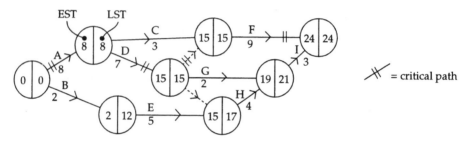

Using the given activity durations, the minimum project duration is 24 months. The critical activities are A, D and F. The associated cost is made up from (in £'000)

Cost of activities + Overheads

$$= (100 + 75 + 135 + 70 + 160 + 255 + 30 + 90 + 55) + 24(5)$$
$$= 970 + 120 = 1,090 \ (\text{i.e. } £1,090,000)$$

b) The only activities which should be considered, are ADF, i.e. on the critical path. If all these are crashed the project time reduces to 18 months and the costs are

$$(125 + 75 + 135 + 85 + 160 + 275 + 30 + 90 + 55) + 18(5) − 25$$
$$= 1,095 \ (\text{i.e. } £1,095,000)$$

However crashing all the 3 critical activities results in a time of 18 months but 20 months are allowed to collect the bonus so A need not be crashed, only D + F. This results in a 20 month duration and costs of

$$1,005 + 20(5) − 25 = 1,080 \ (\text{i.e. } £1,080,000)$$
which is a £10,000 saving over the original position.

e) With a bonus of only £15,000 there would be no direct financial benefit in any crashing. However the firm might wish to complete earlier at slightly higher cost, if it enabled them to gain another contract.

A3. a)   Network

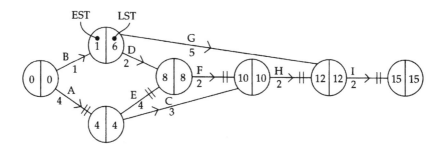

The critical path is A, E, F, H, I.

The duration of the critical path is 15 days.

b)   Assuming all activites start as soon as possible, the following chart shows when activities will start and finish.

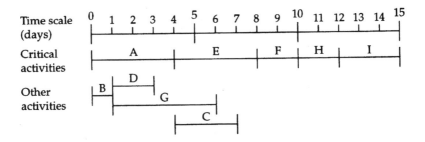

Consequently the following resource allocation is required on a day by day basis.

Day	Activities	Number of increased staff	
1	A, B	1 + 3	= 4
2, 3	A, D, G	1 + 1 + 1	= 3
4	A, G	1 + 1	= 2
5, 6	C, E, G	2 + 2 + 1	= 5
7	C, E	2 + 2	= 4
8	E	2	= 2
9, 10	F	2	= 2
11, 12	H	1	= 1
13, 14, 15	I	1	= 1

c)   Current costs with 5 staff = 15 days $\times$ 5 $\times$ £500 = **£37,500**

If Activity C is delayed to start on day 7 only 4 staff are required and the project duration unchanged. The cost is

$$15 \text{ days} \times 4 \times £500 = £30,000$$

However if the duration is increased by starting B on Day 1 and delaying activities A, E, F, H, I by 1 day and C until day 12 only a maximum of 3 staff are required thus:

549

Day	Activities	Total staffing
1	B	3
2, 3	A, D, G	3
4, 5	A, G	2
6	E, G	3
7, 8, 9	E	2
10, 11	F	2
12, 13, 14	C	2
15, 16	H	1
17, 18, 19	I	1

Consequently the duration is 19 days, with consequent cost

$$19 \times 3 \times £500 = £28,500$$

As Activity B requires 3 men then we cannot reduce resources further.

The above analysis indicates that staff are fully interchangeable for all tasks, which is unlikely.

A4. a)

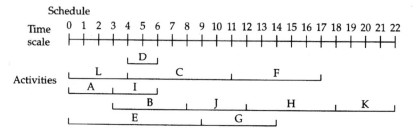

Critical path L C F K with a 21 day duration.

b) Total Floats of each activity:

A:3	E:7	I:11
B:3	F:0	J:11
C:0	G:7	K:0
D:11	H:4	L:0

Total Float = Latest Head Time − Earliest Tail Time − Activity Duration

e.g. Activity G = $21 - 9 - 5 = \underline{7}$

The uses of float can be taken from the text.

c) If A, G and I start at their ESTs there is no overlap Activities BHJK can be started at time 3, 12, 8, 18 (or 3, 8, 13, 18) with only 1 day increase in project duration.

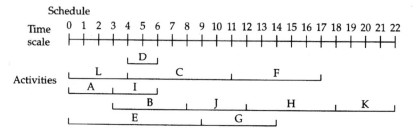

Assessment and revision Chapters 27 to 31

A1. a) From Table VIII the annuity factor for 6 years @ 10% = 4.355

∴ Maximum sum = £10,000 × 4.355 = **£43,550**

b) Net inflow = 25,000 − 5,000 = 20,000

Year end savings = 3,000

$$\therefore \frac{20,000}{3,000} = 6.667$$

Looking down the 10% column in Table VIII the time is between 11 and 12 years, i.e.

11 years Factor 6.495
12 years Factor 6.814

$$\therefore \text{Linear interpolation} = 11 + 1\left(\frac{20,000}{3,000}\right) \approx \textbf{11.54 years}$$

A2. a)

			£
Present heating cost			200,000
In 1 year cost will be	200,000 × 1.03	=	206,000
In 2 year cost will be	200,000 × (1.03)^2	=	212,180
In 3 year cost will be	200,000 × (1.03)^3	=	218,545
In 4 year cost will be	200,000 × (1.03)^4	=	225,102
In 5 year cost will be	200,000 × (1.03)^5	=	231,855
In 6 year cost will be	200,000 × (1.03)^6	=	238,810

b)

			£
In year 1 saving will be	5% × 206,000	=	10,300
In year 2 saving will be	5% × 212,180	=	10,609
In year 3 saving will be	5% × 218,545	=	10,927
In year 4 saving will be	5% × 225,102	=	11,255
In year 5 saving will be	5% × 231,855	=	11,593
In year 6 saving will be	5% × 238,810	=	11,941

c) Discounted cash flow calculation at 12% applied to savings:

Year	Cash flow	Discount factor	Present value
1	10,300	0.893	9,198
2	10,609	0.797	8,455
3	10,927	0.712	7,780
4	11,255	0.636	7,158
5	11,593	0.567	6,573
6	11,941	0.507	6,054
			45,218

The maximum amount the company will pay is £45,218

d) The present value of the savings is less than the cost of the double-glazing (£50,000) therefore the investment is not recommended on purely finance grounds.

However, other factors such as noise reduction and draft reduction may improve staff motivation. The resulting increase in value of the buildings and lower maintenance costs for the new windows may also be taken into account in deciding whether to install double-glazing.

ital investment

Year 0	(£150,000) × 1,000	=	(£150,000)	PV
Year 6	£10,000 × 0.432	=	4,320	PV
			(£145,680)	PV

Cash inflows – high demand

Year 1	£60,000 × 0.870	=	£52,200	PV
2	£62,000 × 0.756	=	46,872	PV
3	£65,000 × 0.658	=	42,770	PV
4	£70,000 × 0.572	=	40,040	PV
5	£70,000 × 0.497	=	34,790	PV
	3.353		£216,672	PV

Cash inflows – low demand

Years 1–5   £50,000 × 3.353   =   **£167,650**   PV

Thus net present value =

High demand £216,672 − 145,680 = £70,992

Low demand £167,650 − 145,680 = £21,970

Expected net present value = (70,992 × 0.7) + (21,970 × 0.3) = **£56,285**

b) PV of existing product = £12,000 × 3.353 = £40,236

∴ based on expected value there is a potential value of £56,285 − 40,236 = £16,049 from accepting the new product.

Usual caveats regarding Expected Value.

A4. a) The 'late payments' for each country are the difference between average credits days and average debt days:

Country	Late Payment (days):
Denmark	20
Finland	26
France	50
Germany	18
Holland	22
Ireland	30
Italy	30
Norway	15
Sweden	18
Switzerland	30
United Kingdom	45

Bar Chart showing average late payments, by country

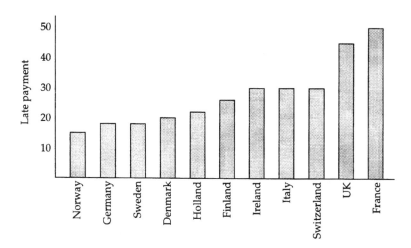

c) The geometric progression is the rate at which the invoice value accumulates per month, i.e. it has a value at the end of the first month of £10,000 × 1.01 and increases by a constant ratio of 1.01 per month.

The cost to the firm is the interest lost by the two month settlement delay, this is found by comparing the interest consequence of two month's delay cf immediate payment.

Interest value with immediate payment

$$S = \frac{A(R^N - 1)}{(R - 1)}$$

$$S = £10,000 \left(\frac{1.01(1.01^{12} - 1)}{1.01 - 1} \right)$$

$$= £10,000 \ (12.8093)$$

$$= £128,093$$

∴ Interest element = £128,093 − 120,000 = £8,093

Interest with 2 months' delay

$$S = £10,000 \left(\frac{1.01(1.01^{10} - 1)}{(1.01 - 1)} \right)$$

$$= £10,000 \ (10.5668)$$

$$= £105,668$$

∴ Interest = £105,668 − 100,000 = £5,668

∴ Cost = £8,093 − 5,668 = **£2,425**

A5. a) i)

M	=	56,000	+	0.05M	+	0.05L	+	0.10C
L	=	250,000	+	0.25M			+	0.15C
C	=	35,000	+	0.10M	+	0.10L	+	0.05C

where **M** = Maintenance, **L** = Laundry and **C** = Cleaning

Transposing gives

0.95M	−	0.05L	−	0.05C	=	56,000
−0.25M	+	L	−	0.15C	=	250,000
−0.10M	−	0.10L	−	0.95C	=	35,000

The LHS values are the matrix coefficients.

ii) Let $\mathbf{X}$ = the matrix coefficients in (1)
      $\mathbf{S}$ = the service department vector
      $\mathbf{T}$ = total cost

$$\text{Thus } \mathbf{T} = \mathbf{XS}$$

$$\therefore \mathbf{S} = \mathbf{X}^{-1}\mathbf{T}$$

Algebraically the matrix of coefficients can be found using the row by row operation method thus:

$$\begin{bmatrix} 0.95 & -0.05 & -0.10 \\ -0.25 & 1 & -0.15 \\ -0.10 & -0.10 & 0.95 \end{bmatrix} \begin{bmatrix} 1 & 0 & 0 \\ 0 & 1 & 0 \\ 0 & 0 & 1 \end{bmatrix}$$

Working on each row will transform the identity matrix on the right into the required matrix inversion.

b)

$$\begin{bmatrix} 1.083 & 0.067 & 0.125 \\ 0.293 & 1.034 & 0.194 \\ 0.145 & 0.116 & 1.086 \end{bmatrix} \times \begin{bmatrix} 56,000 \\ 250,000 \\ 35,000 \end{bmatrix} = \begin{bmatrix} \mathbf{M} \\ \mathbf{L} \\ \mathbf{C} \end{bmatrix}$$

Giving:

$$\begin{aligned} \mathbf{M} &= (1.083 \times 56{,}000 + 1.034 \times 250{,}000 + 0.125 \times 35{,}000) \\ &= \underline{\mathbf{£81{,}773}} \\ \mathbf{L} &= (0.293 \times 56{,}000 + 1.034 \times 250{,}000 + 0.194 \times 35{,}000) \\ &= \underline{\mathbf{£281{,}698}} \\ \mathbf{C} &= (0.145 \times 56{,}000 + 0.116 \times 250{,}000 + 1.086 \times 35{,}000) \\ &= \underline{\mathbf{£75{,}130}} \end{aligned}$$

A6. (a) Timeline showing structure of the repayment scheme:

End of Quarter	Repayment
1	x
2	x
. . .	. . .
12	x

We need to find the equal quarterly repayments (x) with a present value (quarter 0) of £40,000. From tables, the cumulative discount factor for twelve quarters, at 3 per cent per quarter = 9.954.

$$x = \frac{40{,}000}{9.954} = \underline{\mathbf{£4{,}018.49}}$$

(b) Time line showing structure of the savings:

Start of Quarter	Saving
1	x
2	x
. . .	. . .
12	x

We need to find the equal quarterly instalments with a terminal (future) value equal to the expected cost of the work.

Expected cost of work	=	$£40,000 \times 1.04 \times 1.04 \times 1.04$
	=	**£44,995**
Cumulative compound factor	=	$\dfrac{1.02(1.02^{12} - 1)}{1.02 - 1} = \mathbf{13.680}$
Quarterly savings	=	$\dfrac{44,995}{13.680} = \mathbf{£3,289.11}$

(c) In the current economic climate it has already been decided that a modernisation programme needs to go ahead if the business is to survive. What needs to be considered are the alternative ways of financing this investment.

Loan

If it is financed using a loan, the modernisation may be carried out immediately. As there is a danger of losing business without modernisation, there is clearly a major advantage in borrowing the money now. If the modernisation is carried out well it could not only stop a decline in business, but also increase the business. The other benefit of proceeding immediately is that the cost is known to be £40,000. Further costs can only be forecasts or estimates.

Obviously there should be a search to see whether the cost of the loan could be reduced.

Saving

The saving option is less risky but it all depends on the amount of business lost by not modernising immediately.

A7. (a) Pipeline capacity constraint $1.4Q_o + Q_g \leq 140,000$

       Processing capacity constraints $Q_g \leq 70,000$

                                              $Q_o \leq 150,000$

Rail capacity constraints:

              For $O \leq Q_g \leq 20,000$

                         $Q_o + Q_g \leq 120,000$

              For $Q_g \geq 20,000$ and $Q_o \leq 100,000$

                         $0.65\,Q_o + Q_g \leq 85,000$

Where Q_o represents barrels of oil and Q_g represents barrels of LPG.

Constraints on the production of oil and gas

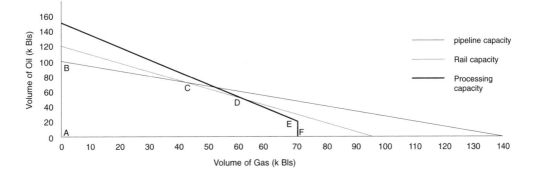

The optimum annual output combination is to be found on one of the extreme points marked A to F. The Net Present Value from 1,000 barrels of each product is:

	Oil	LPG
Capital cost (£)	−40,000	−60,000
Contribution (£)	8,000	11,500
DCF 5.65	45,200	64,975
NPV (£)	5,200	4,975

The contribution is for ten years at 1,000 barrels per annum. The DCF is the cumulative figure for ten years at 12%

Point A	Q_o = 0	Q_g = 0	NPV = 0
Point B	Q_o = 100,000	Q_g = 0	NPV = £520,000
Point C	Q_o = 73,333	Q_g = 37,334	NPV = £567,068
Point D	Q_o = 42,735	Q_g = 57,222	NPV = £506,901
Point E	Q_o = 18,751	Q_g = 70,000	NPV = £445,755

Thus, point C is the optimal output combination. The coordinates of the extreme points may be read off the graph or found by solving simultaneous constraint equations.

Workings for NPV for point C:

$$\text{NPV} = 73,333 \times \frac{£5,200}{1,000} + 37,334 \times \frac{£4,975}{1,000} = \underline{\mathbf{£567,068}}$$

(b) The optimum point is limited by pipeline capacity so the option is worth exploring. The new constraint for pipeline capacity would be:

$$1.4Q_o + Q_g \leq 154,000$$

and this intersects with the Rail Capacity Constraint:

$$0.65Q_o + Q_g \leq 85,000$$

subtracting the latter from the former gives:

$$0.75Q_O = 69,000$$

$$Q_o = \underline{\mathbf{92,000}}$$

so

$$Q_g = 85,000 - (0.65 \times 92,000)$$

$$= \underline{\mathbf{25,200}}$$

The new NPV would be £603,770 – an increase of £36,702 – which is less than the cost of the new pump, therefore the option should not be pursued.

Index

Index